THE
FRIENDLY GUIDE
TO THE
UNIVERSE

NANCY HATHAWAY

VIKING
A Winokur/Boates Book

VIKING
Published by the Penguin Group
Penguin Books USA Inc., 375 Hudson Street, New York, New York 10014, U.S.A.
Penguin Books Ltd, 27 Wrights Lane, London W8 5TZ, England
Penguin Books Australia Ltd, Ringwood, Victoria, Australia
Penguin Books Canada Ltd, 10 Alcorn Avenue, Toronto, Ontario, Canada M4V 3B2
Penguin Books (N.Z.) Ltd, 182–190 Wairau Road, Auckland 10, New Zealand

Penguin Books Ltd, Registered Offices: Harmondsworth, Middlesex, England

First published in 1994 by Viking Penguin, a division of Penguin Books USA Inc.

1 3 5 7 9 10 8 6 4 2

A Winokur / Boates Book

Grateful acknowledgment is made for permission to reprint the following copyrighted works:
Excerpts from *Ulysses: The Corrected Text* by James Joyce. Copyright © 1986 by Random House,
Inc. Reprinted by permission of Random House, Inc., The Bodley Head and the estate of the
author. Excerpt from speech by George Bernard Shaw appearing in *Thirty Years with G.B.S.* by
Blanche Patch (Gollancz, London). By permission of the Society of Authors on behalf of the
Bernard Shaw Estate. Excerpts from "Planetarium" and "The Spirit of Place" from *The Fact of the
Doorframe, Poems Selected and New, 1950–1984* by Adrienne Rich. Reprinted by permission of
the author and W. W. Norton & Company, Inc. Copyright © 1981 by Adrienne Rich.
Excerpt from "Monkey of Stars" from *Good Morning, America* by Carl Sandburg. Copyright
1928 and renewed 1956 by Carl Sandburg. Reprinted by permission of Harcourt Brace & Company.
"Perils of Modern Living" by H. P. Furth appearing in *The New Yorker*, November 10, 1956.
Reprinted by permission. © 1956, 1984 The New Yorker Magazine, Inc.
Excerpt from "I Will Sing You One-O" from *The Poetry of Robert Frost*, edited by Edward
Connery Lathem. Copyright 1923, © 1969 by Henry Holt and Company, Inc. Reprinted by
permission of Henry Holt and Company, Inc. Excerpt from "Iron Horse" from *Collected
Poems, 1947–1980* by Allen Ginsberg. Copyright © 1966 by Allen Ginsberg.
Reprinted by permission of HarperCollins Publishers, Inc.

Box ornaments by Nancy Hathaway. Copyright © Nancy Hathaway, 1994.

LIBRARY OF CONGRESS CATALOGING-IN-PUBLICATION DATA
Hathaway, Nancy, 1946–
The friendly guide to the universe / by Nancy Hathaway.
p. cm.
"A Winokur/Boates book."
Includes bibliographical references and index.
ISBN 0-670-83944-2
1. Astronomy. I. Title.
QB43.2.H373   1994
520—dc20   93-28189

Printed in the United States of America
Set in Century Old Style
Designed by Amy Hill

In memory of my father, Alan Berman, who stood in the front yard with us to see Mars at favorable opposition and *Sputnik* somewhere overhead.

Nothing is too wonderful to be true.

—Michael Faraday (1791–1867)

Acknowledgments

f the many people who deserve acknowledgment, I am particularly grateful to Jon Winokur, who has been helpful, entertaining, and encouraging from beginning to end. Reid Boates has also been tremendously supportive. I am grateful to my mother, Hannah Berman, who provided the drawings on pages 58, 68, 132, 197, 312, 397; my late aunt Beatrice Wachtler; Sharon Bronte; Fernando Tesón; Sandra Kitt and David Roth of the Hayden Planetarium; photo researcher Gillian Speeth; Ashton Applewhite, Pamela Dorman, Susan Elia, Amy Hill, and Paris Wald; and Bob Cornet, Michael Goth, Ava Guss, Margo Kaufman, Barry Kerner, Diana Rico, Bill Scharfman, RitaSue Siegal, Karen Shannon, and Frank Wechsler. Finally, although many scientists have been helpful, I especially wish to thank Dr. Michael Rich of Columbia University, who has generously shared his time and expertise. Any errors that remain are entirely mine.

✳

Contents

PART III
The Milky Way and Beyond

227

PART IV

An Album of Stars and Constellations

363

A Friendly Preface to the Friendly Guide

My eyes are full of star-dust
—Edmond Rostand,
Cyrano de Bergerac, Act 3

n astronomy, as in no other science, important observations and discoveries have been made time and again by curious amateurs. Hermann Goldschmidt, a nineteenth-century German painter living in Paris, took up astronomy as a way to combat depression and discovered fourteen asteroids. Will Hay, an English comedian and box-office star of the 1930s and 1940s, discovered the Great White Spot on Saturn. An officer in the Austrian army discovered Biela's Comet; a Dutch minister first observed the variable star Mira; a pharmacist detected the sunspot cycle. The astronomical tradition formed by these and many other amateurs continues to be upheld today by a community of backyard stargazers so active and knowledgeable that in 1986 NASA called upon them to suggest experiments for the much-beleaguered Hubble Space Telescope. Astronomy inspires something akin to love, which is why it can boast of popular magazines, sea cruises, weekend-long seminars at extension universities, consumer catalogues filled with things that glow in the dark, hundreds of clubs, and a multitude of enthralled fans. No other science—not even paleontology—can make a similar claim.

In February 1885, Simon Newcomb, a professor of astronomy and mathematics at Johns Hopkins University known for his insistence that no machine heavier than air could possibly fly, remarked upon this phenomenon. Astronomy, he wrote in *Harper's* magazine, "seems to have the strongest hold on minds which are not intimately acquainted with its work. The view taken by such minds is not distracted by the technical details which trouble the investigator, and its great outlines are seen through an atmosphere of sentiment, which softens out the algebraic formulae with which the astronomer is concerned into those magnificent conceptions of creation which are the delight of all minds, trained or untrained."

The attitude he describes is precisely mine. I have written this book in the spirit of delight, balancing that which is important and known against that which fills me with wonder, tickles my fancy, and makes me happy. If this sounds like an unscientific way to conduct a tour of the vast, violent, nuclear-powered beast that is the universe, so be it. This book is neither an observational guide nor a history nor an introduction to astronomy, and it is certainly not a textbook. It is a *friendly* guide, mixing science, history, and mythology (along with bits and pieces of art and literary history) according to a formula entirely my own.

Considering the topic, this book is not and cannot be all-encompassing. The universe is too big. In the solar system alone there are 9 planets and dozens of moons, not to mention trillions of comets in the icy reaches of the Oort cloud. On a clear dark night in the country, it is possible to see 1,500 stars with the naked eye. The ancient Greeks classified around 6,000. With the invention of the telescope in 1608, the number of visible stars grew exponentially. In the glittering Pleiades, a cluster of stars known as the Seven Sisters, the average observer can detect only 6 or 7 stars. But look at this tiny patch of sky through a portable telescope, even in the middle of a city, and dozens of stars appear; Tennyson described them as "a swarm of fireflies tangled in a silver braid." All told, thousands of stars have been counted in the Pleiades—and there are more. It's the same way everywhere you look. Thanks to Carl Sagan, the word "billions" has become a cliché, but it's impossible to talk about astronomy without it; in our galaxy alone, there are about 200 billion stars.

Astronomy is more than a description of the heavens. It is the history of humanity's attempt to understand on the largest possible scale, and as such it is fraught with genius and with error. Every astronomer whom we revere for important breakthroughs—William Herschel, for example, the professional musician who discovered Uranus—might also be mocked (unfairly) for theories that in retrospect seem addlebrained, such as Herschel's idea that the Sun was a solid body beneath whose fiery rim lived a race of beings not unlike ourselves.

As astronomy has evolved, our conception of the universe has been transformed, increasing in complexity, variety, and size. Less than a hundred years ago, the Milky Way Galaxy *was* the universe. Then a smudge in the constellation Andromeda was recognized as another galaxy—what Immanuel Kant called an "island universe." Today we know that galaxies number in the, yes, billions. And our knowledge continues to grow. Aside from any errors I may have made, elements in this book are surely already dated or proven incorrect; something unexpected has been discovered, something astounding has been proposed, and without question a few of the superlatives listed herein—the biggest, the farthest, the hottest, the brightest—have been demoted to runner-up.

When I was in college, there were two sorts of science courses: courses for those who wished to become scientists, and courses for the rest of us. This book is for the rest of us.

☆

PART I

A CHRONOLOGY:

Meditations of evolution increasingly vaster: of the moon invisible in incipient lunation, approaching perigee: of the infinite lattiginous scintillating uncondensed milky way . . . of our system plunging towards the constellation of Hercules: of the parallax or parallactic drift of socalled fixed stars, in reality evermoving wanderers from immeasurably remote eons to infinitely remote futures in comparison with which the years, threescore and ten, of allotted human life formed a parenthesis of infinitesimal brevity.

—James Joyce, *Ulysses*

AN ABBREVIATED HISTORY OF THE UNIVERSE

A Chronology:
An Abbreviated History
of the Universe

15–20 billion years ago: The universe begins with a bang. During the first incomprehensibly small fraction of a second, the universe is an infinitely dense, hot fireball; space and time are scrambled; and the laws of physics do not apply.

> **W**e think of the Big Bang as a fraction of time, but in this fraction the most incredible things may have happened; there may have flourished and disappeared the most evolved civilizations, which looked at us as at a distant future.
>
> —Tullio Regge

10^{-43} second—or 0.0000000000000000000000000000000000000000001 second—after the Big Bang: At this moment, known as Planck Time, gravity becomes a force in its own right. Physics as we know it begins.

10^{-36} second after the Big Bang: The universe inflates, doubling and redoubling many times over until it swells from the size of a proton to the size of a honeydew melon. The strong force, which holds together the nuclei of atoms, appears.

10^{-34} second after the Big Bang: The inflationary era ends. Quarks, leptons, photons, and neutrinos flood the universe.

10^{-10} second after the Big Bang: The electromagnetic and weak forces appear.

10^{-6} second after the Big Bang: Quarks slam into one another, forming protons and neutrons.

3 minutes after the Big Bang: Protons and neutrons form nuclei of hydrogen, helium, and lithium.

500,000 years after the Big Bang: Electrons swing into orbit around nuclei.

1 million years after the Big Bang: The universe becomes transparent. It is now possible to see.

300 million years after the Big Bang: Stars and galaxies begin to form.

1 billion years after the Big Bang: Quasars start to shine.

Many years pass.

4.5 billion years ago: The solar system is formed from a cloud of dust and gas.

4 billion years ago: The early atmosphere of Earth, zapped by lightning, produces amino acids, the building blocks of proteins and the foundation of life.

3.6 billion years ago: One-celled organisms are fruitful and multiply.

1 billion years ago: Worms, jellyfish, and algae flourish.

570 million years ago: Large numbers of creatures with hard skeletons suddenly appear in the fossil record, a burst of evolutionary activity known as the Cambrian Explosion.

360 million years ago: Vertebrates paddle onto land.

248 million years ago: Early dinosaurs roam the Earth.

65 million years ago: An asteroid slams into the northern edge of the Yucatán Peninsula in Mexico. Dinosaurs all around the world die.

35,000 years ago: Homo sapiens inherits the Earth.

4977 B.C.: Sunday, April 27. The Creation occurs, according to Johannes Kepler (1571–1630), court astrologer to the emperor Rudolf II and discoverer of the laws of motion.

4713 B.C.: January 1. Day 1, according to Joseph Justus Scaliger (1540–1609), who names his simple system after his abusive father, Julius Caesar Scaliger. January 1, 2000, is Julian Day 2,451,545.

4004 B.C.: October 22 (evening). The world is created, according to Irish archbishop James Ussher (1581–1656).

Johannes Hevelius (1611–87) and his wife.

Yerkes Observatory photograph, University of Chicago

3936 B.C.: October 24 (6:00 P.M.). The world is created, according to Polish astronomer Johannes Hevelius (1611–87).

3761 B.C.: The Jewish calendar begins.

3500 B.C.: The world is created, according to Isaac Newton (1642–1727).

2283 B.C.: Babylonian observers in the city of Ur record a lunar eclipse—the first such account on record.

2136 B.C.: October 22. Chinese astronomers Hsi and Ho, failing to predict the solar eclipse that occurred on this date, are executed by order of the emperor Chung K'ong.

1300 B.C.: A new star appears near Antares in Scorpius and is duly noted by Chinese astronomers, who record the event on an oracle bone.

c. 450 B.C.: Democritus and his teacher Leucippus propose that all matter is made of atoms. "Nothing exists except atoms and empty space; everything else is opinion," Democritus states.

352 B.C.: Chinese observers record a supernova.

270 B.C.: Aratus of Soli writes an influential poem, *Phaenomena,* describing forty-five constellations.

260 B.C.: Aristarchus of Samos suggests that the Earth revolves around the Sun.

240 B.C.: The summer solstice. Eratosthenes, vacationing in Syene (now Aswan), notes that at noon on this date the Sun is directly overhead and hence no shadows are cast, while at Alexandria, the Sun is a little south of the zenith, creating shadows. After measuring the angle of the Sun and the distance between the two cities, he applies the principles of high school mathematics and calculates the circumference of our planet. Although his figure is close to correct, his results are ignored by his contemporaries, who can't believe the Earth is that big.

134 B.C.: Hipparchus notices a star he has not seen before in Scorpius. It inspires him to chart the positions of about a thousand stars and to rank them according to brightness. His system, modified, is still in use.

46 B.C.: Julius Caesar revises the calendar, inserting a leap day every four years and, on a one-time-only basis, adding almost three months to the year, thus making 46 B.C. the longest year ever.

44 B.C.: March. A comet appearing shortly after the assassination of Julius Caesar is assumed to be his soul.

c. (A.D.) 100: Alexandrian astronomer Claudius Ptolemy, relying heavily on the work of Hipparchus, describes a geocentric model of the universe that holds sway for over 1,400 years.

150: Lucian of Samosata writes the first work of science fiction, *Icaromenippus,* in which our hero, Menippus, flies from Mount Olympus to the Moon and sees that the Earth is round.

497: Indian astronomer Aryabhata I suggests that the Earth rotates.

538: February 15. The *Anglo-Saxon Chronicle* describes a solar eclipse, the first so noted in Britain.

570: Isidorus, bishop of Seville, distinguishes between astronomy and astrology. Ever after, scientists will feel compelled to reiterate the point.

773: Astronomy comes to Baghdad when a visitor from India describes the prediction of eclipses to the caliph.

This woodcut, possibly by Albrecht Dürer, shows the eighth-century Jewish astronomer Mash'allah sitting beneath a starry dome as he measures the Earth.

827: Claudius Ptolemy's *Megale mathematike syntaxis* is translated into Arabic; it becomes known as the *Almagest*.

1006: May 1. Egyptian astrologer Ali ibn Ridwan observes a supernova in the constellation Lupus the Wolf, near Scorpius. Chinese and Japanese accounts confirm the sighting, as do the records of a Swiss monk at the monastery of Saint Gall. By the twentieth century, nothing but radio waves will remain of what is probably the most luminous supernova in recorded history.

1054: July. A new star in Taurus outshines Venus. Chinese, Japanese, and Native American observers record the supernova, but in Europe, as Nobel Prize winner Sheldon Glashow will point out, "People did not see the great supernova of 1054, for they were too busy arguing how many angels could dance on the head of a pin!" The remnants of this explosion are visible as the Crab Nebula.

1066: William the Conqueror, interpreting Halley's Comet as "a wonderful sign from heaven," defeats the Saxons in the Norman Conquest.

1175: The *Almagest,* written by Ptolemy in the second century A.D., is translated into Latin.

1229: Frederick Barbarossa's son returns to Italy from a Crusade with the first planetarium ever seen in Europe: an Arabic tent whose cupola-shaped roof is marked with constellations.

1252: Alfonso X, known as the Wise, is crowned king of León and Castile. He is known for overseeing the compilation of astronomical treatises, star catalogues, and tables of planetary motion, and for the possibly aprocryphal remark "If the Lord Almighty had consulted me before embarking on the Creation, I should have recommended something simpler."

1408: October 24. Chinese observers note a "guest star" in Cygnus the Swan. Over five centuries later, scientists will discover a powerful source of X-rays in Cygnus that they believe to be a black hole—the remnant of the supernova explosion seen on this date.

1433: Tatar astronomer Ulugh Beg publishes the first new star map since Hipparchus and establishes an observatory in Samarkand (the ruins will be found in 1908). He also banishes his son Abdallatif on the advice of an astrologer who predicts that his son will murder him.

1449: October 27. Ulugh Beg is assassinated by order of his son.

1456: Pope Calixtus III excommunicates Halley's Comet (or so the story goes).

1473: February 19. Nicolaus Copernicus is born in Prussian Poland.

1483: The Latin translation of the Alfonsine Tables of planetary positions, compiled in 1252 under the auspices of Alfonso X, is printed in Toledo, Spain.

Albrecht Dürer's *Northern Hemisphere of the Celestial Globe* (1515). Dürer's maps of the northern and southern constellations, drawn after someone else indicated the positions of the stars, were the first ever to be printed. Note the picture of Ptolemy (Ptolemaeus Aegyptius) in the upper right.

1543: May 24. Copernicus dies. On his deathbed, he receives the first copy of his book *De revolutionibus orbium coelestium,* which states that the Earth and the other planets revolve around the Sun.

1564: February 15. Galileo Galilei is born.

1566: December 29. The great observer Tycho Brahe and Manderup Parsbjerg duel to decide who is the superior mathematician. Parsbjerg proves himself the better swordsman by slicing off a portion of his opponent's nose. From then on, Tycho wears a metal prosthesis.

The Earth at the center of the universe in a page from a sixteenth-century mathematical book. Urania is the Muse of Astronomy.

1571: December 27. Johannes Kepler is born in Germany.

1572: Tycho Brahe sees a supernova in Cassiopeia.

1576: In an addendum to an English translation of Copernicus' *De revolutionibus,* Thomas Digges theorizes that space is infinite.

1596: August 13. Protestant minister David Fabricius discovers a variable star in the constellation Cetus the Whale. Half a century later, the astronomer Hevelius will name it Mira, the wonderful star.

1599: Shakespeare's tragedy *Julius Caesar* is performed for the first time. "When beggars die, there are no comets seen; The heavens themselves blaze forth the death of princes" (II.ii.30–32).

Galileo Galilei as pictured in his book *Siderus Nuncius* (1610).

1600: February 17. Giordano Bruno writes: "Innumerable suns exist; innumerable earths revolve about these suns in a manner similar to the way the seven planets revolve around our sun. Living beings inhabit these worlds." For these and other ideas, he is burned at the stake.

1604: October. Kepler and Galileo see a supernova in Ophiuchus.

1605: May 29. Kepler's wife forces him, "by her pesterings," to take a bath. He finds the experience unpleasant.

1608: October 2. Hans Lippershey, a Dutch maker of spectacles, applies for a license to manufacture telescopes. His request is denied.

1609: Galileo, having read about Lippershey's invention, makes his own telescope and demonstrates its military uses to the doge and the Senate of Venice. His salary is doubled.

1610: Galileo sees the moons of Jupiter, Saturn's rings, the phases of Venus, and the stars in the Milky Way. He publishes the news in *The Starry Messenger.*

1616: Copernicus' *De revolutionibus* is banned by the Roman Catholic Church. Galileo is warned not to defend the heliocentric doctrine.

1617: May 7. David Fabricius, who discovered the variable star Mira, is murdered by a member of his congregation.

1619: Kepler publishes *Harmonice mundi,* explaining his third law of motion.

1632: The pope is insulted by Galileo's *Dialogue Concerning the Two Chief World Systems.* Galileo is ordered to appear before the Inquisition.

1633: Galileo stands trial in Rome. He is put under house arrest, where he will remain for the rest of his life.

1635: Arcturus becomes the first star to be seen with a telescope during the day.

1642: January 8. Galileo Galilei dies.

December 25. Isaac Newton is born prematurely.

1643: Kepler's unfinished manuscript, *Somnium,* a description of a journey to the Moon accomplished with the assistance of heavenly spirits and opiates, is published posthumously.

1656: Christiaan Huygens recognizes that the odd structure on either side of Saturn is a continuous ring surrounding the planet.

1664: Robert Hooke sees the Great Red Spot of Jupiter.

1665: The bubonic plague strikes London. Cambridge University closes, forcing Isaac Newton to go home. During this plague year he realizes that the force that makes the apple fall from the tree and the force that keeps the Moon in orbit around the Earth are both obeying the inverse-square law of gravity. "All this was in the two plague years of 1665–1666," Newton will write. "For in those days I was in the prime of my age for invention and minded Mathematicks & Philosophy more than at any time since." At this same time, Samuel Pepys, later president of the Royal Society, takes a series of private lessons in order to learn division.

Danish astronomer Olaus Roemer's telescope. Roemer (1644–1710) made the first reasonable estimate of the speed of light by timing eclipses of the moons of Jupiter.

Yerkes Observatory photograph, University of Chicago

1672: Giovanni Domenico Cassini observes Mars at opposition and determines, with 90 percent accuracy, its distance from the Earth.

1675: Olaus Roemer establishes that light has a finite velocity. His estimate of 141,000 miles per second is about 76 percent of the actual value.

1682: November 22. Edmond Halley sees the comet that will later bear his name.

1684: Edmond Halley convinces Isaac Newton to write the *Principia,* describing the law of universal gravitation.

1718: Halley discovers that Sirius, Aldebaran, and Arcturus are no longer in the positions reported by Ptolemy and Hipparchus, thus proving that they move.

1752: Great Britain (and its colonies) abandon the Julian calendar. An act of Parliament authorizing the "New Style" calendar moves New Year's Day from March 25 to January 1 and drops eleven days from September 1752.

1755: Immanuel Kant suggests that the universe is filled with other island universes similar to the Milky Way.

1758: September 12. Charles Messier sees a patch of light—the Crab Nebula—in Taurus. It becomes the first object in the *Messier Catalogue.*

Christmas night. Saxon amateur astronomer Johann Georg Palitzsch sights a comet. A month later it will be identified as the comet whose appearance had been predicted by Edmond Halley.

1761: By observing the transit of Venus across the Sun, a phenomenon that happens twice in eight years and then not again for over a century, Russian chemist Mikhail Vasilyevich Lomonosov detects the atmosphere of Venus.

1766: Henry Cavendish discovers hydrogen.

1772: Bode's law predicting the positions of the planets, proposed by Johann Daniel Titius in 1766, is publicized by Johann Elert Bode.

1781: March 13. Thinking he has found a comet, William Herschel discovers Uranus.

1782: November 12. Deaf-mute astronomer John Goodricke notices that the star Algol in the constellation Perseus looks dimmer than usual. Later he will hypothesize correctly that it is regularly eclipsed by an unseen companion.

1783: English geologist John Michell, noted for his study of earth-quakes, suggests that if a star were big enough, its gravity would be so strong that light could not escape. The concept of a black hole is born, although it will not be given that name until the late 1960s, when the term is coined by John A. Wheeler of Princeton.

November 21. Two Frenchmen, riding in a wallpaper-and-linen hot-air balloon designed by the Montgolfier brothers, Joseph-Michel and Jacques-Etienne, ascend into the atmosphere from a garden in the Bois de Boulogne and fly for twenty-six minutes.

1800: William Herschel ascertains that the part of the Sun's spectrum that heats a thermometer the most is beyond the red—literally off the chart. Thus he discovers infrared radiation.

1801: January 1. Giuseppi Piazzi detects Ceres, the first known asteroid.

1814: Joseph von Fraunhofer charts the dark lines in the solar spectrum.

1835: A French gunboat is ordered to shoot down a hostile hot-air balloon. It turns out to be the planet Venus.

1838: April 22. Haxti, a young Sioux captured by the Pawnee, is painted red and black, branded, and shot through the heart as a ritual sacrifice to the Morning Star, identified as either Venus or Mars. Due to bad publicity, this is the last such sacrifice to be performed.

German astronomer Friedrich Wilhelm Bessel, applying the principle of parallax, calculates the distance to 61 Cygni, the first star to be so plotted.

1839: January 8. Thomas Henderson calculates the distance to Alpha Centauri, the closest star system.

1842: Christian Johann Doppler explains the wavelength velocity shift that bears his name. Dutch meteorologist Christoph Buys Ballot demonstrates it with an experiment in which a group of musicians sit by some railroad tracks and record what they hear as a band of trumpet players rides past them on a flatcar. As the trumpeters approach the listeners, the notes sound higher, and as they are pulled away, the notes sound lower.

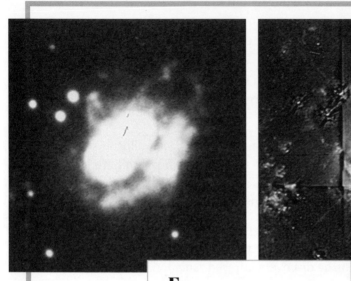

NASA

Eta Carinae and its surrounding nebula, ejected in 1843. A ground-based telescope provided the picture on the left. The more detailed image on the right was taken by the Hubble Space Telescope.

1843: April. Eta Carinae, recorded by Edmond Halley in 1677 as a fourth-magnitude star, brightens so markedly that it becomes the second brightest star in the sky. (By 1868 it will no longer be visible to the naked eye.)

1845: John Couch Adams calculates the position of an undiscovered planet beyond Uranus. In France, Urbain Jean Joseph Leverrier makes a similar prediction.

1846: Frédéric Petit announces that three Frenchmen have seen the Earth's second moon—a small body with a long elliptical orbit. (Seventy-five years later, Harvard professor William H. Pickering recommends looking for this body as "an opportunity for the amateur.")

September 23. Astronomers at the Berlin Observatory discover Neptune in its predicted place.

1847: October 1. In Nantucket, Maria Mitchell becomes the first person ever to discover a new comet not visible to the naked eye, thereby winning an ornate gold medal offered by the king of Denmark for just such an accomplishment.

1850: July 16. Vega becomes the first star to be photographed when a daguerreotype of it is made at the Harvard Observatory.

1862: Using the largest refractor telescope in the world, an instrument he had just constructed, Alvan Graham Clark discovers Sirius B, the white dwarf companion to the Dog Star.

1877: Asaph Hall discovers Deimos and Phobos, the moons of Mars.

Giovanni Schiaparelli discovers *canali* on the surface of Mars.

1878: Thomas Alva Edison sets up an infrared detector in a Wyoming chicken coop and prepares to watch a solar eclipse. The chickens disrupt his plans.

1879: March 14. Albert Einstein is born.

1885: August 31. Ernst Hartwig discovers a bright new star in the Andromeda Nebula. It later turns out to be a supernova.

1892: September 9. Edward Emerson Barnard discovers a fifth moon of Jupiter, which will be named Almalthea.

1895: November. Wilhelm Konrad Roentgen discovers a mysterious, penetrating form of radiation he calls X-rays.

1897: Joseph John Thomson discovers the electron, a particle Sir Arthur Eddington will later describe by saying, "Something unknown is doing we don't know what."

Admiral Robert Peary discovers a thirty-one-ton meteorite in Greenland. It is transported to the American Museum of Natural History in New York, where it becomes the largest meteorite on display in a museum.

1899: After spending a night in a hot-air balloon watching the Leonid meteor showers, Dorothea Klumpke becomes the first astronomer to make observations above the surface of the Earth.

1900: Max Planck hypothesizes that energy comes not in waves but in tiny bundles he calls quanta.

1901: June. Tycho Brahe's body is exhumed. Analysis of a green stain on his skull suggests that his metal nose, long thought to be silver or gold, was partially copper.

1902: William Thomson, Lord Kelvin, becomes convinced that Martians are using radio waves in an attempt to establish contact with New York City.

1903: December 17. Kitty Hawk, North Carolina. With Orville as pilot, the Wright *Flyer I* travels 852 feet through the air.

1905: Percival Lowell begins his search for a planet beyond Neptune.

Ejnar Hertzsprung classifies stars according to color and luminosity.

Albert Einstein, age twenty-six, presents the special theory of relativity, showing that $E = mc^2$.

1908: June 30. A huge fireball plows into the atmosphere above Siberia near the Tunguska River and explodes in midair, incinerating trees for miles around and killing 1,500 reindeer.

Percival Lowell publishes *Mars as the Abode of Life*.

1909: "I came in with Halley's Comet in 1835," Mark Twain reports. "It is coming again next year, and I expect to go out with it. It will be the greatest disappointment of my life if I don't."

1910: April 20. Halley's Comet reaches perihelion, its closest approach to the Sun.

April 21. Mark Twain dies.

1911: A meteorite from Mars kills a dog in Egypt.

1912: Henrietta Swan Leavitt discovers that the rate at which a Cepheid variable changes is related to its absolute magnitude. This makes it possible to determine distances beyond the Milky Way.

1914: Albert Einstein publishes his general theory of relativity, which postulates that gravitation and acceleration are equivalent and that light is bent by gravity.

1915: Australian astronomer R. T. Innes discovers a pipsqueak companion star to Alpha Centauri. Known as Proxima Centauri, the tiny red dwarf is the closest of all stars to planet Earth.

1916: E. E. Barnard discovers a nearby red dwarf moving faster than any star in the sky. Every 180 years, Barnard's "runaway star," as it is sometimes known, covers a distance in the sky equivalent to the width of the full moon.

1917: Willem de Sitter proves that according to the general theory of relativity, the universe must be expanding. Einstein dislikes the idea and adds the "cosmological constant" to halt the expansion.

1918: Harlow Shapley estimates the size of the Milky Way and determines that we are not in the middle.

June 8. The brightest nova in 300 years appears in the constellation Aquila the Eagle.

1919: March 29. Observation of stars during a solar eclipse shows that their light has been bent by gravity in accordance with Einstein's figures, thereby proving his theory. Asked how he would have reacted if the observations had indicated otherwise, Einstein responds, "I would have had to pity our dear Lord. The theory is correct."

June 8, 1918. A total eclipse of the sun.

Yerkes Observatory photograph, University of Chicago

1920: April 26. "The Great Debate" pits Heber D. Curtis against Harlow Shapley in a discussion about whether so-called spiral nebulae are within the Milky Way, as Shapley contends, or are distant galaxies, as Curtis believes.

Albert A. Michelson determines the diameter of Betelgeuse, the first star to be so measured.

1922: Alexander A. Friedmann uses Einstein's theories to show that the universe has been expanding and becoming less dense.

George Ellery Hale visits Egypt, watches the excavation of the tomb of King Tut, resigns as director of the Mount Wilson Observatory, and goes into seclusion.

1923: October 6. Edwin P. Hubble detects Cepheid variables in M31, the so-called Andromeda nebula. The discovery enables him to show that M31 is not a nebula at all but another galaxy entirely outside the Milky Way.

1925: The *Soviet Encyclopedia* declares that relativity is unacceptable in the light of dialectical materialism.

1927: May 20–21. Charles Lindbergh flies the *Spirit of St. Louis* across the Atlantic.

Georges Lemaître theorizes that the universe was once a dense concentration of matter and energy—a cosmic egg that exploded and has been expanding ever since. Twenty years later, Fred Hoyle will mock this idea with the term "Big Bang."

Werner Heisenberg presents his uncertainty principle, which states that you can know the location or the momentum of a particle, but not both. The more precise one measurement is, the more indeterminate the other.

1929: Edwin Hubble announces that the more distant a galaxy is, the faster it is slipping away from us. Hubble's law indicates that the universe is expanding.

1930: February. Clyde W. Tombaugh, age twenty-four, discovers Pluto.

The Harvard College Observatory

Although *The Observatory Pinafore* was written in the nineteenth century, it was not performed until members of the Harvard College Observatory staged a production on December 31, 1929. Cecilia Payne-Gaposchkin, the first woman to become a full professor and department chair at Harvard, is second from the left. Fourth from the left is Mildred Shapley, wearing a starry belt buckle; 878 Mildred, an asteroid discovered in 1916 by her father, Harlow Shapley, was named after her.

1931: Paul A. M. Dirac predicts the discovery of anti-matter.

1932: Carl Anderson discovers anti-matter.

Karl Jansky, studying radio interference for Bell Telephone Laboratories, discovers radio waves emanating from the center of the Milky Way, in Sagittarius.

1935: May 4. "Einstein Attacks Quantum Theory," reads a headline in *The New York Times.* He never does make peace with this theory.

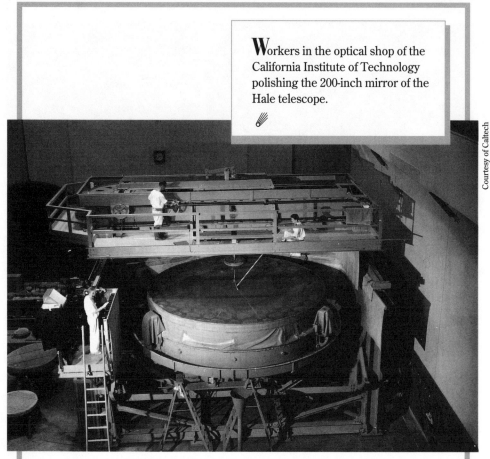

Workers in the optical shop of the California Institute of Technology polishing the 200-inch mirror of the Hale telescope.

Courtesy of Caltech

1936: April 10. After a two-week railway journey across the country during which enormous crowds cheer its progress, the two-hundred-inch glass disk of the Hale Telescope arrives in Pasadena, California.

1938: Hans Bethe and Carl Friedrich von Weizsäcker determine separately that the energy in stars comes from nuclear fusion.

1939: Ham radio operator Grote Reber builds the first radio telescope in his backyard in Wheaton, Illinois, and creates the first radio map of the galaxy.

August 2. A letter from Albert Einstein prompts Franklin Delano Roosevelt to authorize the Manhattan Project.

1940: October 1. Albert Einstein becomes a citizen of the United States.

1943: Carl Seyfert discovers extremely luminous active galaxies with bright, starlike nuclei.

1944: During the blackouts of World War II, Walter Baade, precluded from doing sensitive wartime work because of his German background, uses the hundred-inch Mount Wilson telescope to distinguish between young, blue, Population I stars found in the spiral arms of the Andromeda Galaxy, and older, redder, Population II stars in the nucleus.

1945: July 16. The first atom bomb is exploded near Alamogordo, New Mexico. J. Robert Oppenheimer quotes the *Bhagavad Gita:* "I am become Death, the destroyer of worlds."

1946: Cygnus A, a powerful source of radio waves, is discovered.

1947: February 12. An iron meteorite hits Siberia near Vladivostok, producing 106 craters and twenty-three tons of fragments.

June 24. Kenneth Arnold, piloting a plane near Mount Rainier, sees nine objects flying through the air. He calls them saucers. Members of the press, after a brief flirtation with the terms "flying disks" and "sky widgets," follow Arnold's lead and settle on "flying saucers."

October 14. Chuck Yeager, flying in a U.S. Bell X-1 rocket plane, exceeds the speed of sound.

December. The Hale Telescope, with its 200-inch mirror, sees first light on Mount Palomar. "I had never seen so many stars in my life," Byron Hill will recall in Richard Preston's *First Light.* "It was like pollen on a fish pond."

1948–49: Hermann Bondi, Thomas Gold, and Fred Hoyle set forth the Steady State theory, in which the universe has neither a beginning nor an end and matter is continually created.

1948: George Gamow and Ralph Alpher predict that the universe is filled with cosmic background radiation left over from the Big Bang.

1949: Fred L. Whipple likens comets to dirty snowballs made of ice and dust.

1950: Dutch astronomer Jan Hendrik Oort describes a spherical cloud of comets surrounding the solar system.

1951: The Roman Catholic Church throws its support behind the Big Bang.

1952: Walter Baade discovers two kinds of Cepheid variables. New calculations based on this discovery show that the galaxies are more distant than had been thought.

In a famous experiment, Stanley Miller and Harold Urey show that electricity—i.e., lightning—striking a mixture of compounds similar to the early atmosphere of Earth creates organic molecules.

1956: The neutrino is discovered.

1957: October 4. The Soviet Union launches a small metal sphere called *Sputnik 1* into orbit around the Earth. The Space Age begins.

November. The USSR sends a mutt named Laika into space and ten days later, puts it to sleep, making Laika the first—but not the last—dog to die in orbit.

1958: March 17. *Vanguard 1* is launched into orbit by the United States.

1959: Soviet spacecraft *Luna 1* misses its target, goes into orbit around the Sun, and is dubbed the first artificial planet; *Luna 2* crash-lands; *Luna 3* fulfills an age-old dream and photographs the far side of the Moon.

1960: Frank D. Drake initiates Project Ozma. Named after the princess in L. Frank Baum's *Wizard of Oz* books, its purpose is to scan two nearby stars in search of radio signals that might indicate conscious intelligence. The search fails.

1961: April 12. Yuri Gagarin, a twenty-seven-year-old Russian pilot, completes one orbit of the Earth, thereby becoming the first human in space and the first person ever to see the Sun rise twice in two hours.

May 5. Alan B. Shepard, Jr., becomes the first American in space. His pay for the fifteen-minute flight totals $14.38.

November. A chimpanzee named Enos orbits the Earth twice.

1962: February 20. John Glenn becomes the first American to orbit the Earth. "The view is tremendous," he states.

1963: June. Former textile worker Valentina Tereshkova becomes the first woman in space. "It is our girl who is first in space," says Soviet premier Nikita Khrushchev. In 1967, she will be elected to the Supreme Soviet.

December. Maarten Schmidt realizes that the spectral lines of quasars reveal an extraordinarily high redshift, indicating that they are receding rapidly and hence must be far away. Quasars prove to be the most distant objects in the universe—beacons from the edge of time.

1964: X-rays are discovered in the Crab Nebula.

Murray Gell-Mann and George Zweig suggest separately that protons and neutrons are formed from even smaller components. Zweig recommends calling these tiny particles "aces." Gell-Mann thinks of them as "kworks." They receive their name after Gell-Mann thumbs through a copy of James Joyce's *Finnegans Wake* and comes upon the line "Three quarks for Muster Mark!"

1965: Astronomers at the Arecibo Observatory in Puerto Rico discover that Venus rotates backward, with the Sun rising in the west.

Arno Penzias and Robert Woodrow Wilson, radio astronomers for Bell Laboratories, detect a low hiss streaming from all parts of the sky. It is the background radiation predicted by George Gamow and, later, Robert Dicke, and it is considered a major confirmation of the Big Bang theory.

August. NASA introduces Tang to a grateful world.

1966: September 8. *Star Trek* premieres on nationwide TV, ranking fifty-second out of fifty-four shows.

November 17. The Leonid meteors flash across the sky in the western United States at rates approaching 150,000 shooting stars an hour.

1967: January 27. Astronauts Virgil I. "Gus" Grissom, Edward H. White II, and Roger B. Chaffee die aboard the *Apollo* space capsule during a ground-based fire at Cape Canaveral.

Graduate student Jocelyn Bell discovers pulsars.

An underground vat containing 100,000 gallons of cleaning fluid is set up in a South Dakota mine to capture solar neutrinos.

Neil Armstrong took this picture of Buzz Aldrin on the Moon.

NASA

1969: July 20. Neil Armstrong becomes the first human to walk on the Moon. Armstrong and Edwin "Buzz" Aldrin, Jr.—whose mother's maiden name was Marian Moon—take pictures, collect moon rocks, eat cookies, and leave behind a plaque signed by President Richard M. Nixon.

A satellite designed to detect unauthorized Soviet thermonuclear tests conducted in space discovers instead the first evidence of gamma ray bursts—mysterious, violent explosions of unknown origin scattered randomly across the cosmos.

1970: Bennett's Comet, visible in April and May, is taken by Arab observers to be an Israeli weapon.

December 12. On the seventh anniversary of Kenya's independence, NASA goes to that country to launch the X-ray satellite *Uhuru* (Swahili for "Freedom").

1971: June 6. "You look down there and you get homesick," says Soviet cosmonaut Vladislav Volkov after twenty-three days in space. "You want some sunshine, fresh air, you want to wander in the woods." When air leaks from their capsule during reentry, Volkov and two other cosmonauts die.

1972: December. Eugene Cernan becomes the last man—for now—to walk on the Moon.

1973: November. Photographs of Jupiter, taken by *Pioneer 10,* appear a slice at a time over television sets at the Ames Research Center in California, which is later awarded an Emmy by the National Academy of Television Arts and Sciences.

1973: December 28. Comet Kohoutek reaches perihelion. Although scientists had predicted a dramatic appearance and publicity surrounding the event is enormous, the comet is barely visible.

1974: *Pioneer 11* passes Jupiter, receiving a gravitational boost toward Saturn.

November 16. A three-minute message leaves the radio telescope in Arecibo, Puerto Rico, on its way to a globular cluster in Hercules, where it is expected to arrive in 24,000 years.

1975: June. A shower of meteorites batters the Moon.

July 17. About 140 miles above the Earth, American astronauts present Soviet cosmonauts with a recording in Russian of a Conway Twitty song.

Two pictures of Venus are sent to Earth by Russian space probes.

1976: March. Comet West makes a spectacular, tail-splitting appearance, but hardly anyone sees it because the media, stung by Kohoutek, give it no publicity.

1976: The first space shuttle is named *Enterprise* after *Star Trek* fans deluge President Gerald R. Ford with letters.

1977: March 10. James Elliot and his colleagues, sighting Uranus from NASA's flying observatory, discover rings around the planet.

August 20. *Voyager 2* is launched on a grand tour of the solar system to include Jupiter, Saturn, Uranus, and Neptune. *Voyager 1* follows in September, with planned encounters at Saturn and Jupiter.

1978: June 22. Charon, moon of Pluto, is discovered by James W. Christy.

An unexplained explosion over the southern Atlantic Ocean is thought to be a nuclear test performed by South Africa and Israel. Fourteen years later, scientists will decide that a small asteroid may have been to blame.

NASA's SETI—Search for Extraterrestrial Intelligence—program receives a Golden Fleece award from Senator William Proxmire.

1979: January 21. Pluto crosses inside the orbit of Neptune. For twenty years, Neptune, not Pluto, will be the outermost planet.

March 5. *Voyager 1* discovers the rings of Jupiter. "Our sense of novelty could not have been greater had we explored a different solar system," a member of the scientific team later states.

September 1. *Pioneer 11* is the first spacecraft to reach Saturn.

December 6–7. On this night, Alan Guth calculates the evolution of the universe and concludes that in the earliest moments after the Big Bang, the universe inflated wildly at a rate that exceeded that of the Big Bang itself.

NASA

Voyager 1 was 20 million miles away from Jupiter when it took this picture on February 1, 1979.

1981: April 12. The first space shuttle is launched.

July. Senator William Proxmire adds an amendment to NASA's appropriations budget that forbids spending money on SETI.

Robert Kirshner and colleagues discover a great void 250 million light-years wide in the constellation Boötes.

1982: Following a meeting with Carl Sagan, Senator Proxmire withdraws his objections to SETI funding. On October 1, Congress approves funding for NASA's attempts to detect radio signals from other worlds.

October 16. Almost four years before its reappearance, Halley's Comet is photographed. It is over a billion miles from the Sun.

1983: IRAS, a satellite designed to pick up infrared radiation, discovers a protostar in Perseus so young it has not yet begun to emit visible light.

1984: A partial ring system is detected around Neptune.

A disk of gas and dust that looks like a solar system in the making is discovered around the star Beta Pictoris.

1986: January 24. *Voyager 2* reaches Uranus and continues on to Neptune.

January 28. The Space shuttle *Challenger* explodes seventy-three seconds after liftoff, killing astronauts Francis R. Scobee, Michael Smith, Judith Resnik, Ellison S. Onizuka, Gregory B. Jarvis, Ronald E. McNair, and teacher-in-space Christa McAuliffe.

August. Egyptian astronomer Nahed Youssef announces that there are nine atoms of gold for every trillion atoms of hydrogen in the Sun. Thus, the Sun contains 10,000,000,000,000,000 (10 quadrillion) tons of gold.

1987: February 23. Sanduleak 69°202, a blue supergiant in the Large Magellanic Cloud, explodes, becoming the first supernova in 383 years visible to the naked eye.

1988: Quasars are discovered at a distance of about 17 billion light-years.

November 15. The radio telescope at Green Bank, West Virginia, where almost three decades earlier Frank Drake began a search for extraterrestrials, collapses.

1989: May 4. The *Magellan* spacecraft is launched on its journey to Venus, where it will use radar signals to map the surface of the planet.

August 24. *Voyager 2* passes Neptune.

September 12. Pluto swings as close to the Sun as it ever gets.

November 17. Margaret Geller and John P. Huchra announce the discovery of a huge concentration of galaxies known as the Great Wall.

November 18. NASA's Cosmic Background Explorer (COBE) satellite is launched.

1990: April 24. The twelve-ton Hubble Space Telescope, equipped with a 94-inch mirror, is sent into orbit from the space shuttle *Discovery.* Within two months, a flaw in the mirror will become apparent.

The bright white spot in the center of this nebula is one of the hottest stars on record. Its picture was captured by the Hubble Space Telescope Planetary Camera and improved by computer image restoration.

S. Heap, NASA/Goddard Space Flight Center

September. "Let's not mince words: Hubble truly is crippled, and years of planning for HST operations had to be thrown out," writes planetary scientist Clark R. Chapman.

December 5. The first photograph taken with Hawaii's Keck Telescope, which will be the world's largest once its thirty-six hexagonal mirrors are in place, is displayed in the *Los Angeles Times*. It shows spiral galaxy NGC 1232 in Eridanus, 65 million light-years away.

December 7. A computer on the space shuttle *Columbia* overheats and fails. The failure is blamed on lint from the astronauts' uniforms—a problem, according to a representative from the Johnson Space Center, that has plagued the shuttle program since its inception.

December 7. *Galileo* becomes the first interplanetary spacecraft ever to visit Earth when it flies from Venus back to its home planet. Somewhere over the Bermuda Triangle, it receives a gravitational boost which helps propel it toward Jupiter.

1991: February 7. Eight hundred miles south of Buenos Aires, debris from *Salyut 7* crashes through the atmosphere. Some of it lands in a garbage dump. Another part, according to two tramps, splashes into the Atlantic Ocean.

April 5. The Compton Gamma Ray Observatory (GRO) satellite is launched.

May. *Life* magazine features on its cover a picture of Mars and the headline "Our Next Home."

July 11. A total eclipse of the Sun, visible from Hawaii and Mexico, is seen by more tourists than any other eclipse in history.

October. The *Galileo* spacecraft—impaired by an inoperable antenna—flies past the asteroid Gaspra, revealing craters, grooves, ridges, indentations, and clear evidence that Gaspra was once part of a larger body.

December. Discovery of a vast blob of hydrogen near the edge of the observable universe supports the idea that galaxies are formed "top down" from clouds of dust and gas that collapse into giant pancakes, which, in turn, form galaxies.

1992: January. Noting that every few hundred thousand years the Earth is hit by an asteroid large enough to wipe out the species, Edward Teller and other scientists at the Los Alamos National Laboratory discuss creating a fleet of missiles armed with nuclear weapons whose first-strike role would be to get it before it gets us.

April. A star in the Large Magellanic Cloud is photographed by the Hubble Space Telescope. At 360,000°F, it is the hottest star on record.

April 24. The Big Bang theory receives major confirmation when George Smoot announces that the Cosmic Background Explorer (COBE) satellite has detected minute temperature fluctuations in the background radiation. These ripples indicate enough variation in the early universe to give rise to structure. This addresses a major criticism of the Big Bang theory, which featured a missing episode between the seemingly smooth background radiation of the early universe and the lumpy universe of today. The discovery is hailed by Michael Turner as "the Holy Grail of cosmology."

May. On the 100th space walk of the Space Age, astronauts aboard the space shuttle *Endeavour* save a communications satellite in a too-low orbit. After hardware designed for the purpose fails to work, they capture the 4.5-ton satellite with their thinly gloved hands and wrestle it into position.

September 16. The International Astronomical Union announces the discovery of a reddish object with a 120-mile diameter orbiting the Sun beyond Pluto. This is the first evidence of the existence of the Kuiper Belt, a band of minor planets thought to be the source of short-period comets.

September 25. NASA's Mars Observer spacecraft is launched. Its goal: a detailed study of the atmosphere and surface of the red planet.

October 31. Pope John Paul II announces that the Roman Catholic Church erred in condemning Galileo for his belief that the Earth revolved around the Sun.

1993: January 31. Two hundred and fifty miles above Earth, the Gamma Ray Observatory (GRO) registers the brightest burst of gamma rays ever detected, now known as the Super Bowl Burst.

March 28. A star explodes in the M81 galaxy in Ursa Major. Although not visible with the naked eye, the supernova is the brightest one in the Northern Hemisphere since 1937.

May 26. Scientists announce that *Voyagers 1* and *2,* respectively 4.9 billion miles and 3.7 billion miles from Earth, have detected intense low-frequency radio waves thought to come from the edge of the solar system, where the hot solar wind slams into the cold, interstellar wind of outer space. This boundary, known as the heliopause, is estimated to be between 7.6 and 12 billion miles from the Sun.

August 21. Three days away from its planned entry into orbit around the red planet, Mars Observer falls ominously silent, causing astronomers to recall the Great Galactic Ghoul, often blamed for the failure of missions in space.

December. Astronauts aboard the space shuttle *Endeavour* succeed in mission to correct the defective vision of the Hubble Space Telescope.

1994: July 20. Comet Shoemaker-Levy 9, a fragmented comet with so many nuclei it is often compared to a string of pearls, crashes into Jupiter.

1995: December 7. *Galileo* arrives at Jupiter.

1999: March 14. Pluto crosses over Neptune's orbit and becomes once again the most distant planet in the solar system. It will remain there for 228 years.

August 11. A solar eclipse sweeps across Europe.

November 15. Mercury's rare transit across the Sun is visible from the west coasts of North and South America.

2004: June 8. Venus travels across the face of the Sun for the first time since 1882, when, according to astronomer William Harkness, "that wondrous scientific activity which has led to our present advanced knowledge was just beginning. What will be the state of science when the next transit season arrives God only knows."

2012: June 5. For the second time in eight years, Venus transits across the Sun.

2045: August 12. A solar eclipse is visible in the United States on a path that extends from coast to coast.

2061: Halley's Comet returns, looking far more dramatic than it did in 1986.

2126: August 14. Comet Swift-Tuttle, first seen in 1862 during the Civil War, has a 1-in-10,000 chance of hitting the Earth on this day and obliterating our species.

2137: Halley's Comet puts on the best show in several thousand years. With luck, someone is here to see it.

c. 2500: Because the Earth wobbles like a top, its shifting axis is now oriented so that on the vernal equinox, when the Sun crosses the celestial equator and the night and the day are equal, the Sun appears against the background stars of the constellation Aquarius, the age of which now begins.

1.5–2 billion years from now: The Sun heats up, becoming 15 percent more luminous. The weather in Iceland becomes temperate.

4–5 billion years from now: The Sun exhausts its hydrogen and balloons into a red giant. The Earth's atmosphere evaporates, the oceans boil away, and our planet starts to spiral into the Sun.

1 trillion (10^{12}) years after the Big Bang: Hydrogen and helium are used up. Stars and galaxies die. Even neutron stars and white dwarfs wink out. Only black holes remain.

10^{27} years after the Big Bang: The galaxy is a black hole.

10^{31} years after the Big Bang: Supergalactic black holes form.

10^{36} years after the Big Bang: "I believe there are 15,747,724,136,275,002,577, 605,653,961,181,555,468, 044,717,914,527,116,709, 366,231,425,076,185,631, 031,296 protons in the universe, and the same number of electrons," said Sir Arthur Eddington (1882–1944). Around this time, they all decay.

10^{67} years after the Big Bang: Ordinary black holes disintegrate.

10^{97} years after the Big Bang: Galactic black holes disappear.

10^{106} years after the Big Bang: Supergalactic black holes evaporate. Nothing is left but a thin wash of radiation and, every once in a while, a lonely particle.

THE BIG CRUNCH

• ➵

What if there's a lot more matter than scientists think? In that case, the universe will eventually contract. Galaxies will stream toward each other. Redshifts, whose long wavelengths indicate that a source of light, be it a galaxy or a quasar, is receding, will change to blueshifts, signaling approach. The temperature of the background radiation will rise.

* 1 billion years before the Big Crunch: Galaxy clusters merge.

* 100 million years before the Big Crunch: Galaxies merge.

* 70 million years before the Big Crunch: Stars are so close together that on Earth the sky is never dark.

* 1 million years before the Big Crunch: The temperature of the background radiation rises to thousands of degrees Fahrenheit.

* 3 weeks before the Big Crunch: Temperature rises into the millions of degrees. Matter dissolves.

* 3 minutes before the Big Crunch: Temperature rises into the billions of degrees. Atomic nuclei break up.

* The Big Crunch: Matter and radiation are reduced to a point of infinite density. Space and time cease to exist, and the laws of physics no longer apply. Does it start all over again with a Big Bounce? Maybe.

PART II

THE

Gasballs spinning about, crossing each other, passing. Same old dingdong always. Gas: then solid: then world: then cold: then dead shell drifting around, frozen rock, like that pineapple rock. The moon. Must be a new moon out, she said. I believe there is.

—James Joyce, *Ulysses*

SOLAR SYSTEM

Crystal Balls
and Other Spheres:
The Cosmology of the
Ancient Greeks

he philosopher Thales (624–546 B.C.), considered the first Greek scientist, thought that the Earth was essentially flat, an island riding on an infinite sea. His student Anaximander (610–547 B.C.) disagreed, believing instead that the Earth was cylindrical, curved on top and surrounded by a hollow sphere of stars. The cylindrical image of the world never attracted a huge following. But the notion of spheres took hold, and for almost 2,000 years it infected cosmology.

Sometimes conceived of as metaphorical devices and sometimes as solid objects made of crystal, spheres were intended to hold heavenly objects in place and to describe the complicated choreography of the skies. But because the planets meander across the heavens and upon occasion seem to stop and reel backward, a single celestial sphere was not enough.

So Anaximander's sphere soon multiplied. The mathematician Pythagoras (c. 582–c. 497 B.C.), who is credited with suggesting that the Earth is round, added spheres for the Sun, the Moon, and the planets. In the fourth century B.C., Plato's student Eudoxus of Cnidus increased the number to 26 nested, rotating spheres; Eudoxus' student Callippus upped the ante to 34; and Plato's most famous student, Aristotle

(384–322 B.C.), who believed that the Earth was imperfect and ever-changing while the heavenly bodies were eternal and without flaw, amended the model to include 54 crystalline spheres, which Aristotle evidently accepted as real, not metaphorical.

This system, based on two incorrect assumptions (that the planets revolve around the Earth, and that their orbital paths are circular rather than elliptical), was by now thoroughly unworkable. Consequently, Hipparchus (c. 190–120 B.C.) reduced the number of large spheres to seven (for the Sun, the Moon, Mercury, Venus, Mars, Jupiter, and Saturn) and added smaller spheres called epicycles, which veered off the main orbital paths to form little loops. The large spheres, called deferents, revolved around an imaginary point that revolved around the Earth. The little epicycles were invented to account for the troubling backward motion of the planets.

All told, the geocentric mechanism created by Hipparchus was an awkward-looking machine. Its purpose was to "save the appearances," to make theory and reality agree, and in this it partially succeeded. Astronomers could make reasonable predictions about planetary positions using this model, even though it was unbearably complicated and intrinsically wrong. Yet because the stars really do look as if they're moving around the Earth, the Hipparchan system maintained dominance for centuries.

Not that other proposals didn't come along from time to time. Philolaus (c. 480 B.C.–?) argued that the Earth moved, but because he was a member of the persecuted Pythagorean sect, which believed in the mystical primacy of numbers, his insight was dismissed. In his nearly heliocentric system, nine known celestial bodies—the six planets, the Sun, the Moon, and the starry sphere—revolved in circular orbits around an unseen central fire of which the Sun was a mere reflection. In addition, because Philolaus believed ten to be a magical number of wholeness, he added a tenth body—a parallel world, opposite the Earth and hence invisible to us.

Two centuries later, Aristarchus of Samos calculated that the Sun was larger than the Earth and concluded that the Earth, as the smaller body, probably revolved around the Sun. For a hundred years, astronomers grappled with this concept before ultimately discarding it.

By the second century A.D., the Earth-centered universe was dogma. Its catechism was written between 140 and 149 by Claudius Ptolemy, about whom so little is known that it is uncertain whether he was Greek

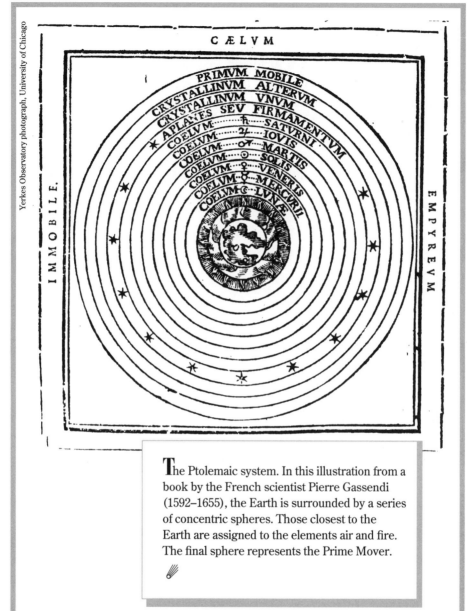

The Ptolemaic system. In this illustration from a book by the French scientist Pierre Gassendi (1592–1655), the Earth is surrounded by a series of concentric spheres. Those closest to the Earth are assigned to the elements air and fire. The final sphere represents the Prime Mover.

or Egyptian. His model of the universe—so similar to that of Hipparchus that Ptolemy has been accused of plagiarism—was a swaying, off-center, Rube Goldberg contraption. His thirteen-volume work, which included a star catalogue and a section on trigonometry, was known as the *Megale mathematike syntaxis* (*Great Mathematical Composition*), or the *Megiste* (*Greatest*) for short. Its importance can hardly be overestimated, for it paralyzed cosmology for almost 1,400 years. In the first seven centuries following its composition, the library at Alexandria was destroyed, the Roman Empire fell, Plato's Academy, in existence for over 900 years, closed forever, Muhammad went to Mecca, Charlemagne was crowned Roman emperor by the pope, and, still, the Sun was thought to revolve around the Earth. In 827, Ptolemy's book was translated into Arabic and became known as the *Almagest.* For another 700 years after its translation, its authority went essentially unchallenged.

And then, in 1543, another book was published. Its author was Nicolaus Copernicus.

❖

Copernicus and the
Center of the Universe

A fool who went against the Holy Writ.

—Martin Luther

eginning virtually at the moment of his death, Nicolaus Copernicus (1473–1543) revolutionized astronomy with his book, *De revolutionibus orbium coelestium* (*Concerning the Revolution of the Heavenly Spheres*), which stated that the Sun, not the Earth, was the center of the universe. Yet he was a reluctant prophet, for he was withdrawn and secretive, an essentially conservative man who worked most of his life as a church canon.

The son of a copper merchant who died when Nicolaus was ten years old, Copernicus was adopted by his uncle Lucas Waczenrode, who made certain that the shy boy and his rakish older brother received an education. Copernicus (originally Mikolaj Kopernik—he latinized his name, as was the fashion) studied mathematics and art at the University of Cracow, astronomy in Bologna, medicine at the University of Padua (a three-year course of study), and canon law in Ferrara. By the time he finished his education in 1506, his uncle—a man who reputedly never laughed—was the bishop of Ermeland, a politically volatile region in East Prussia. Copernicus, who had already been appointed canon to the Catholic church, became his uncle's assistant as well as his personal physician. He devoted himself to issues of currency reform and worked to maintain Ermeland's independence from its powerful neighbors, Poland and the Knights of the Teutonic Order, but he had few friends and never married. Late in his life, it became known that he was romantically involved

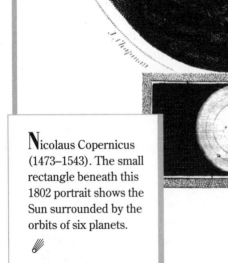

Nicolaus Copernicus (1473–1543). The small rectangle beneath this 1802 portrait shows the Sun surrounded by the orbits of six planets.

with his housekeeper, Anna; church authorities asked him to end the affair and he complied.

But if his personal and public life involved upholding the status quo, his intellectual life was aimed at overthrowing it. At the start of his career as church canon, he spent many hours contemplating the geocentric Ptolemaic system, which seemed inadequate to him because it required complicated explanations for ordinary phenomena such as retrograde motion or the constant proximity of Mercury and Venus to the Sun. It occurred to Copernicus that if the Earth was indeed the center of the system, no planet should ever travel backward. Similarly, if Venus and Mercury traveled around the Earth, they should occasionally move away from the Sun—which they never do.

On the other hand, if one used the Sun-centered system suggested by Aristarchus of Samos, these inherent difficulties were easily overcome. Venus and Mercury would appear close to the Sun because they actually *are* close to the Sun. The planets would occasionally seem to swim backward because the Earth would sometimes overtake them on its endless lap around the Sun. All this made obvious sense to Copernicus, but he kept it to himself.

Then, in 1512, after Copernicus and his uncle attended the wedding reception of the king of Poland in Cracow, Waczenrode came down with a bad case of food poisoning and died. (The death was so sudden that murder was considered a possibility.) Copernicus moved to Frauenburg, where he took up duties as canon of the cathedral and established lifelong residence in a square crenellated tower overlooking a lake whose waters flowed into the Baltic Sea. There he wrote a brief summary of his ideas, explaining that the Sun is the center of the universe, that the Earth rotates on its axis and revolves around the Sun, and that this motion accounts for the retrograde motion of the planets.

Although this revolutionary treatise was only privately circulated, the news got around. For three decades after his theory's appearance, Copernicus neither published nor taught, yet his system was talked about wherever astronomers congregated.

Copernicus did not participate in these discussions. He did, however, refine his theory. In the margins of books he was reading, he often made astronomical jottings, along with notes regarding cures for toothache, kidney stones, corns, and rabies—ailments whose medications involved ingredients such as cinnamon, dittany, rust, pearls, bone from a deer's heart, and unicorn horn. He worked on new tables of planetary motion, and he wrote at length. But like many other writers, he kept his manuscript in his room. His inclination—based on his retiring nature, his awareness that his theory would stir up ecclesiastical controversy, and perhaps his sympathy with the Pythagorean cult of secrecy—was never to publish.

He probably never would have, either, except that toward the end of his life he fortuitously and unexpectedly acquired a disciple, a young professor of mathematics and astronomy who came to Frauenburg to study with the great man. Georg Joachim Iserin, known as Rheticus (he took the Latin name to avoid being linked to his father, a physician who was beheaded for sorcery), immediately urged Copernicus to publish. This threw Copernicus into a tizzy of indecision. Anxious because his theory ran counter to the accepted wisdom of the day, Copernicus nevertheless *did* want to publish his tables of planetary motion—he just didn't want to mention the theory behind them. When he took Rheticus, a Lutheran, to the home of his one true friend, Tiedemann Giese, bishop of a neighboring diocese, his friend and his disciple tried to convince him of the importance of publishing both his tables *and* his views. Finally, a compromise was reached; Rheticus would write a book in which he would explain Copernicus' ideas but refer to their author merely by his first name and his birthplace.

Rheticus thus wrote a "letter" to one of his teachers in which he described the theory of "the Reverend Father, Dr. Nicolaus of Torun, Canon of Ermeland." He had the letter, which included some astrological and biblical commentary of his own, printed up and sent to a few people. With the theory now out in the open, pressure on Copernicus to publish his findings in full increased. At last, he relented.

Rheticus did the work, meticulously copying (and making minor corrections to) Copernicus' bulky manuscript. When he was done, he began the process of printing the book, but having already spent two years away from his university, he set it aside in order to take up his professorial duties again. He returned to the University of Wittenberg and was promptly elected dean. When the term ended in May 1542, he traveled to Nuremberg, manuscript in hand, to complete the task.

Shortly thereafter, Rheticus got a new job at the University of Leipzig and dropped the project. He may have felt estranged from Copernicus, for in the acknowledgments to the book, which Rheticus surely saw, Copernicus omitted all mention of the man who had helped him the

THE RIDDLE OF RETROGRADE

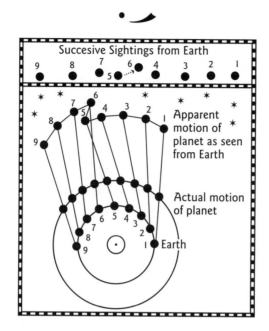

Succesive Sightings from Earth

Apparent motion of planet as seen from Earth

Actual motion of planet

Earth

Of all the astronomical anomalies that confused pre-Copernican observers, the most puzzling was the way the planets, which usually travel across the sky from west to east, occasionally seem to reverse direction. Early astronomers thought that this apparently backward, or retrograde, motion was real, and they invented complicated schemes to account for it.

Copernicus showed that the backward motion of the planets is an illusion. It occurs because the planets revolve around the Sun at varying distances. As a result, every so often the Earth "overtakes" a more distant planet, which then appears to be going backward, just as a local train chugging forward into the country nonetheless seems to be sliding back into the city if the express train you're riding on, on an adjacent track, is moving faster.

Similarly, an interior, faster-moving planet, such as Mercury, which runs around a shorter track, may look as if it's moving backward because, during a single Earth year, Mercury races around the Sun several times. Thus its direction seems to change repeatedly as it overtakes the Earth and then moves away from us. In fact, its direction is constant.

most. So Rheticus handed responsibility for the printing over to someone else.

Enter Andreas Osiander. A Lutheran priest, he had suggested two years earlier that if Copernicus were to publish the book, he would be wise to say that the hypotheses therein were not "articles of faith" but merely computational devices. By making this disclaimer, Osiander thought, Copernicus would deflect criticism from "the Aristotelians and the theologians whose contradictions you fear." With this notion still in mind, Osiander took it upon himself to protect Copernicus by adding an equivocating preface, famous in the history of astronomy, that discounted the importance of the book. "These hypotheses need not be true nor even probable; if they provide a calculus consistent with the observations, that alone is sufficient," Osiander wrote. "So far as hypotheses are concerned, let no one expect anything certain from astronomy, which cannot furnish it, lest he accept as the truth ideas conceived for another purpose, and depart from this study a greater fool than when he entered it. Farewell." The unsigned preface, which everyone assumed Copernicus had written, cast doubt upon the book's ideas by implying that even their author didn't believe them.

It took a year to finish printing the book, during which time Copernicus had a stroke and was partly paralyzed. The first printed copy of the book, which was dedicated to the pope, arrived at Frauenburg Castle on May 24, 1543.

Later that day, Copernicus died.

His system prevailed. It was not the elegant model we imagine it to be, because Copernicus, like Aristotle, was wedded to the idea of the perfect, circular orbit. (Only after Johannes Kepler announced that orbits are elliptical did a truly accurate system become possible.) As a result, some details needed adjustment. Copernicus did what everyone else had done: he added epicycles, wheels within wheels upon which various bodies revolved. Nine different wheels, for example, accounted for the various motions of the Earth. In addition, they revolved, as they did for Philolaus, not around the Sun but around a point close to the Sun. And according to Kepler, who uncovered Osiander's authorship of the infamous preface, Copernicus did not even discard the notion that the heavenly bodies might be embedded in solid crystal. In short, the Copernican system was a mess.

It didn't matter, just as it didn't matter that the entire Christian world officially rejected the thesis. Martin Luther called Copernicus an "upstart astrologer" and complained, "This fool wishes to reverse the entire science of astronomy." Luther was right. During Galileo's lifetime, the pope put Copernicus' book on the Index of Prohibited Books (where it stayed until 1835, the year Charles Darwin sailed to the Galápagos Islands aboard the *Beagle*). Nonetheless, the Copernican system, unlike the Ptolemaic system, was grounded in reality. The primacy of the Earth had been forever shaken. The Sun was king.

⊙

Immanuel Kant and the Birth of the Solar System

Two things fill the mind with ever-increasing wonder and awe . . .: the starry heavens above me and the moral law within me.

—Immanuel Kant

 nown primarily as a moral philosopher, Immanuel Kant (1724–1804) was a lifelong bachelor famous for the rigidity of his habits. According to a much-repeated story, Kant took his afternoon walk up and down the same path so punctually that his neighbors in the Prussian city of Könisgsberg, where he spent his entire life, could determine the time by the moment of his passing. Only once, when he became engrossed in reading Jean Jacques Rousseau's *Émile,* did he fail to take his constitutional.

Kant's deliberations into such abstractions as freedom, peace, experience, knowledge, and the limits of the mind were matched by a lively interest in the natural world. As a young man, he wrote about earthquakes, and one of his last papers considered the moon's influence on the weather. But without doubt his most important scientific contribution, published twenty-six years before *Critique of Pure Reason* in 1781, was a book entitled *Universal Natural History and Theory of the Heavens.*

The ideas in this book were inspired by the musings of Thomas Wright (1711–86), a clockmaker who wrote that the Deity could be found in the center of the universe, around which the Sun and the stars in the Milky Way all moved. Wright speculated that the Milky Way is a hollow ball of stars and that the blurry nebulae seen here and there are other "abodes of the blessed." Although Kant did not actually read Wright's book, he read a newspaper summary from which he received the incorrect impression that Wright thought the Milky Way was a disk of stars.

The theory Kant developed to explain this, elaborated upon by the Marquis de Laplace four decades later, forms the core of our contemporary understanding of the creation of the solar system, which, Kant suggested, began with a spinning cloud of dust and gas known as the primordial nebula. For billions of years it floated around in space until it eventually collapsed inward under the crush of its own gravity.

Why that happened is uncertain. A current notion is that a nearby star exploded, sending shock waves through the nebula. The center bulged; the edges flattened; it began to contract and whirl. The more it contracted, the faster it rotated. Eventually, the atoms and molecules inside the central ball of gas were colliding so frequently that they created a kind of pressure that slowed the process of contraction but increased the heat. When the temperature reached about 18,000,000°F, nuclear explosions rocked the core. The ball of gas became a star.

Meanwhile, out in the suburbs, rings of gas and dust on the flattened edges of the cloud were orbiting around the newborn star. The atoms combined and condensed into microscopic grains of aluminum, titanium, iron, and other solids. Every so often, they tumbled into each other. At high speeds, the force of the encounter shattered them into spray. But when the velocity of the encounter was relatively low, the grains stuck together like clumps of sticky rice, gradually forming odd-size chunks of rock, metal, and ice known as planetesimals—the building blocks of the solar system. The smaller planetesimals pummeled the larger ones, which grew by accretion, gradually forming planets. The stream of impacts also generated heat; inside the planets, heavy elements such as iron melted and sank to the middle while the lighter ones rose to the surface.

Several theories attempt to account for the differing compositions of the planets in our solar system. One explanation is that although the planets were formed more or less simultaneously, different elements condensed at different distances from the Sun. Consequently, each planet differs from the others according to its distance from the Sun. Another theory suggests that originally all the planets were gaseous balls.

The ferocious wind blowing from the Sun stripped the gas from the nearby planets, but the outer planets were able to maintain something akin to their original form. Or perhaps the heat of the Sun melted and evaporated most of the icy particles on the inner planets. In the cold outer solar system, the ices—methane, water, ammonia—remained frozen, which meant that the planets retained more mass. This greater mass gave them the gravitational force needed to hold on to the thick blankets of gas that cling to them still.

As for the solid moons that populate the entire solar system, they may have been planetesimals captured into the gravitational field of the more powerful planets, or they may have formed from scratch, accreting slowly from the dust and bits of matter orbiting the planets when the solar system was young.

Scientists believe that planets are being formed this way even now, possibly around the bright star Fomalhaut, in Piscis Austrinus (the Southern Fish), or around Beta Pictoris, the second-brightest star in the humdrum southern constellation Pictor (the Painter). Infrared analysis of a disk of gas and dust surrounding the star indicates that the inner portion is filled with rocks and comet dust. Kant wouldn't have been surprised. He believed that stars and planets were constantly being formed. "The creation is never finished or complete," he wrote. "It has indeed once begun, but it will never cease. It is always busy producing new scenes of nature, new objects, and new worlds."

Kant was certain that life existed on these new worlds. Indeed, although he is credited with recognizing that the universe is filled with other galaxies he called "island universes" (an idea not confirmed until the twentieth century), he did not find it necessary to look beyond our own small system for other forms of life. He was convinced that "most of

> *. . . [Bloom] had conjectured as a working hypothesis which could not be proved impossible that a more adaptable and differently anatomically constructed race of beings might subsist otherwise under Martian, Mercurial, Veneral, Jovian, Saturnian, Neptunian or Uranian sufficient and equivalent conditions, though an apogean humanity of beings created in varying forms with finite differences resulting similar to the whole and to one another would probably there as here remain inalterably and inalienably attached to vanities, to vanities of vanities and to all that is vanity.*
>
> —James Joyce, *Ulysses*

the planets are certainly inhabited, and those which are not will be at some time." And since he was also a moral philosopher, he thought about the creatures on those unknown worlds and decided that although the Martians were undoubtedly less highly evolved than we are, the citizens of Jupiter were our moral superiors.

✳

Classical Mythology and the Planets

But still the heart doth need a language; still
Doth the old instinct bring back the old names.

—Samuel Taylor Coleridge

n the beginning, wrote the Roman poet Ovid, Chaos and Nyx, the goddess of Night, ruled a realm where "the jarring seeds of things confusedly roll'd." They oversaw this gloomy kingdom for many years until they grew tired and asked their son Erebus—Darkness—to help bring order to things. This he did in Oedipal fashion, overthrowing his father and marrying his mother. Night and Darkness gave birth to Aether (Air) and Hermera (Day), who in turn overthrew their parents and begat Eros. They also created the Sea and the Earth, who was called Gaia.

Gaia was not beautiful, however, until Eros loosed his famous arrows at her, whereupon the Earth bloomed with seedlings and flowering plants. Gaia was moved to create the Sky, who was named Uranus and who became her mate. After overthrowing Aether and Hermera, Uranus and Gaia ruled in their stead and had many children, including the twelve Titans.

But all was not domestic bliss. Uranus, knowing that the past is the best predictor of the future, was afraid his children might overthrow him. To forestall this event, he imprisoned them—along with his subsequent offspring, the Cyclopes—in Tartarus, a dark and cavernous abyss.

Gaia begged Uranus to release her children, but he ignored her pleas. So one day she set free her son Cronus, the Titan of Time, whose Roman name was Saturn. Saturn hated his father. So the next time Uranus and Gaia were in bed together, Saturn grabbed a sickle, castrated his father, and tossed his genitals into the ocean. Uranus cursed his son, screaming that one day Saturn's children would overthrow him. But Saturn found that impossible to imagine. He claimed the throne, set free the other Titans, giving them each an area to rule, and married his sister Rhea (whose name today belongs to one of the satellites of Saturn).

Diana, goddess of the Moon, in an illustration by the sixteenth-century artist Jost Amman.

But when they had a baby, his father's curse rang in Saturn's memory. He didn't want to imprison the child, but he didn't want him around either. So he swallowed his son. Every time Rhea had a baby—and it happened frequently—he repeated the act. Like his father before him, Saturn ignored his wife's pleas on behalf of her children. Finally, Rhea came up with a scheme. The next time she gave birth, she handed the infant to some nymphs and wrapped a large stone in baby clothes. Then she gave the swaddled stone to her none-too-bright husband, cried piteously, begged him to let her keep her child, and watched while he swallowed it whole. Thus she saved her baby, who was called Zeus, or Jupiter.

Apollo, the Sun God.

When Saturn learned that his son was alive, he felt afraid. Sure enough, Jupiter eventually attacked his father and forced him to regurgitate his progeny. Out came Neptune and Pluto along with three goddesses whose Roman names would be given to asteroids in the nineteenth century: Vesta, Ceres, and Juno, whom Jupiter married.

Jupiter then claimed the throne but the Titans were dissatisfied. War broke out. Looking for assistance, Jupiter freed the Cyclopes in exchange for the weapons of thunder and lightning. After ten years of fighting, Jupiter won and Saturn retired to Italy. Exhausted, Jupiter gave Neptune dominion over the sea and Pluto power over the underworld, and he set forth on a long series of adventures, including many of the extramarital sort. The sky is filled with stars, planets, and moons that attest to Jupiter's infidelity. Among his out-of-wedlock children were Phoebus Apollo, the god of the Sun, and Diana, the goddess of the Moon, both of whom were the children of Latona; Mercury, the son of Maia; and, according to some accounts, Venus, whom the Greeks called Aphrodite. One story says that when Saturn threw Uranus' genitals into the sea, Venus was born from the bloody foam—a story much prettified in Botticelli's famous painting. Another story says that she is the daughter of Jupiter and Dione, the goddess of Misfortune. According to Greek legend, only Mars was Jupiter's legitimate son. Even that is in dispute, though, for according to Roman legend, Mars was linked to Jupiter only indirectly. After Jupiter gave birth to Minerva out of his forehead, this story goes, Juno got angry and visited the goddess Flora, who held out a handful of herbs. The moment Juno touched them, she became pregnant and subsequently gave birth to Mars—the god of War, the child of spite.

How Big Is It?

When the heavens were a little blue arch, stuck with stars, methought the universe was too straight and close: I was almost stifled for want of air: but now it is enlarged in height and breadth, and a thousand vortices taken in. I begin to breathe with more freedom, and I think the Universe to be incomparably more magnificent than it was before.

—Bernard de Fontenelle

o adjectives are impressive enough to convey the vastness of space. Everything astronomical is bigger—or denser, hotter, more diffuse, or weirder—than you can possibly imagine. (Throw in astrophysics, which deals with heavenly objects on the subatomic level, and it's also unimaginably small.) Even the solar system, our own tiny corner of the universe, is so enormous that when scientists try to describe our place in the celestial scheme of things, they are forced to shrink things down and use everyday analogies.

Consider these comparisons.

If the Sun were an orange, as Robert Jastrow imagines in his book *Red Giants and White Dwarfs,* the Earth would be a grain of sand thirty feet away; a block away would be a cherry pit, representing Jupiter, and a block beyond that, another cherry pit, representing Saturn. Pluto would be a grain of sand ten blocks away. "Two thousand miles away is another orange, perhaps with a few specks of planetary matter circling around it," Jastrow writes. "That is the void of space."

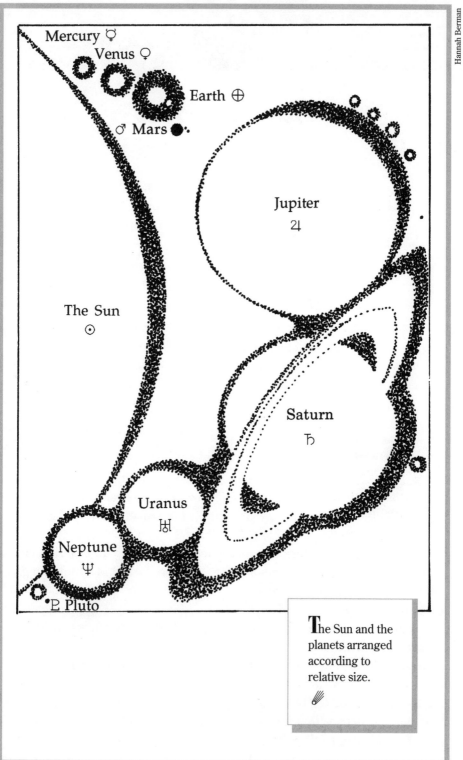

Hannah Berman

The Sun and the planets arranged according to relative size.

The solar system is so insignificant that on a map of the universe the size of a city block it would not appear at all. Harvey H. Nininger, the great collector of meteorites, visualized such a map. "An ordinary punctuation mark," he wrote, "a period such as is used on this page, would be more than fifty thousand times too large to represent even the outmost limits of our solar system."

Even on a map of our own Milky Way the planets would not appear. "I have often constructed a model of the Milky Way Galaxy on a classroom floor by pouring a box of salt into a pinwheel pattern. The demonstration is impressive, but the scale is wrong," writes Chet Raymo in *The Soul of the Night.* "If a grain of salt were to accurately represent a typical star, then the separate grains should be thousands of feet apart; a numerically and dimensionally precise model of the Galaxy would require 10,000 boxes of salt scattered in a flat circle larger than the cross-section of the Earth."

And where does our galaxy fit into the universe? Astronomer James Gunn, quoted in Richard Preston's *First Light,* describes how big the universe would be if our galaxy were the size of a dime. On that scale, he said, "This universe—or at least the universe we can see—is a small object, something like a cloud of dimes four miles in radius."

Maarten Schmidt, known for his explication of quasars, the most distant objects in the universe, has also been linked with the concept of a "small" universe ever since Alan Lightman and Roberta Brawer, authors of *Origins,* asked him how he would have designed the universe. After objecting strenuously to the question, he admitted the truth. "I would have constructed a bigger universe," he said. "I think the universe is small. There we go. If I'd had my rathers, I would do that. I find the universe too confined. I find it amazing that it is so small."

How small is it? Here are some of the numbers. Beginning in our own solar system, the most common measurement is the astronomical unit (AU), which represents the average distance between the Sun and the Earth: approximately 93 million miles. The Sun and Jupiter, the largest planet in the solar system, are 5.2 AU apart, and the average distance between the Sun and Pluto, the most distant planet in the solar system (as far as we know), is 39.4 AU—about 3.65 billion miles.

Outside the solar system, the astronomical unit is too tiny to matter. The next useful measurement is the light-year, the distance that light, which travels at 186,282 miles per second, covers in a year: almost 5.9 trillion miles. Because light takes time to reach our eyes, everything we

THE SOLAR SYSTEM

	Distance from the Sun (miles)	Diameter (miles)	Diameter if Earth = 1
Sun		865,000	109.13
Mercury	36,000,000	3,031	0.38
Venus	67,200,000	7,521	0.95
Earth	93,000,000	7,926	1.00
Mars	141,500,000	4,217	0.53
Jupiter	483,300,000	88,700	11.19
Saturn	886,200,000	74,600	9.41
Uranus	1,782,000,000	31,800	4.01
Neptune	2,792,400,000	30,770	3.89
Pluto	3,663,800,000	1,457	0.18

see is an image from the past. We see the Sun as it was eight minutes ago; that is, the Earth is eight light-minutes from the Sun. (Or to look at it another way, an astronomical unit is equivalent to eight light-minutes.) The nearest star in our galaxy is 4.1 light-years away. One of the brightest stars, Rigel in Orion, is 910 light-years away. When we look at Rigel, we see it as it was shortly after the Norman Conquest.

Yet compared to the universe as a whole, the light-year is a small unit of measurement. Our galaxy, one among many, measures approximately 100,000 light-years across. Andromeda, the nearest major galaxy to our own, is 2.2 million light-years away. Beyond that, the numbers rise so rapidly that astronomers commonly use the parsec, a distance equal to 3.26 light-years; the kiloparsec, which equals a thousand parsecs; or the megaparsec, which equals a million parsecs. At the farthest edges of the observable universe, even these measurements fade into meaninglessness. One of the most distant objects known at the moment is a quasar approximately 14 billion light-years away. To ordinary mortals, if not to astronomers, that's inconceivable.

❈

Sunspots

The Sun, with all the planets revolving around it and depending on it, can still ripen a bunch of grapes as though it had nothing else in the universe to do.

—Galileo Galilei

Give me the splendid silent sun with all his beams full-dazzling.

—Walt Whitman

n 1611, Christoph Scheiner, a Jesuit priest and mathematician, became convinced that sunspots were *on* the Sun and not merely near it. When he described his telescopic observations to his superior in the church, he was admonished that since Aristotle had stated that everything in the universe with the exception of the Earth was perfect and without flaw, dark spots could not conceivably mar the surface of the glorious Sol. Scheiner revised his opinion and decided that perhaps sunspots rotated around the Sun, close to but not actually on its surface.

Galileo Galilei, however, trusted his eyes and wanted credit for it. After reading Scheiner's anonymously published letters about sunspots, he announced that he had seen them first. It wasn't true, but he accurately described their duration and their irregular, constantly shifting shape, and he realized that they were part of the Sun. Consequently, he is often credited with the discovery of sunspots.

The nature of sunspots remained a mystery. Giovanni Domenico Cassini, famous for his discovery of the eponymous gap in the rings of Saturn, thought sunspots might be mountains whose tops were revealed

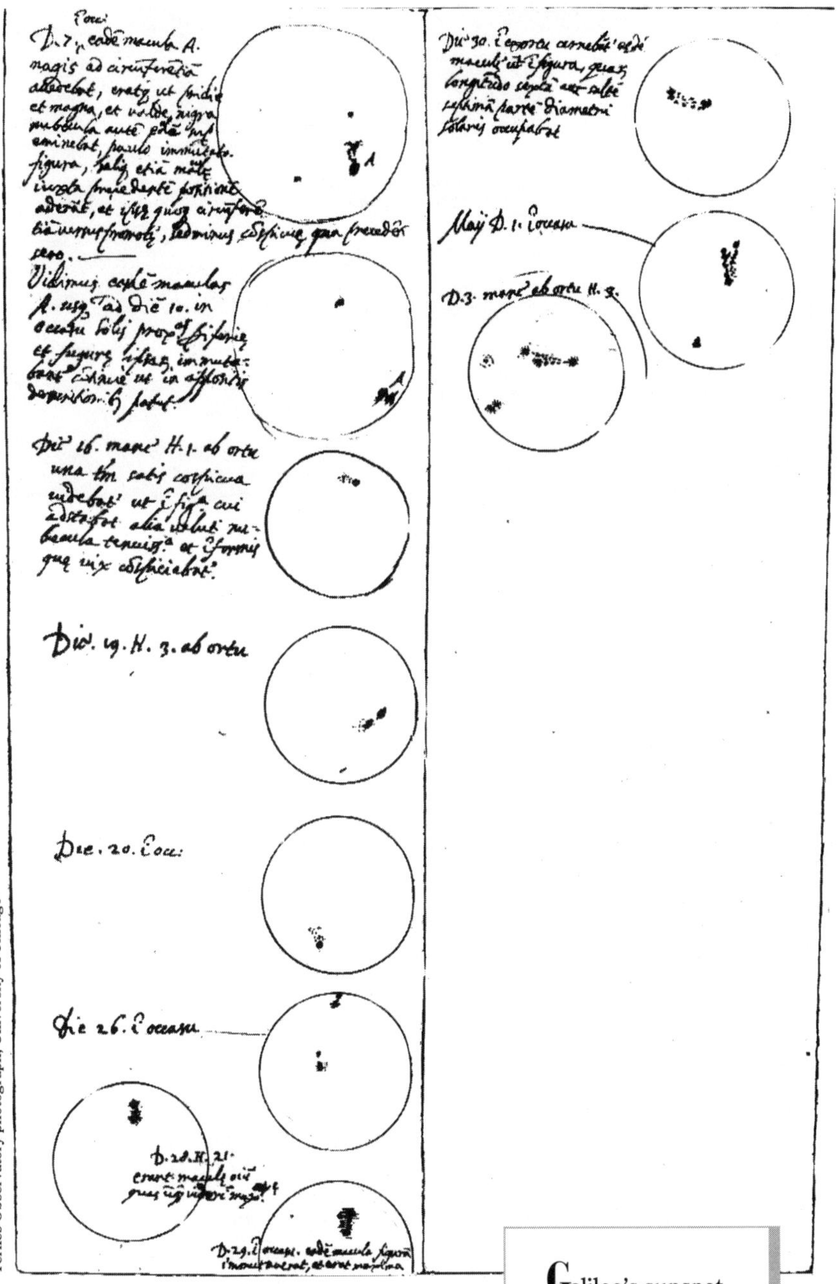

Yerkes Observatory photograph, University of Chicago

Galileo's sunspot notes and drawings.

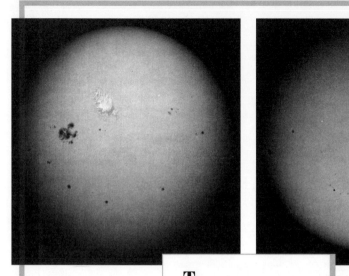

The changing positions of the sunspots on these pictures reveal the rotation of the Sun.

only when they briefly peaked above the ebb and flow of the photosphere, the visible surface of the Sun. In 1769, the Scottish astronomer Alexander Wilson suggested that sunspots were holes in the photosphere. This idea is incorrect but not unreasonable; sunspots are darker than the rest of the Sun, and as they approach the perimeter they look hollow.

Sir William Herschel, whose great contributions to astronomy have eclipsed some of his more muddled ideas, noticed this and wrote in 1795 that if sunspots were glimpses of the darker surface seen through a break in the photosphere, the Sun might be "nothing else than a very eminent, large and lucid planet. Its similarity to the other globes of the solar system . . . leads us on to suppose that it is most probably also inhabited, like the rest of the planets, by beings whose organs are adapted to the peculiar circumstances of that vast globe."

The most important addition to the study of sunspots, however, was made by an obscure nineteenth-century pharmacist named Samuel Heinrich Schwabe, who began systematically observing the Sun with the intention of finding a new planet within the orbit of Mercury. In his futile search for the nonexistent planet known as Vulcan, he became fascinated with sunspots. From 1826 to 1850, he doggedly recorded sunspot groups, and as a result was the first observer to see that they occur in cycles. This was new information; previously, scientists thought sunspot activity was random and unpredictable. Nonetheless, Schwabe's findings went unheralded.

At least one man, however, read the unknown pharmacist's paper: Baron Alexander von Humboldt, the explorer and scientist, who traveled the world, climbed volcanoes, met President Thomas Jefferson, explored Russia at the behest of Czar Nicholas I, became good friends with the king of France, and was appointed to a diplomatic post by the king of Prussia. In *Kosmos,* a five-volume work written while Humboldt was in his seventies, he announced Schwabe's discovery of the sunspot cycle. People paid attention.

Today we know that sunspots are vast magnetic storms on the agitated surface of the Sun. Although they look dark, even Galileo knew that that was merely a comparative effect. At 6,700° to 7,600°F, they are cooler than the rest of the surface, radiate less light, and consequently look darker. They are formed by the rotation of the Sun, a process that takes 25.4 days at the equator and 11 days longer at the poles. These varying speeds cause the magnetic lines of force within the Sun to become so twisted that eventually they burst through the photosphere and loop back down again. Sunspots are the dark swirls formed at each foot of the loop. They travel in pairs, with the eastern spot having a magnetic polarity opposite to that of the western spot.

THE SUN

Mean Distance from Earth
93,000,000 miles

Diameter at Equator
865,000 miles

Diameter if Earth = 1
109.13

Mass if Earth = 1
332,946

Surface Gravity if Earth = 1
27.9

Sunspots follow a 22-year cycle divided into two 11-year periods. At the beginning of the first period, a few sunspots cluster in the higher latitudes, 30° to 35° north and south of the equator, while the corona, made of thin, invisible gases in the Sun's outer atmosphere, appears as a re-

THE GREEN FLASH

•⟿

Green is the night, green kindled and apparelled.
It is she that walks among astronomers.
—Wallace Stevens

Plan your life correctly and it's possible to see a solar eclipse (assuming the sky is clear) simply by making arrangements with your travel agent. But you can't buy the elusive green flash, although you can look for it. It appears rarely, either at dusk, just as the topmost curve of the Sun dips below the horizon, or at dawn, just before the Sun begins its climb. On those occasions, a cap of Day-Glo green shimmers for a moment on the rim of the Sun and is gone.

It happens because for an instant the atmosphere acts like a prism, separating the light of the Sun into its component colors and overlapping them. As the Sun sinks beneath the horizon, the red, orange, and yellow at the top of the Sun—the usual colors of the sunset—slip away one by one, leaving, for a moment, green. (Blue and violet are scattered so easily by the atmosphere that they are the hardest colors to see in the Sun.) Sometimes, at the end of the sunset, the green can cling for a few seconds.

The Egyptians evidently observed this phenomenon, as did the Celts. Jules Verne wrote a novel (*Le Rayon vert*) about it, Admiral Richard E. Byrd spotted it in Antarctica, and the author of this book saw it from a dock on Fire Island. Most easily glimpsed at sea or in tropical latitudes, the green flash isn't something you can count on seeing. But it's worth watching the sunset every night for the rest of your life, preferably from sea level, where the horizon is low and flat. Just in case.

strained halo around the orb. The sunspots gradually multiply and float toward the equator until their population peaks at what is called sunspot maximum, at which point the corona arcs wildly from the Sun in every direction, crowning it with incandescent plumes of gas so huge and hot they dwarf the Sun in size and exceed it in temperature. A few years later, another 11-year period begins. This time, the magnetic field of the Sun flips along with the polarities of the sunspots. In a complete reversal, positive becomes negative, north becomes south, and the process begins anew.

You'd think that since sunspots are dark, more sunspots would mean a darker Sun. But in fact, the more sunspots there are, the brighter the Sun is. One explanation is that sunspots are usually accompanied by bright patches called faculae, which appear at 70° to 80° latitude north and south of the Sun's equator and drift to within 3° of it. First discovered by Christoph Scheiner, faculae are so bright that they more than make up for the darkness of the sunspots.

In contrast, the intensity of high-energy particles hurtling toward the Earth from all directions varies inversely to the sunspot cycle. Why this should be so is uncertain; after all, these speeding particles, known as cosmic rays, do not primarily come from the Sun. They may be affected in some way by the magnetic field of the solar wind.

With so much activity keyed to sunspots, it's no surprise that scientists (and others) have looked for ways sunspots influence our planet and our lives. The most famous instance is the Maunder Minimum, the years from 1645 to 1715 in which few sunspots were seen. Although many scientists remain skeptical about cause and effect, it is nonetheless true that during the Maunder Minimum (named after E. Walter Maunder, who noticed it in 1890), eclipses almost entirely lacked the great flares that usually unfurl from the Sun on those occasions, aurorae such as the northern

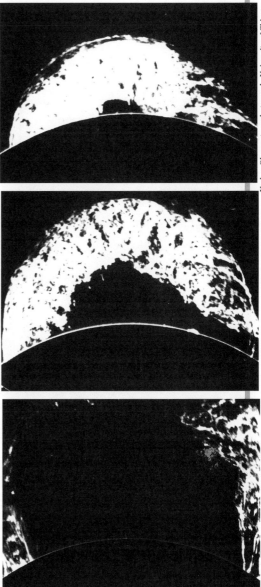

Yerkes Observatory photograph, University of Chicago

The arch prominence of June 4, 1946, lifting from the Sun in a series of photographs taken during a single hour. The picture in the middle shows a small bead of light on the sun's rim. At this scale, the diameter of the Earth would be smaller than that spot.

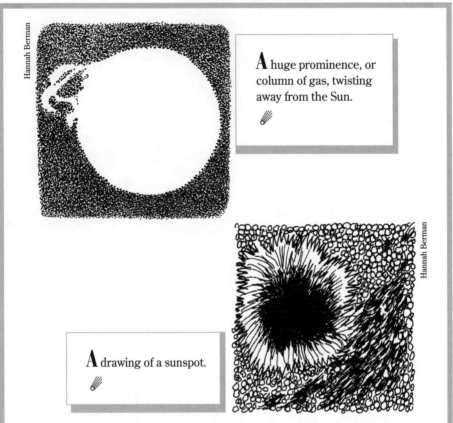

A huge prominence, or column of gas, twisting away from the Sun.

A drawing of a sunspot.

lights—a phenomenon directly associated with sunspots—were seldom seen, and temperatures in Europe were so low that the period was known as the Little Ice Age. (A time in which sunspots were especially numerous occurred between 1100 and 1387. Known as the Medieval Maximum, it has been associated with an epidemic of skin cancer.)

Other parallels have been suggested. Jonathan Swift, among others, wondered if sunspots affected the weather, William Herschel tried to connect them with the price of wheat, and Fred Hoyle wondered if they might be somehow implicated in outbreaks of flu. Sunspot cycles have also been linked to droughts, rock formation, patterns in the sediment of glacial lakes, and the amount of carbon 14 in the rings of trees. They also may affect human activity. According to Noel W. Hinners, a former associate deputy administrator of NASA, surges of sunspot activity correlate "rather well" with what he calls "NASA's peaks in frenzy." But as for the influence of sunspots on hemlines and the Dow-Jones average, the verdict is in.

The Moon

The moon—yes, that would be the place for me—
My kind of paradise! I shall find there
Those other souls who should be friends of mine—
Socrates—Galileo—

—Edmond Rostand, *Cyrano de Bergerac*, Act 5

he only celestial body ever visited by Earthlings, the Moon is by all accounts a dead world. It has little or no magnetic field, and no signs of life have been found on its charcoal-colored crust: not the river Giovanni Cassini thought he had detected, not the creatures William Herschel imagined made their home there, not even the fossilized shells of organic molecules twentieth-century scientists searched for in moon rocks brought back from the great space expeditions. Although volcanic eruptions helped shape the surface of the Moon in the primordial past, throughout human history the barren surface appears not to have changed. Early maps, however crude, and recent maps, however detailed, indicate identical landscapes.

Yet from time to time, something happens there. In 1783, William Herschel claimed to have seen a spot like a burning ember on the unlit portion of the Moon. In 1963, moon watchers at Lowell Observatory reported another red glow, possibly caused by an emission of gas near Aristarchus, the satellite's brightest crater. Six years later, astronauts of *Apollo 11* and other observers reported that the same area seemed to have "brightened."

But perhaps the most exciting lunar extravaganza occurred one summer night shortly after sunset in the year 1178, when a group of monks at Canterbury Cathedral, where Thomas à Becket had been murdered

Galileo's drawings of the Moon.

less than eight years before, saw light blazing from the tip of the crescent moon. Astronomers believe the conflagration may have been caused when a meteor struck the Moon's surface, and that the monks did not see the impact because it occurred on the far side. Rather, they saw the cloud of dust, sparks, and partly melted rocks that spewed into space as a result of the crash. It may be that some moon rocks, ejected from the Moon's gravitational field on that long-ago day, even fell to Earth. Scientists believe that a meteorite discovered in Antarctica and known as Allen Hills 81005 (after the location where it was found) may have come from the crater born in that crash—a crater later named after Giordano Bruno, who suffered a church-sponsored execution for, among other heresies, his belief that the universe was filled with other suns and other worlds.

Impacts such as the one in 1178 cause the Moon to ring like a bell. We know this thanks to a solar-powered instrument collection called the Apollo Lunar Surface Experiment Package (ALSEP), which Neil Armstrong and Buzz Aldrin lowered to the surface of the Moon in 1969, mere moments after receiving a message from President Richard Nixon. Since then, seismic readings have recorded thousands of small moonquakes as well as the

NASA

Geologist Harrison Schmitt standing next to a lunar boulder during the *Apollo 17* mission, the last manned flight to the Moon.

impacts of meteorites and spacecraft (some of which have been dropped onto the Moon for the precise purpose of monitoring the vibrations). When one of these objects plows into the lunar crust, it throws up clouds of debris and carves out a crater. Such collisions are small events reminiscent of the way the Moon reached its current shape, and they cause the Moon to shudder. These impacts vaporize bits of the lunar surface, releasing into the Moon's whisper-thin atmosphere a cloud of gases that trails behind it like the tail of a comet, blown by the solar wind. The cloud on the side of the Moon facing away from the Sun is fully five times longer than the one on the other side. Even as the mist drifts into space, meteoric impacts, when they occur, continue to feed it. The Moon may be lifeless—but it isn't static.

THE MOON

Mean Distance from Earth
238,900 miles

Diameter at Equator
2,160 miles

Diameter if Earth = 1
0.27

Mass if Earth = 1
0.012

Surface Gravity if Earth = 1
0.17

The Birth of the Moon

Although most planets have satellites, the relationship between the Earth and the Moon is unique. Only one other planet—Pluto, the planet about which we know the least—has a moon so relatively like it in size. Most planets are far larger than their satellites, whereas the Earth in diameter is only 3.6 times the size of the Moon. For that reason, the Moon and the Earth are sometimes considered a double planet. How they got that way 4.5 billion years ago is a mystery. Here are some scenarios:

* *Simultaneous creation.* Maybe the Moon and the Earth formed at the same time, coalescing from the same planetesimal stew, orbiting around each other, influencing each other's shape and rotation. It's a nice theory except that the Earth and the Moon are compositionally different, with the Earth holding three times as much iron as the Moon, which may not even have a metallic core.

* *Fission.* George Darwin, son of Charles, hypothesized that the Moon and the Earth were originally one body. As it revolved around the Sun, it cooled, contracted, and began to rotate faster. Eventually, when it was whipping around its axis in two or three hours, the flattening at the poles and the bulge at the equator became so extreme that a giant bubble of still-molten material pulled away and escaped. Caught in the orbit of the Earth, it became the Moon.

  One might think that the chemical difference between the Earth and the Moon would argue against this idea. Oddly, it does just the opposite, because the fission theory suggests that the Moon would have been torn out of the mantle of the Earth—not out of its heavy metal innards. Hence, it would be—and is—less dense than the Earth.

  A variant on this theory suggests that rather than being torn out of the Earth in a single surgery, globules of material may have spun off the Earth and formed a ring not unlike one of Saturn's. That material would gradually have merged into the Moon.

* *Capture.* It could be that the Moon, formed elsewhere in the solar system, was pulled from its original orbit and captured by the Earth's gravitational field. Astronomers believe that this happened with Neptune's moon Triton. Unfortunately for the analogy, Neptune is a giant planet, with a mass seventeen times that of Earth, while Triton is smaller than our Moon. The Earth and the Moon are far closer in size, which makes capture a lot more difficult. The capture theory could work—but only under a very specific set of circumstances.

* *Collision.* Maybe the young Earth was revolving around the Sun when an errant asteroid or a wandering comet the size of Mars slammed into it, carving out a monstrous crater and spewing into space a huge ring of matter that gradually formed the Moon. This theory explains the chemical differences, because even an enormous impact would not have pulverized the core of the Earth, which then as now had far more iron than does the Moon. The collision theory is a favorite at the moment—although, as the history of astronomy amply shows, a popular answer is not always a correct one, and disturbing ideas are not necessarily wrong. The German astrophysicist Rudolf Kippenhahn has quoted Groucho Marx on this point: "They claimed Galileo was mad when he said the Earth orbited the Sun, but he was right," said Groucho. "They said the Wright brothers were crazy when they tried to fly, but they did. They said my uncle Wilbur was mad as a hatter—and he was."

The Face of the Moon

"Beautiful! . . . magnificent desolation!"
—Buzz Aldrin, July 20, 1969

If the manner of its birth is still in question, the way in which the Moon acquired its pockmarked face is known. For close to 700 million years after its formation, the Moon was viciously bombarded with meteors. Large meteors blasted out impressive impact craters such as Copernicus, which features a central peak, terraced cliffs around its rim, and long fingers of debris pointing from the center, as well as smaller secondary craters formed when the lunar crust, blown away by the initial impact, crashed back into the surface some distance away. Small meteors scooped out circular basins such as Galilaei, a tiny crater named by Father Giovanni Battista Riccioli (1598–1671), a Jesuit astronomer whose lunar map inaugurated the system of nomenclature used today and reflected his opinion that Galileo, as a heretic, was undeserving of a grander crater.

The meteoric blitz and radioactive breakdown of elements within the Moon caused the crust to melt. Heavy elements sank to the bottom, while lighter feldspars bubbled to the top like foam in a glass of beer. Embedded within them were potassium (K on the periodic table of elements), a collection of rare earth elements (given the abbreviation REE), and phosphorus (P). The compound they formed, known by the acronym KREEP, is virtually a moon-rock signature.

Eventually, the crust cooled and lava from the interior gushed into cracks and crevices. Basalt flooded the plains and filled the basins of many of the craters, thereby creating the dark patterns clearly visible on the Moon's face. Aristotle, who believed the Moon was a perfect sphere unmarred by topography, thought that these dark smudges might be reflections of the Earth (and it is easy to imagine this). But Galileo proved Aristotle wrong when he looked through his spyglass and saw that the mountains on the Moon cast shadows, which indicated that they were actual objects. The Moon was not flawless, and the dark areas were not simply reflections. Making what E. M. Forster called a "gracious error," Galileo named the dark plains *maria,* or seas, although they have nothing to do with water, as he undoubtedly realized. There seems to be no water on the Moon, and evidently there never has been. Even the sinuous curves known as rilles, which resemble dried-out riverbeds, are the cooled-off remains of flowing streams of lava. (There is, however, a distant possibility that frozen water vapor, perhaps from long-ago volcanic eruptions or ancient comet crashes, could lie frozen at the lunar poles.)

Some craters on the Moon. The large crater on the left is Copernicus, with Eratosthenes to the right and slightly above. At the top of the picture, partially out of the frame, is Archimedes.

Hale Observatories

COSMIC CODEPENDENCY

Like other satellites in the solar system, the Moon is deformed by the proximity of its planet, but the distortion cuts both ways. The Earth's gravity pulls the Moon into a slightly flattened egglike shape, with the length of the egg pointing toward us, and the Moon's gravity likewise causes the Earth to bulge. As they tug on each other, tides swell either side of each body, the tides create friction, and the Earth and the Moon slow each other down. Consequently, the Earth is rotating on its axis less rapidly than it used to, and the Moon is pulling away at the rate of four centimeters a year. Four hundred million years ago, in the Devonian era, well before the dinosaurs were even a gleam in a thecodont's eye, the Moon looked larger in the sky (to any simple invertebrate who might have been looking), the tides were higher, the day was only 21.9 hours long, and a year had approximately four hundred days. Some 4.5 billion years from now, the Moon will be much smaller in the sky, total solar eclipses will be a thing of the past, and the day will be almost twice as long as it is now. Although the system seems to be in balance, it is in fact forever changing.

The *maria,* as they are still known, are far more abundant on the side of the Moon facing the Earth. The Sea of Tranquillity and the Ocean of Storms, the Marsh of Epidemics, the Bay of Dews, and other dark areas form the shape variously seen as the man in the Moon, the goddess of the Moon, Coyote, or a hare, an image prominent in many cultures. Although these areas are smooth compared to the lighter, more heavily cratered areas, they are nonetheless dotted with thousands of smaller craters, proof that meteoric bombardment did not stop even after the deluge of basalt. Indeed, it is going on to this day, albeit slowly: every 50 million years or so, three inches of lunar soil are added to the surface of the Moon. Given the lack of an atmosphere and the absence of weather, that coating of dust should affect the Moon very little. Millions of years from now, the Moon will probably look much as it did when Father Riccioli, whose name belongs to a small, dark crater near the rim, first published his map.

About Eclipses

Whenhen the Earth, the Moon, and the Sun are precisely aligned, the sky darkens in an eclipse. Solar eclipses occur at new moon, when the Moon passes between the two larger bodies, blotting the Sun from view and casting a shadow on the Earth.

There are three types of solar eclipse. In a partial eclipse, the Moon eats into the Sun but does not devour it. The day dims slightly, and the Sun, viewed through protective eye gear or with a pinhole camera, looks like a cookie with a bite taken out of it. In a total eclipse, the face of the Sun disappears behind the Moon, the normally invisible corona blooms, and fortunate observers within the Moon's shadow can experience darkness at noon. The third type of eclipse, called annular, occurs when the Moon is at its greatest distance from the Earth and consequently looks smaller than usual. Even at the height of such an eclipse, the rim of the Sun surrounds the Moon, like this: ◉.

The full moon is the time for lunar eclipses, when the Earth is caught between the luminaries and its shadow falls across the surface of the Moon. Like solar eclipses, lunar eclipses don't happen every month; they occur only when the three-way lineup is exact. That only happens occasionally, because the Moon's orbit, which rotates, is tilted at a 5° angle to the plane of the Earth's orbit around the Sun.

Real eclipse lovers will stop at nothing to see one. For instance, on October 3, 1986, Glenn Schneider of Baltimore and eight other intrepid souls watched an eclipse of the Sun from a small airplane 40,000 feet above Iceland. Writing in *Sky & Telescope* magazine, Schneider described what he saw as the Moon moved in front of the Sun and the sunlight filtered through the valleys and mountaintops on the Moon's edge, producing a phenomenon known as Baily's Beads. "For a full six seconds," Schneider recalled, "the blinking dance of beads flashed along the limb. . . . A minute after 'totality,' we looked out the left windows to judge the increasing sky brightness. The Moon's shadow was projected on the cloudtops like a blob of spilled ink! For a full minute we watched this dark stain, shaped like a squashed cigar, speeding away from us toward the horizon."

Eclipses occur every year—seven at the most, two at the least. So why don't we see more of them? Lunar eclipses are visible only at night. Half the time, they happen during the day and can be seen only from the other side of the Earth. Solar eclipses are even more elusive because they are visible for only a few minutes and only within the path cast by the Moon's shadow. That path is so narrow that in all of England over the last thousand years, only four total solar eclipses have been visible.

So eclipses are not easy to see. Take a four-eclipse year like, say, 1997. In March, a lunar eclipse will be visible from anywhere in the United States—but it will be only a partial eclipse. That same month, you'll be able to see a total eclipse of the Sun—but only from a thin strip of land in China or a ship in the Arctic Ocean. In September, a solar eclipse will be visible as long as you happen to be watching from Australia or New Zealand, but even down under, it's only a partial eclipse. Two weeks later, you can catch a total eclipse of the Moon—but not from North America. That's what eclipse hunting is like. The idea of chartering a plane for the purpose of seeing one begins to seem not at all unreasonable.

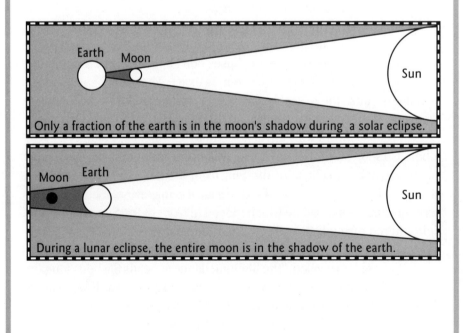

Only a fraction of the earth is in the moon's shadow during a solar eclipse.

During a lunar eclipse, the entire moon is in the shadow of the earth.

Six Noteworthy Eclipses

> ... the attendant phenomena of eclipses, solar and
> lunar, from immersion to emersion, abatement of wind,
> transit of shadow, taciturnity of winged creatures,
> emergence of nocturnal or crepuscular animals,
> persistence of infernal light, obscurity of terrestrial
> waters, pallor of human beings.
>
> —James Joyce, *Ulysses*

In Dahomey mythology, the Moon, whose name is Mawu, and her twin brother the Sun, called Lisa, made love during an eclipse. The seven sets of twins thus conceived became the stars and the planets.

But in most mythologies, eclipses have more frightening associations. The ancient Chinese and the Bolivians imagined that during an eclipse mad dogs tore at the Sun or the Moon with their teeth. In Yugoslavia, vampires were said to ravage the luminaries. Egyptians believed that occasionally the serpent Apep, ruler of the Underworld and lord of the dead, rose up and swallowed the boat in which the Sun god Ra was riding across the sky. At those moments, the Sun disappeared.

Historical accounts tend to be semi-mythological. Often they tell of a superior being—a conqueror or a scientist—who is able to predict an eclipse, thereby averting disaster while simultaneously illustrating that knowledge is power. Two examples:

✻ May 28, 585 B.C. Despite his belief in a flat Earth, Thales of Miletus is considered the first Greek scientist. He linked mathematics and logic and was the first to articulate various mathematical truths most of us learn in high school. The ancients revered him for his ability to stop a battle, a feat he accomplished with the help of Babylonian tables. According to Herodotus, the Medes and the Lydians were in the midst of a battle when

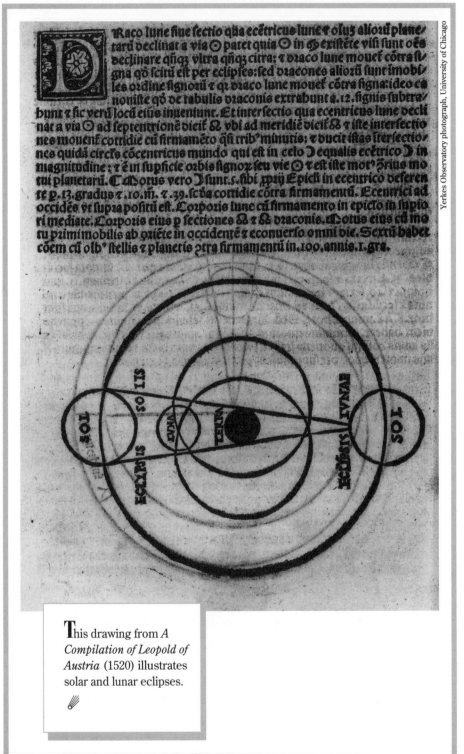

This drawing from *A Compilation of Leopold of Austria* (1520) illustrates solar and lunar eclipses.

"the day changed into night. And this change had been predicted to the Ionians by Thales of Miletus and he had given the year in which it took place." Even though Thales had not specified the day, his prediction was sufficiently awe-inspiring to result in an immediate peace.

* February 29, 1504. Christopher Columbus had been marooned for months with his disgruntled crew on the coast of Jamaica. Legend says that he arranged a meeting with the native inhabitants for a date on which he knew a total eclipse of the Moon would occur. He based his prediction on the navigational tables of the astronomer Johann Müller, who is better known by his Latin name, Regiomontanus. The eclipse occurred on schedule, the Indians were impressed, and the explorers were temporarily able to recover a little of their fading influence.

A few eclipses are memorable for scientific reasons:

* June 21, 1629. The Chinese were able to predict eclipses, but not well. Imperial astronomers, who had failed to anticipate an eclipse in 1610, predicted a solar eclipse for this date in 1629. Jesuit missionaries, however, insisted that the prediction was an hour early and that rather than lingering for two hours the eclipse would last only two minutes. The Jesuits were correct. As a result, the emperor ordered that the Chinese calendar be revised, and the Jesuits were encouraged to build telescopes and to begin translating books on optics, music, and mathematics into Chinese.

* July 8, 1842. During this solar eclipse, European scientists deduced that the rose-colored prominences and opalescent streamers of light surrounding the Moon at totality were neither emissions from the Moon's atmosphere nor an optical illusion, but part of the Sun.

* August 18, 1868. Pierre Jules César Janssen, a French banker turned astronomer, took a spectroscopic reading of the solar corona during this eclipse, which enabled scientists to analyze the composition of the solar atmosphere. The corona was so dramatic that Janssen was convinced it should be detectable under ordinary conditions. The next day, he located the protuberances

and recorded a spectrum. Another scientist, J. Norman Lockyer, had been doing similar work. Between them, they proved that the corona, which was present at all times albeit visible only during a solar eclipse, was part of the Sun although in composition it differed slightly from the body of the Sun. They also identified in the yellow part of the spectrum an element that would be named after the Greek word for "sun" and would not be found on Earth for another quarter of a century: helium.

Janssen was so thrilled with these results that in 1870, when an eclipse was due to be visible in Algeria, he let nothing stand in his way. Getting out of Paris was a problem, however, because the city was surrounded by hostile Prussian troops. Mobs swarmed through the streets, hungry citizens ate cats and rats, fancy restaurants raided the zoo and served dishes made from the two elephants, Castor and Pollux, and the only way out of the city was via hydrogen balloon. Janssen wafted out of Paris and reached Algeria in time. Unfortunately, as totality approached, the temperature dipped, clouds covered the Moon, and he couldn't see a thing.

* March 29, 1919. Albert Einstein had predicted that light passing a heavy object such as the Sun would be bent a specific amount by the object's gravitational field. This had yet to be demonstrated, but the solar eclipse of March 1919, when the Sun would be silhouetted against the tightly packed stars of the Hyades cluster, offered a perfect opportunity to compare the usual positions of those well-known stars with their positions during the eclipse. With that in mind, Sir Arthur Eddington traveled to an island off the western coast of Africa, and a group of British scientists went to Brazil. During the eclipse, the observers measured the positions of several stars in the Hyades, and discovered that the light from these stars was bent by the gravity of the Sun and hence shifted from its usual position precisely in accord with Einstein's predictions, thereby confirming his theory—and changing his life. As soon as Einstein heard the news, he sent a postcard to his mother, announcing "Joyous news today." A *New York Times* headline proclaimed, "Lights All Askew in the Heavens/Einstein Theory Triumphs," and Einstein was propelled into permanent, worldwide celebrity.

Edgar Allan Poe
and the Man in the Moon

Two nineteenth-century astronomical fantasies exemplify the human desire not to be alone in the universe. The notion that Mars was criss-crossed by canals indicative of intelligent life is one. Less well known is the story of the civilization on the Moon—a hoax perpetrated in 1835 by an enterprising newspaperman who attributed the discovery to Sir John Herschel, whose *Treatise on Astronomy* had just been published in the United States.

Edgar Allan Poe, a knowledgeable amateur, read the *Treatise* and was inspired by Herschel's speculations. "This theme excited my fancy, and I longed to give free rein to it in depicting my day-dreams about the scenery of the moon—in short, I longed to write a story embodying these dreams," he wrote in an article published twelve years later in *Godey's Lady's Book*. He originally planned a story of "very close verisimilitude," but instead, "I fell back upon a style half plausible half bantering, and resolved to give what interest I could to an actual passage from the earth to the moon, describing the lunar scenery as if surveyed and personally examined by the narrator. In this view I wrote a story which I called 'Hans Pfaall.' "

Three weeks after the story's publication in the *Southern Literary Manager,* Poe read about some astounding discoveries in a series of articles in the *New York Sun.* The articles reported that John Herschel, using a tele-scope "of vast dimensions and an entirely new principle," had seen lunar trees similar to "the largest class of yews in the English church-yards," a blue lake or inland sea, dark red poppies, "monstrous amethysts," and a chain of lilac-colored pyramids. Roaming across this remarkable moon-scape were herds of miniature bison and an animal "of bluish lead-color, about the size of a goat, with a head and beard like him, and a single horn."

It got better. The next installment of the series described flocks of "large winged creatures" who could fly and walk. "Whenever we saw them, these creatures were evidently engaged in conversation; their gesticulation, more particularly the varied action of their hands and arms, appeared impassioned and emphatic. We hence inferred that they were rational beings, . . . capable of producing works of art and contrivance."

Written by Richard Adams Locke, the articles, which according to Poe "not for a moment could I doubt had been suggested by my own *jeu d'esprit,*" were presented as reprints from the *Edinburgh Journal of Science.* Called "probable and plausible" by *The New York Times,* the articles became immensely popular. While Poe saw that Locke was perpetrating a hoax, most people evidently believed that there were on the Moon huge purple crystals, unicorns, and people who could fly. The circulation of the *Sun* jumped, and the articles, reprinted in pamphlet form, were read throughout the United States and Europe. Poe wrote a piece denouncing the series and explaining its many errors. (For example, he pointed out, the temperature of the Moon "is rather above that of boiling water, and Mr. Locke, consequently, has committed a serious oversight in not representing his man-bats, his bisons, his game of all kinds—to say nothing of his vegetables—as each and all done to a turn.")

The response to his rebuttal was not what Poe expected. He "was astonished at finding few listeners, so really eager were all to be deceived. . . . Not one person in ten discredited it, and (strangest point of all!) the doubters were chiefly those who doubted without being able to say why—the ignorant, the uninformed in astronomy, people who would not believe because the thing was so novel, so entirely 'out of the usual way.' A grave professor of mathematics at a Virginian college told me seriously that he had *no doubt* of the truth of the whole affair!"

Eventually, it became known that the *Edinburgh Journal of Science* did not exist. In New York, the *Journal of Commerce* declared the story a hoax. In France, François Arago, director of the Paris Observatory, read the series aloud to an assembly of scientists who pronounced it "utterly incredible." In Cape Town, Sir John Herschel finally heard about his so-called discoveries. To his credit, he found the brouhaha funny.

As for Poe, Locke's series had been so widely read and discussed that he decided against continuing the adventures of Hans Pfaall. "The chief design in carrying my hero to the moon was to afford him an opportunity of describing the lunar scenery, but I found that he could add very little to the minute and authentic account of Sir John Herschel," Poe wrote. "I did not think it advisable even to bring my voyager back to his parent Earth. He remains where I left him, and is still, I believe, 'the man in the moon.'"

How the Moon Got Its Spots

A young Eskimo woman was visited in the dark, night after night, by a man who made love to her but whom she could not see. One night, longing to know his identity, she rubbed ashes on her hands so that later she could look for the man with dark palm prints on his back. When the lamps were lit, she found him. It was her brother. This upset her so much that she cut off her breasts, threw them at him, and ran away. She became the Sun while her brother, with his sister's sooty palm prints still on him, became the Moon. And that is why the moon looks so patchy, with white and dark spots all over its surface.

Other stories about the Sun and the Moon reflect the connection between the Moon's cycle and menstruation. In a story from Guiana, the Sun and the Moon were friends until the Moon began to have sex with the Sun's daughter. When the Sun found out, he instructed his daughter to wipe her menstrual blood on her lover's face. This she did. The face of the Moon is still marked by the evidence of the alliance; and the Sun and the Moon still avoid each other.

Sometimes the dark spots on the Moon are animals. In *The Origin of Table Manners,* Claude Lévi-Strauss describes an Arapaho story, variants of which are told by other Plains Indian tribes. It seems that the Sun and the Moon were brothers, both of whom wanted wives. The Moon said he hoped to find a human wife because women were so beautiful. The Sun, however, didn't like human women because their eyes squinted and their faces wrinkled when they looked at him. He preferred a water wife. So the Sun and the Moon went down to Earth, where the Moon immediately attracted a human wife by

> L et's get this straight once and for all:
> is that a face up there or is it a rabbit, and if
> it's a face, then why does it hold itself back,
> why doesn't it take control and say, Who made
> this mess, who's responsible?
> —Stephen Dobyns, "Missed Chances"

Hale Observatories

The full Moon as photographed with the 100-inch telescope. The dark round spot near the three o'clock position is Mare Crisium, the Sea of Crises. To its left is Mare Tranquillitatis, the Sea of Tranquillity, where the astronauts of *Apollo 11* first walked on the Moon.

sitting in a tree and turning into a porcupine. Two women saw him, and one climbed after him. The tree grew higher, and the woman kept climbing. When the porcupine changed back into the Moon, the woman agreed to marry him.

In the meantime, the Sun also found a wife. She was a frog.

Back home, the celestial brothers' parents noticed that the woman gracefully chewed her food in the regular fashion, while the frog held a piece of charcoal in her mouth, which caused black juice to dribble down her chin. Her table manners made a poor impression on her in-laws, as did her habit of stopping frequently to urinate. When the Moon laughed at her, the frog jumped on the Moon's face, where she can still be seen today.

Another common lunar image is that of the rabbit, which appears in stories from India, Europe, Africa, Japan, and South America. This repetition of images may seem peculiar, but Dr. E. C. Krupp, a pioneer in the field of archaeo-astronomy and author of *Beyond the Blue Horizon,* believes that the images are chosen for a simple reason: They really do resemble the hazy shapes in the full moon. This is particularly true in the case of the rabbit. Its head is the Sea of Tranquillity (Mare Tranquillitatis), where the astronauts landed in July 1969; its floppy ears are the Sea of Fertility (Mare Fecunditatis) and the Sea of Nectar (Mare Nectaris), and its body curls around the edge of the full moon like a peacefully sleeping pet. It's a distinctive and memorable shape.

Space Trash

You wish to know by what mysterious means
I reached the moon?—well—confidentially—
It was a new invention of my own.
—Edmond Rostand, *Cyrano de Bergerac,* Act 3

If visitors from another galaxy ever land on the Moon, they'll know someone else was there first. Millions of years from now, the astronauts' footsteps will still be sunken into the dusty lunar soil, and over twenty tons of high-ticket, high-tech junk left behind by the American and Soviet space programs will still litter the lunar landscape. Visitors from distant worlds will see the jetsam of the space program: crashed satellites, rocket parts, robot explorers, barely used moon buggies, seismometers, laser reflectors, and assorted tools and pieces of equipment left behind simply to lighten the payload. In addition, they might find various medals, a commemorative statue of astronauts who died on duty, an American flag, three cameras, two golf balls, one photograph encased in plastic of an astronaut's family, a pin, and a feather from a falcon. Should they visit Venus or Mars they'll find those distant territories similarly marked with discarded pieces of earthly technology.

And then there's the junk in orbit. During the very first American space walk, Astronaut Ed White dropped a glove into eternity. In 1966, Mike Collins lost a Hasselblad camera during a space walk; and during a 1971 Apollo mission, a toothbrush was sucked into space. A comb and a screwdriver once circled the Earth, and bags full of garbage are frequently ejected from Soviet space stations. Like many other objects launched into orbit since the dawn of the Space Age in 1957, most of these objects have reentered the atmosphere and been incinerated. But literally thousands of satellites and spacecraft, operational and otherwise, whole or in parts, are still spinning around the Earth, including rocket boosters the size of small apartment buildings and a portion of the *Apollo 10* lunar module the size of a truck. As of September 30, 1988, a minimum of 7,122 objects, launched by the United States, the Soviet Union, the European Space Agency, China, Japan, and Israel, were still up there, endangering other flights.

And then there are fragments, tiny shards of the space program no bigger than a centimeter but large enough to inflict damage—like the paint chip that crashed into a 1983 *Challenger* flight and left a quarter-inch crater in a window. Two 31-inch telescopes at MIT searching the skies for orbiting scraps discovered nearly 48,000 teensy satellites larger than a centimeter. At the right velocity, such an object crashing into a burned-out rocket stage—or, say, a space suit—could pulverize it. "The resulting debris cloud, which scatters into new orbits, might easily include 40,000 centimeter-size and 10 million millimeter-size objects," writes NASA project scientist Donald J. Kessler. "These pieces could then collide with other spacecraft producing even more fragments at an exponential rate."

As for fragments smaller than a millimeter, they're out there in numbers too large to be imagined. Tiny impact pits discovered on returned satellites suggest that "we have created anywhere from 10 billion to thousands of *trillions* of orbiting objects in the size range of 1 to 100 microns," Dr. Kessler believes. These particles, evidence of our attempt to explore the universe, surround our blue planet like a halo of junk.

Mercury:
A Heavy Metal World

Then at last I saw it, little Mercury,
a pinprick of light, a mote of dust in
the gathering day.

—Chet Raymo

egend has it that on his deathbed Nicolaus Copernicus, a great theoretical astronomer but an indifferent star-gazer, regretted one omission in particular: he had never seen the planet Mercury. In his meager writings he complained that other people—i.e., "the ancients"—had better viewing opportunities than he because "they were helped by clearer skies," and that all in all Mercury had "inflicted many perplexities and labors on us."

The reason is that Mercury is so tiny and close to the Sun that it is easy to miss; a seventeenth-century writer called it "a squirting lacquey of the sun, who seldom shows his face in these parts, as if he were in debt." Visible only at sunrise and sunset, always low in the sky, Mercury never makes a dramatic appearance. Even mythologically, it often plays a secondary role. To the Skidi Pawnee, it was known as "the Little Brother of the Morning Star." The early Greeks gave it three names: Stilbon, as a general name; Apollo, for its morning appearance; and Hermes, in the evening, after the god of magic, medicine, commerce, thievery, oratory, and the occult. With his winged sandals and cap (familiar to us from the FTD florist's delivery van), Hermes carried messages between gods and mortals and conveyed souls to the Underworld. In the Roman pantheon of the Gods, Hermes was known as Mercury. Neither as sexy as Venus and Mars nor as powerful as Jupiter and Saturn, Mercury was simply the divine messenger—always a thankless role.

The association between the god and the celestial body was undoubtedly made because the planet whizzes around the Sun in only 88 days, faster than any other member of the solar system. For every two turns around the Sun (i.e., every two mercurial years), Mercury rotates on its axis exactly three times. Thus its day is bizarrely long, equaling 58.65 Earth days (a fact determined by radar transmissions picked up by the Arecibo radio telescope in Puerto Rico). The Italian astronomer Giovanni Virginio Schiaparelli (1835–1910), best known for the discovery of dark streaks on Mars he called *canali,* believed that the mercurial day and year were the same length, and that Mercury always kept the same face to the Sun. Although this isn't true, it isn't unreasonable either, for among the many oddities of its cycle, Mercury faces the same direction during the times when it is most easily visible from Earth.

Compared to the orbits of most of the other planets, Mercury's orbit is both extremely eccentric (i.e., elliptical) and steeply tilted. When Mercury is nearest the Sun, the Sun's gravitational force is so strong that the little planet turns in sync with it, facing it in the same way for about sixty days. To an observer on the surface of Mercury, the enormous Sun would appear to stand still in the sky for that entire time. Finally, as Mercury swung away in its orbit, the Sun would set.

Like the Moon, Mercury is a cratered, inhospitable place that shows no signs of recent geological activity. The 45 percent of the planet

MERCURY

Mean Distance from the Sun
36,000,000 miles

Diameter at the Equator
3,031 miles

Diameter if Earth = 1
0.38

Mass if Earth = 1
0.055

Surface Gravity if Earth = 1
0.38

Moons
none

Length of Year
(revolution around the Sun)
87.97 Earth days

Length of Day
(rotation at equator)
58.65 Earth days

NASA

Eighteen computer-enhanced pictures were pieced together to form this photomosaic of Mercury.

mapped by *Mariner 10* is covered with fault lines, rolling plains, banks of cliffs that crisscross the rocky surface like giant wrinkles, and ancient craters, the most impressive of which is the Texas-size Caloris Basin. Ringed by mountains, the basin was created by an impact so powerful it sent seismic shock waves to the opposite side of the planet.

Although Mercury is closest to the Sun, it is not the hottest planet because its thin atmosphere is incapable of retaining heat. Venus, which is farther away, is hotter because its thick, soupy atmosphere traps heat, causing the planet to suffer from the greenhouse effect. On Mercury, where heat dissipates rapidly, the temperature ranges from a high of 800° F at the equator to around 235° below zero at the poles—which, despite the planet's proximity to the Sun, may be covered with ice.

Odder still is Mercury's interior. Imagine a huge metal ball dipped in mud, and you've got a good image of the place. It is overwhelmingly metal. Its iron-nickel core, the densest in the solar system, takes up almost four fifths of the diameter of the entire planet.

SIXTEEN CRATERS

Never let it be said that planetary scientists lack appreciation for the arts. While it is true that the International Astronomical Union committee on names wanted to give the craters of Mercury the names of birds, this did not occur, thanks to the persuasive interference of Carl Sagan. Excluding the enormous Caloris Basin, the sixteen largest craters on Mercury bear the following names (listed here in descending order of size):

1. Beethoven	9. Haydn
2. Tolstoy	10. Mozart
3. Raphael	11. Bach
4. Goethe	12. Välmiki
5. Homer	13. Renoir
6. Vyasa	14. Wren
7. Rodin	15. Vivaldi
8. Monet	16. Matisse

The question of whether the core is entirely solid has yet to be answered. Scientists presumed so, but had to reconsider when *Mariner 10* proved that Mercury, like the Earth but unlike Venus, Mars, and the Moon, has a magnetic field. Since magnetic fields are usually produced by the movement of fluid material inside a body's core, the outer rim of Mercury's core may be partially molten. Or maybe the magnetic field is generated from two "hot spots" on the planet's surface located exactly where you'd expect—on either side of the globe, right where the surface receives maximum heat from the Sun. In any case, the magnetic field acts as a shield, protecting the tiny planet from the harsh particle wind that blows from the Sun in a never-ending torrent.

Venus: Toxic Twin

Star light, star bright,
First star I see tonight,
I wish I may, I wish I might,
Have the wish I wish tonight.

wo aspects of Venus impressed early civilizations: its luminous beauty and its cycle of evening and morning appearances. The first attribute led them to correlate the planet with goddesses rather than gods (the only planet other than Earth for which this is true). But these stories often cast an ominous shadow, especially for men. For instance, the Greek goddess of love, Aphrodite—Venus to the Romans—was born from the sea into which Saturn had tossed his father's severed genitals. Unfaithful to her husband, dangerous to her mortal lover, the indirect cause of the Trojan War, she combined allure with threat.

Similar qualities exist in a tale told by the Tumupasa Indians of South America. The Sun caught two sisters, Venus and the Moon, stealing fruit from his garden. He nonetheless made the Moon his mistress, but his penis grew so enormous that he was forced to cart it around in a basket. One day he pointed it at the other sister, Venus. Thinking it was a snake, she cut it in half, causing the Sun to die and soar into the sky, where he can still be seen.

The Mayans paid particular attention to the planet's predictable pattern of appearances. For months on end, Venus shines only after sunset in the western sky. During this period, it is usually the first "star" visible (although unlike actual stars, it shines with a constant rather than a twinkling light). Finally, it draws so close to the Sun that it is lost to sight. After a brief intermission, it reappears in the east, where it is visible for about nine months at dawn. Then it disappears behind the Sun. Seven weeks later, it reappears as the evening star once more. This 584-day cycle formed the basis of the Mayan calendar, a system so accurate that in 481 years it was off by only about two hours.

Many cultures, inspired by this cycle, endowed Venus with a double identity. The early Egyptians, the Greeks of Homer's time, and the Romans all gave it two names. The Pueblos associated it with twin war gods, sons of the Sun. And some tribes in western Africa thought of Venus as the two wives of the Moon.

Venus and the Moon are often linked mythologically, but to astronomers the real analogy has been between Venus and the Earth. The diameter of Venus (7,521 miles) is 95 percent of the Earth's, its gravity is 91 percent and its mass 82 percent of the Earth's. But in other ways the planets are completely different. "If Venus is our sister," declares space scientist Andrew Fraknoi, "it is a twisted sister." For instance, its orbit is nearly a perfect circle—not an ellipse, as are the orbits of the other plan-

NASA

The cloudy cover of Venus as photographed by the *Pioneer-Venus* spacecraft.

ets. It has no magnetic field. And it is the slowest planet in the solar system. So leisurely is its rotation that although the planet requires 225 Earth days to complete its yearly journey around the Sun, it takes 243 days to turn on its axis a single time. On Venus, the day is longer than the year.

In addition, Venus rotates backward. As a result, the long Venusian day begins with sunrise in the west and ends when the Sun sets in the east. Not that anyone could see it: Thick layers of hydrochloric and sulfuric acid, including one layer twelve miles thick, blanket the planet and give it a bland face. Through a telescope, Venus is a featureless wash of yellowish clouds. These clouds, slashed by lightning as often as twenty-five times a second, whip around the planet every four days and reflect huge amounts of sunlight, which is why the planet shines so brightly. But from the surface, the clouds completely obscure the sky. On Venus, you can't see the Sun or the stars.

Astronomers once thought that Venus was a natural garden—perhaps even a spectacular one. Figuring that the cloud cover and proximity to the Sun indicated a warm, wet environment, Svante Arrhenius (1859–1927), a Nobel Prize–winning Swedish chemist, proposed that Venus was a fertile, swampy place not unlike the Earth during the Mesozoic era. Other scientists early in this century thought it might be a world of water, teeming with life.

VENUS

Mean Distance from the Sun
67,200,000 miles

Diameter at the Equator
7,521 miles

Diameter if Earth = 1
0.95

Mass if Earth = 1
0.815

Surface Gravity if Earth = 1
0.91

Moons
none

Length of Year
(revolution around the Sun)
224.70 Earth days

Length of Day
(rotation at equator)
243.01 Earth days

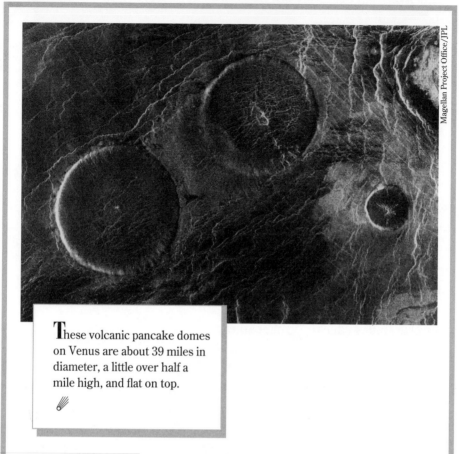

Magellan Project Office/JPL

These volcanic pancake domes on Venus are about 39 miles in diameter, a little over half a mile high, and flat on top.

But no giant ferns sway in its sulfurous yellow wind, and no primeval organisms swim in its seas. It has no seas. A Soviet spacecraft of the 1970s discovered that beneath its thick acid clouds, Venus has a surface temperature of about 900° F—hot enough to melt lead and boil away entire oceans. This scalding heat is caused primarily by the atmosphere, a thick mist of 96 percent carbon dioxide suffused with sulfuric acid. The problem lies in the carbon dioxide. On Earth, CO_2 dissolves in water and is bound in rocks. But on Venus it remains suspended in the atmosphere, which then traps heat and makes Venus the hottest place in the solar system—a greenhouse from hell.

The atmosphere is oppressive in another way, too. Because it is so thick, the surface pressure is almost 100 times that of Earth's. As a result, a series of Venus probes launched by the Soviets in the 1960s and 1970s were crushed to death before they ever hit the ground.

Fortunately, the Soviets eventually succeeded in landing a number of spacecraft on the arid surface of the planet, and these sent back pictures. The American Pioneer missions launched in 1978 sent back radar and gravity maps. And the spacecraft *Magellan,* which began orbiting Venus in August 1990, returned detailed radar images that have enabled astronomers to map most of the surface.

The surprises discovered by *Magellan* include an area dubbed Crater Farm, where the impact craters look like enormous sunflowers; thousands of round, pancake-shaped domes, some of which are twenty miles wide and a mile high; a series of long, deep faults that form a vaguely humanoid shape that many observers have likened to Gumby; mountain ranges that, like the Rockies, form straight lines; and a fractured plain

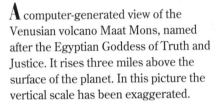

A computer-generated view of the Venusian volcano Maat Mons, named after the Egyptian Goddess of Truth and Justice. It rises three miles above the surface of the planet. In this picture the vertical scale has been exaggerated.

Magellan Project Office / JPL

the size of Rhode Island crisscrossed by lines so straight and parallel that they could easily be city streets. To the best of our knowledge, similar terrain exists nowhere else.

Venus is less heavily cratered than Mercury, and the craters are large because small meteors could not survive the free-fall plummet through the atmospheric soup. The craters have another distinctive quality, too: They look surprisingly young.

The scarcity of craters and their pristine shapes have led scientists to conclude that the surface of Venus was rejuvenated by a series of volcanoes as recently as 400 million years ago. The pancake domes may have been created when thick, pasty lava oozed from the hot interior; enormous circular structures called coronae, 100 to 600 miles across, are thought to have been created when the hot interior bulged upward and then collapsed like a soufflé; and evidence suggests that rivers of lava once snaked across the landscape for hundreds of miles. Whether volcanoes still erupt on Venus is a question yet to be answered.

Almost all the features on Venus have been named after women. They include two main highland areas, Ishtar Terra in the north and Aphrodite Terra nearer the equator, and a multitude of craters named after a formidable group of women including Clara Barton, Amelia Earhart, Florence Nightingale, Anna Pavlova, Sacajawea, Sappho, and Colette, whose name graces a crater two miles deep. Only one major feature is named after a man. It will surprise no one—not the ancient Greeks, not the Tumupasa Indians, not Sigmund Freud—to learn that that feature, named after the physicist James Clerk Maxwell (1831–79), is the highest mountain on the planet.

The Emperor of Venus

Look through a telescope at Venus on a few different occasions and you will notice that it has phases, a fact Galileo discovered in 1610 and announced in a most seventeenth-century fashion: via anagram. Written in Latin, Galileo's message said, "Haec immatura a me iam frustra leguntur oy," which translated meant, "These immature things I am searching for now in vain." Johannes Kepler—who liked to shuffle letters of the alphabet around so much that he once put three anagrams of his own name on the title page of a book—immediately recognized Galileo's wordplay as an anagram, and he came up with eight possible solutions to the puzzle. Unfortunately, none was correct.

Reordered, the letters spelled out another equally cryptic sentence: "Cynthiae figuras aemulatur mater amorum," or, in translation, "The mother of love imitates the figures of Cynthia," which meant that Venus looks like the Moon—sometimes full and round, sometimes gibbous, sometimes a slim crescent.

Unlike the Moon, however, Venus seems to grow and shrink dramatically depending upon its distance from the Earth. As a crescent, it looks seven times larger than when it is full, which suggests that during its crescent phase, it is closer to us, while its full phase occurs when it is far from us and on the other side of the Sun.

In 1643, Giovanni Riccioli noticed another way in which Venus resembles the Moon. Just as the dark part of the Moon is sometimes dimly visible in the reflected glow of the Earth (a phenomenon known as earthshine and first identified by Leonardo da Vinci), the dark part of Venus occasionally seems faintly illuminated. That sheen is called the ashen light.

Franz von Paula Gruithuisen, a physician who became head of the Munich Observatory in 1826, was fascinated by this light. A sharp observer whose interpretations were often tinged with romantic fancy, Gruithuisen was one of the first astronomers to state that craters had been formed when meteors slammed into the surface of the Moon—but he also thought he had seen ancient ruins there and clearings in the lunar forests. After many observations, he concluded that the ashen light of Venus was artificially produced by the inhabitants of a planet he imagined to be a verdant paradise, "incomparably more luxuriant than even the virgin forests of Brazil." He thought the Venusians might be burning their jungles, perhaps to produce arable land or to prevent the migration of large groups of people who might start a war. Alternatively, he considered, the ashen light could be a fireworks display produced in the course of a gala festival. It could even signal a coronation.

This last idea made sense to him. He noted that the ashen light had been observed in 1759 and again in 1806: 47 Earth years—or 76 Venusian years—passed between those two sightings. "If we assume that the ordinary life of an inhabitant of Venus lasts 130 Venus years, which amounts to 80 Earth years, the reign of an emperor of Venus might well last 76 Venus years." The ashen light could easily signal the coronation of "some Alexander or Napoléon."

He was mistaken. The ashen light, like aurorae seen on Earth, is caused by the interaction of atomic particles with solar radiation. It is strictly an atmospheric phenomenon, not a political one.

♀

Gaia: The Once and Future Earth

Talk of mysteries! Think of our life in nature—
daily to be shown matter, to come in contact
with it—rocks, trees, wind on our cheeks! The
solid earth! the actual world! the common
sense! Contact! Contact! Who are we? Where
are we?

—Henry David Thoreau

n early pre-Hellenic mythology, the Earth's name was Gaia. She was the creator, the mother, the central deity originally associated with the Delphic oracle; she ruled the living and the dead and her body was the Earth itself.

But gradually her role was diminished, pushed aside by patriarchal mythmakers including Hesiod and Homer. Gaia was relegated to the status of wife and mother whose male children—Saturn and Jupiter among them—ruled the Heavens and the Earth. Her name fell into obscurity. Even astronomers have often forgotten about the Earth. Our own planet has more than once been excluded from astronomy books or dismissed with a single slur and consigned to the realm of mere geology.

But that attitude has begun to change. Strangely, Gaia has returned. The term is now associated with a movement—part ecological, part scientific, part neopaganist—that considers the Earth a living organism that reacts as a single body to its circumstances. Looking simultaneously backward and forward, the Gaia movement has received its most convincing image from the space program, which allowed us to view the Earth from the outside. The startling photographs from the

Apollo mission showed us our world: small, blue, swirled with clouds, suspended in space, alone. They emphasized as nothing else could that the Earth is a single place with a single atmosphere and a single future.

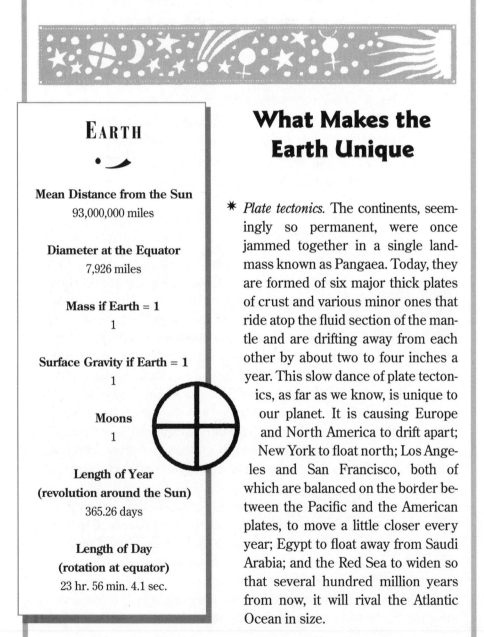

EARTH

Mean Distance from the Sun
93,000,000 miles

Diameter at the Equator
7,926 miles

Mass if Earth = 1
1

Surface Gravity if Earth = 1
1

Moons
1

Length of Year
(revolution around the Sun)
365.26 days

Length of Day
(rotation at equator)
23 hr. 56 min. 4.1 sec.

What Makes the Earth Unique

✳ *Plate tectonics.* The continents, seemingly so permanent, were once jammed together in a single landmass known as Pangaea. Today, they are formed of six major thick plates of crust and various minor ones that ride atop the fluid section of the mantle and are drifting away from each other by about two to four inches a year. This slow dance of plate tectonics, as far as we know, is unique to our planet. It is causing Europe and North America to drift apart; New York to float north; Los Angeles and San Francisco, both of which are balanced on the border between the Pacific and the American plates, to move a little closer every year; Egypt to float away from Saudi Arabia; and the Red Sea to widen so that several hundred million years from now, it will rival the Atlantic Ocean in size.

* *Liquid water.* Oceans like ours, which cover about three quarters of the globe, exist nowhere else in the solar system—although liquid water may lie beneath the surface on other, icier bodies, and rivers may once have flowed on Mars.

* *Location, location, location.* Neither too close to the Sun, like arid Mercury and steaming Venus, nor too far away, like frigid Mars, the Earth's position is the only one to which the word "temperate" could possibly be applied.

* *The atmosphere.* Consisting primarily of nitrogen (78 percent) and oxygen (21 percent), it is the end result of the eons-long interaction of evaporation, precipitation, and photosynthesis; it arises because of, and permits the continued existence of, the Earth's rarest feature:

* *Life.*

NASA

In this view of Earth, taken from the *Apollo 17* spacecraft, Antarctica is at the bottom. The Red Sea and the northeastern portion of Africa are near the top, with Somalia clearly visible beneath a cloudless sky.

WHY THERE ARE SEASONS

Surprisingly, distance is not the determining factor. Indeed, because of the Earth's slightly elliptical orbit around the Sun, the Northern Hemisphere is actually about 3 million miles *closer* to the Sun in winter than in summer. It's the tilt that matters. The Earth is tilted to the plane of the planets at a fairly steep angle of 23° 27′. During the summer, one hemisphere is tilted toward the Sun, which appears higher in the sky and therefore is in the sky for a longer stretch of time. So the day is longer, that hemisphere receives more radiation, and the heat rises. Six months later, the same hemisphere is tilted away from the Sun. The Sun is lower in the sky, the day is shorter, the radiation is less, and the temperature drops.

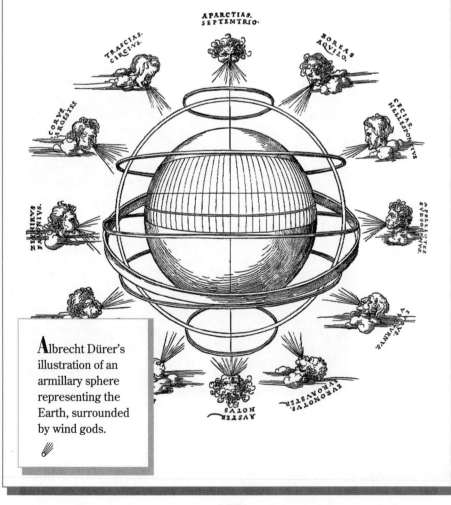

Albrecht Dürer's illustration of an armillary sphere representing the Earth, surrounded by wind gods.

WHY THE SKY IS BLUE

The Sun may look yellow, but the light it pours forth is white—which is to say, a combination of all the colors, as Isaac Newton discovered when he passed a ray of sunlight through a prism and saw a rainbow-colored strip of light. Yet the gas and dust in the atmosphere scatter light waves in such a way that the sky is not white but blue (most of the time). That's because the short blue waves, with their high-energy photons, are scattered sideways, hither and yon, an effect called Rayleigh scattering (after the English physicist who discovered it). So when the Sun is high in the sky, the sky is bathed in blue.

But when the Sun sinks beneath the horizon, the long, slow wavelengths on the red end of the spectrum begin to make an appearance. The reason is that when the Sun is on the horizon, sunlight has to cruise through a larger swath of the atmosphere and hence a greater concentration of dust before it reaches its worshipers. The frantic blue light is not just scattered but dissipated; stolid red forges ever onward on a straight path through the atmospheric muck. The more dust there is, the more likely the sun is to go down in a riot of red. That's why the most spectacular sunsets occur when the dust is densest—as after volcanic eruptions, when people around the world watch the sunset with amazement.

Mars: The Reality

So this is where Elysium lies,
just north of Atlantis,
on the far side of Barsoom.

—Diane Ackerman

ars glows like a burning ember embedded in the night, and as a result it has long been linked with the color red and the god of war, an association made by Aristotle in the fourth century B.C., and before that by the Babyloni-ans, who named the planet after Nergal, a warrior god of death, destruction, and disorder. A fifteenth-century French astrological treatise described a man born under the influence of Mars as "red and angry . . . a great walker, and maker of swords and knives, and shedder of man's blood, a lecher and speaker of ribaldry, red bearded, round vis-aged. . . ." Similar thoughts may have been in the minds of the Melane-sians, who saw Mars as the home of a giant red pig.

Nonetheless, Mars is not red. It is russet and brown, an autumnal, rust-colored world, as *Viking 1* dramatically showed. The rocky soil is the color of clay, thanks to the presence of iron oxides, and the sky, suffused with large particles of dust, is neither blue, as we might imagine, nor black, as some scientists had expected, but apricot, salmon, and peach.

The air on this coral-colored world is 95 percent carbon dioxide, so it's impossible to breathe, and it's so thin you'd be bathed in solar radiation. At night, you'd freeze to death. And without a space suit, your blood would boil, courtesy of the low surface pressure.

But Mars wasn't always like that. Early in its history, after radioactive heat inside the planet caused the underground permafrost to melt, wide rivers meandered across the plains. Active volcanoes poured forth gases

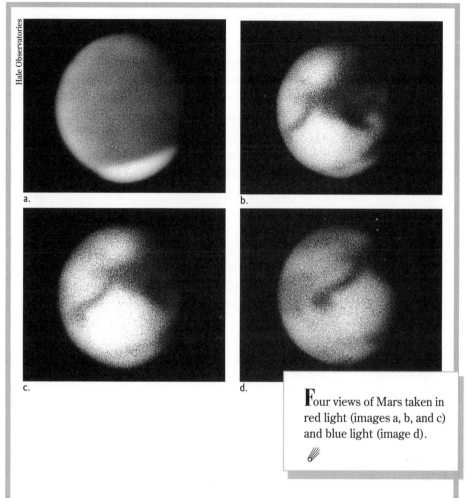

Hale Observatories

a.

b.

c.

d.

Four views of Mars taken in red light (images a, b, and c) and blue light (image d).

that clung to the surface and formed a thick atmosphere. In many ways, Mars resembled the young Earth.

But the Martian gravity—38 percent that of Earth's—was too low to hold on to the atmospheric gases, and as the planet cooled, they leaked away or froze at the poles. Without a warm blanket of gas, the temperature plunged, a process astronomers David Morrison and Tobias Owen have dubbed "a runaway refrigerator effect."

Today, although water may be frozen beneath the poles, there is none on the surface, where water acts like dry ice and an ice cube would be instantly vaporized; like blood, water on Mars would boil because the boiling point of water gets lower as the atmospheric pressure falls.

Yet in certain ways Mars resembles Earth. About half the size of our planet, Mars is tilted at approximately the same angle as the Earth and turns on its axis every 24 hours 37 minutes, a unit of time inexplicably called a "sol" in order to distinguish it from an Earth day. Mars orbits the Sun every 687 days, so the seasons are approximately twice as long as they are here. Cyclones and high-speed winds stir up dust devils and storms, and murky yellow dust sometimes envelops the entire planet. Of the two polar caps, the northern one, made primarily of frozen water, is colder and boasts a larger permafrost, while the southern pole is smaller, warmer, and made of carbon dioxide.

Huge volcanoes dominate Mars. The three large, evenly spaced volcanoes of the Tharsis Montes are spread out in a line that slants across the equator. Nearby is Olympus Mons, three times as high as Mount Everest and the largest volcano in the solar system. Elsewhere the surface is gashed with deep canyons, such as the Valles Marineris, which is twenty-six times longer than the Grand Canyon and almost three times as deep. In the south, where the land is older and more heavily cratered, the most dramatic feature is Hellas, a large, light area where dust carpets the floor of an ancient impact canyon long since filled with lava that welled up from the interior.

Mars is beautiful in its way, even though, compared to the blue and

This Martian boulder, photographed by *Viking 1*, has been nicknamed "Big Joe."

verdant Earth, it is a harsh and unlivable steppe. On the other hand, compared to the blistering hell of Venus, it's not half bad. And as for the other planets—what can you do on a world made of hydrogen? It's no wonder that despite its mythological links with war and disaster, the idea of life on Mars has captured the human imagination. For the intrepid space travler, the red planet is the destination of choice.

Fear and Panic: The Moons of Mars

The existence of moons orbiting Mars was suspected long before it was confirmed. Johannes Kepler, assuming that geometric progressions were a law of nature, figured that since the Earth had one satellite and Jupiter had four (we now know it has many more), Mars must have two; it made a Pythagorean kind of sense. (He also predicted the existence of "six or eight round Saturn, and perhaps one each round Mercury and Venus.") In *Gulliver's Travels,* his masterpiece of 1726, Jonathan Swift posited the existence of "two lesser stars, or satellites, which revolve about Mars," and a quarter century later, Voltaire echoed the sentiment. But it wasn't until 1877 that the Martian moons were found.

The man responsible was Asaph Hall (1829–1907). Apprenticed to a carpenter at age thirteen, he became a professor of astronomy at the Naval Observatory in Washington, D.C. (where he once discussed the cosmos with Abraham Lincoln). In August 1877, when Mars was as close to the Earth as it ever gets, he began a search for the missing moons using the observatory's powerful 26-inch refractor telescope—the largest then in existence.

On August 11, cheered on by his wife, he detected a moon faintly visible in the glare around Mars. But no sooner did he mark the moon's position than fog rolled in from the Potomac River and he was forced to stop. He notified the observatory's scientific chief of his discovery and

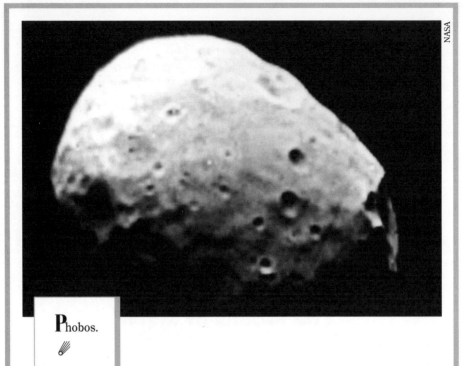

NASA

Phobos.

turned his attention to another, equally pressing matter: how to "get rid of my assistant. It was natural that I should wish to be alone; and by the greatest good luck Dr. Henry Draper invited him to Dobbs Ferry at the very nick of time."

A few days later, after the fog cleared and he no longer had to worry that his assistant would horn in on the discovery, Hall found a second moon. He named the pair Phobos and Deimos—Fear and Panic—after two sons of the god of war mentioned in the *Iliad*.

Phobos and Deimos are strange moons. Deimos, which is smaller and more distant, creeps around Mars every 30 hours 18 minutes. Phobos, the larger satellite (it is nonetheless only about fourteen miles across), lies so close to the planet that it takes only a little over 4 hours to rise and set. In addition, it appears to rise in the west and set in the east. A Martian would see two moons crossing the sky in opposite directions.

Phobos has another peculiar quality: it is gaining momentum. The faster it goes, the lower it goes; the lower it goes, the stronger the pull of gravity. Forty million years from now it will smash into the surface of Mars—a fact that in 1959 gave rise to a new and wonderful theory developed by the Soviet scientist Iosif Shklovskii. It occurred to him that if Phobos had a density similar to Mars, it shouldn't be accelerating. It must therefore be lighter and less dense. "But how can a natural satellite have such a low density?" Shklovskii asked in his book *Intelligent Life in the Universe.* Clearly, Phobos had to be rigid, but rigidity required a fairly high density, in which case, he wrote, "only one possibility remains. Could Phobos be indeed rigid, on the outside—but hollow on the inside? A natural satellite cannot be a hollow object. Therefore, we are led to the possibility that Phobos—and possibly Deimos as well—may be artificial satellites of Mars."

But when *Mariners 7* and *9* took pictures of the moons in 1969, the possibility that the Martians were satellite-launching cousins of ours was quashed. The moons resemble potatoes. The most spectacular feature is a crater on Phobos so massive that it looks as if a sizable portion of the moon had simply been removed with a giant scoop. Named after Chloe Angeline Stickney, Hall's devoted wife, it was produced in a crash so powerful that the moon cracked; grooves radiating from the point of impact meet on the opposite side of the little satellite.

Phobos and Deimos are covered with dark gray dust quite unlike the reddish soil of Mars. Astronomers theorize that the misshapen moons were once asteroids, captured by the Martian gravitational force.

MARS

Mean Distance from the Sun
141,500,000 miles

Diameter at the Equator
4,217 miles

Diameter if Earth = 1
0.53

Mass if Earth = 1
0.107

Surface Gravity if Earth = 1
0.38

Moons
2

Length of Year (revolution around the Sun)
686.98 Earth days

Length of Day (rotation at equator)
24 hr. 37 min. 22.6 sec.

Despite their names, the oddly shaped bodies inspired cooperation between Russian and American scientists. *Phobos 1,* launched on July 7, 1988, was lost for all eternity when a ground controller sent it a computer command that was one character off. *Phobos 2,* launched five days later, went into Mars' orbit and returned thirty-seven images. But radio contact was lost due to a computer malfunction, and the probe never did manage to land its "hopper." This "high-tech pogo stick" was supposed to bounce all over the little, low-gravity moon taking gravitational, magnetic, and chemical measurements. Its failure was an international disappointment.

Life on Mars

The history of humanity's obsession with life on Mars is a long and glorious one, for although optimists have spun fantasies about life throughout the solar system, Mars has always seemed the only probable location. Like Venus, it is relatively close to Earth. But whereas Venus is enshrouded in clouds, the surface of Mars is tantalizingly visible. As early as 1659, the Dutch astronomer Christiaan Huygens, whose discovery of a dark triangular region named Syrtis Major (Large Bog) enabled him to estimate the length of the Martian day, assumed the presence of intelligent life there. Observers soon discovered the changing polar ice caps as well as the dark patches, which shrink in winter and grow in summer. These areas, accumulations of rock and dust that shift when the wind blows, were vaguely reminiscent of vegetation. Vegetation suggested agriculture; agriculture implied civilization. The logic seemed faultless. In 1784, William Herschel was sure that the planet's "inhabitants probably enjoy conditions analogous to ours." But these were just suppositions.

Then, in the summer of 1877, startling evidence emerged. Giovanni Schiaparelli, the color-blind director of the Brera Observatory in Milan (and uncle of the Parisian fashion designer Elsa Schiaparelli), noticed a series of straight lines he called *canali* slicing across the surface. In Italian, the word means "channels," but no one bothered with the transla-

Lowell Observatory

Percival Lowell
(1855–1916).

tion. They were simply called canals, and canals, as everybody knew in those years shortly after the construction of the Suez Canal, were built by intelligent beings. Mars was therefore inhabited.

No one was more excited about this than Percival Lowell (1855–1916). An amateur astronomer from a famous aristocratic American family (it has been said that "the Lowells talk to the Cabots, and the Cabots talk only to God"), Lowell was a brilliant mathematician who turned down an offer to teach at Harvard even though he considered mathematics "the thing most worthy of thought in the world." Instead, he toured Europe, worked for six years as head of his grandfather's cotton mill, and traveled in the Far East, an experience he described in four books written over a four-year period.

During his final visit to the Far East, Lowell heard that Schiaparelli's deteriorating vision was forcing him to give up his observations. Lowell decided to pick up the torch, and in 1894 he built an observatory in Arizona for the express purpose of observing Mars. He found everything Schiaparelli had seen, and more. Eventually, he mapped over 500 canals. "It is the systematic network of the whole that is most amazing," he said. "Each line not only goes with wonderful directness from one point to another, but at this latter spot it contrives to meet, exactly, another line which has come with like directness from quite another direction." He theorized that the water that flowed on Mars in the distant past had largely evaporated; thus the Martians built an elaborate web of canals to transport whatever water still existed at the poles to agricultural sites elsewhere on the planet. He did not believe that the Martians were human because, he knew, the Martian atmosphere wouldn't support creatures such as ourselves. But what about a different sort of being? "To argue that life of an order as high as our own, or higher, is impossible because of less air to breathe than that to which we are locally accustomed is, as [the French astronomer Camille] Flammarion happily expresses it, to argue not as a philosopher, but as a fish."

Alas, there are no canals on Mars and nothing even mildly resembling them. Lowell's and Schiaparelli's certainty regarding the canals is usually blamed on three phenomena: the inadequacies of the human eye, which tends to see lines that don't exist between otherwise unconnected spots; the problem of observing through the Earth's shimmering atmosphere, which makes it difficult to resolve details (that's why astrophotography is so important); and our psychological tendency to see what we want to see. "When certain phenomena can only be seen with great difficulty," wrote Lord Rosse, a nineteenth-century telescope maker, "the eye may imperceptibly be in some degree influenced by the mind." The canals on Mars are an illusion.

In 1908, Lowell published *Mars as the Abode of Life,* married for the first time (at age fifty-three), and turned his attention to other matters: mainly, the search for another planet in our solar system. In this, too, he failed. But when a ninth planet was discovered after his death, it was named Pluto, in part because the first two letters were also Percival Lowell's initials.

Meanwhile, the search—or the hope—for life on Mars went on . . . and on . . . and on. In 1924, the U.S. Navy suspended radio transmissions for three days in the hope of catching signals from Mars. When no signals arrived, Camille Flammarion sighed, "Perhaps the Martians tried before, in the epoch of the iguanodon and dinosaur, and got tired." In 1959, the year Shklovskii announced that the moons of Mars were artificial satellites, another scientist claimed to detect organic matter in the Martian spectrum. Neither claim was true, and in 1965, when *Mariner 4* sent back photographs of the rock-strewn surface, *The New York Times* announced in its front-page article, "A heavy, perhaps fatal, blow was delivered today to the possibility that there is or once was life on Mars."

The blow was far from fatal. It is true that Mars, unprotected by an ozone layer, is bombarded—and sterilized—by solar ultraviolet radiation. It is true that analysis of Martian soil scooped up from the Plains of Gold by *Viking 1* in 1976 revealed no carbon-based organisms and no

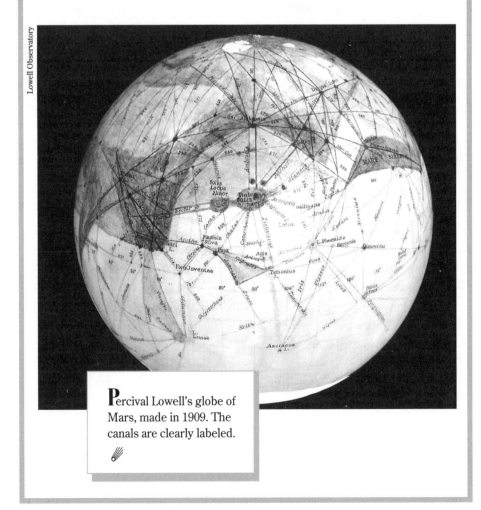

Lowell Observatory

Percival Lowell's globe of Mars, made in 1909. The canals are clearly labeled.

sign that they had ever been there. Nonetheless, the idea did not die. Astronomers—including Carl Sagan, the strongest modern-day proponent of life on Mars—have not given up hope. Some scientists believe that three and a half billion years ago, when rivers ran across the Chryse Planitia, early forms of life may have existed on Mars. All too soon, the atmosphere leaked into the cosmos and the water disappeared. But biological experiments performed by *Viking 1* showed that even now the Martian soil is not completely inert. It's remotely possible that an unimagined life form with a different chemical base might exist there—maybe in mud buried deep beneath the surface. Or perhaps life is lurking in places where our spacecraft have not gone, such as the edge of the polar caps. Perhaps we simply haven't known where to search, or what we were searching for.

But the chances of finding life on Mars are not promising. So scientists are developing a complicated, long-term scenario whereby the red planet could be turned into a habitable ecosystem. The plan, which would take two centuries to complete, begins with the establishment of a permanent space station on Mars complete with greenhouses, computers, and a life-support system. Pioneering Earthlings would eventually construct factories from which gases would be released into the atmosphere, making it thicker and warmer. The increase in temperature would cause the polar caps to melt, and oxygen could be pumped into the atmosphere. First microorganisms and then hardy plants could be introduced to contribute further to the oxygen supply. Creating this Frankensteinian environment would require plenty of money, technology, and courage, especially on the part of those whose mission, should they choose to accept it, would be to build and manage gas factories on a cold, desolate, airless frontier. This plan is so incredible that it has given rise to a debate concerning both its methods and its ethics. Nonetheless, there is a plan, and as the scientist Christopher P. McKay remarked, "A green Mars is better than a red Mars." Sooner or later, one way or another, there will be life on Mars.

Devil Girl from Mars
(and Other Cultural Landmarks)

Although Swift and Voltaire wrote about Mars (and are honored on Deimos, whose two main craters bear their names), it was not until the waning years of the nineteenth century that the Martian landscape (and character) staked a claim in the popular imagination. In 1880, Percy Gregg's novel *Across the Zodiac* featured a journey to Mars, and in 1897, Mars-inspired literature passed beyond the picaresque with the publication of two adventures in which Martians invaded the Earth. The first was a novel called *On Two Planets* by Kurd Lasswitz, a German mathematics professor. The second, a magazine serial by an English author, was published the following year as a book: *The War of the Worlds,* by H. G. Wells.

The boom was on. Edgar Rice Burroughs, the creator of Tarzan, wrote eleven books about Mars—or "Barsoom," as he called it—beginning in 1912 with *A Princess of Mars,* in which a southern gentleman named John Carter travels to the red planet without benefit of technology—he wishes himself there—and falls in love with the Princess of Helium. Often, Mars was depicted as a hostile place, the home of an implacable alien intelligence. In 1938, when Orson Welles broadcast a radio play based on *The War of the Worlds,* the you-are-there, live-broadcast quality of the show convinced panicked listeners that Martians had actually invaded New Jersey. "It's . . . it's . . . ladies and gentlemen, it's indescribable," said the actor playing the radio reporter. "I can hardly force myself to keep looking at it, it's so awful. The eyes are black and gleam like a serpent's; the mouth is . . . kind of V-shaped, with saliva dripping from its rimless lips that seem to quiver and pulsate." Hysterical listeners called the police and the newspapers, fled from their homes, and gazed at the sky in terror. "When the Martians started coming north from Trenton we got really scared," reported a listener. All told, 28 percent of the listeners thought the invasion was real.

MESSAGES FROM
BEYOND

• ✎

In the nineteenth century, observers reported a marking on Mars shaped very much like a Hebrew letter used to represent the name of God. Was this a message? And if so, what did it mean? No one knew. Similar questions arose in this century when photographs of Mars revealed a group of surface features resembling a sphinxlike face; was it a monument, consciously designed and similar in purpose to Stonehenge or the Egyptian pyramids? Might it be a message directed toward Earth by members of a civilization long dead? And were its creators Martians—or Earthlings? In 1988, a Russian satellite allegedly photographed a long, narrow "cigar-shaped" object thought to be either a shadow or a UFO, and a new generation of amateur observers spotted on Mars unmistakable evidence of an intelligence that transcends the bonds of Earth: a statue of Elvis. The unfathomable mystery only deepened in 1991, when the director of the little-known Crypto-Phenomena Museum (yes, it's in California) announced the discovery of a volcanic formation on Mars that bore a startling resemblance to Senator Ted Kennedy. Further reports are expected.

In the decades that followed, the literary alien landscape was often either a reflection of or a sanctuary from the violent civilizations of Earth. In Arthur C. Clarke's *The Sands of Mars,* published in the mid-1960s, Earthlings discover a way to turn Mars into a habitable place by artificially increasing the temperature—much as is being considered now. In Ray Bradbury's immensely popular *The Martian Chronicles,* visiting humans find a place populated by intelligent beings with "fair, brownish skin," "yellow coin eyes," and "soft musical voices," a place where "up and down green wine canals, boats as delicate as bronze flowers drifted." Robert Heinlein's 1961 novel *Stranger in a Strange Land* described the adventures of Valentine Michael Smith, a human raised on Mars who returns to Earth with extraordinary abilities.

And there were many, many more. The denizens of Mars became standard figures in science fiction and in the movies, fueling the fear of alien invasion and the idealistic hope for life elsewhere. Notable cinematic contributions to the exploration of life on Mars include:

✳ *Flight to Mars* (1951), in which an underground Martian civilization plans to use U.S. spacecraft to invade Earth;

✳ *Red Planet Mars* (1952), in which broadcasts from Mars—which turns out to be a Christian planet—inspire fear, panic, and the demise of the Soviet Union;

✳ *Invaders from Mars* (1953), a Cold War allegory;

✳ *Devil Girl from Mars* (1954), a British film in which a nymphomaniac Martian, clad in a black leotard and fresh from a feminist revolution, comes to Earth seeking men for breeding purposes;

✳ *The Angry Red Planet* (1959), in which astronauts on Mars battle giant amoebas;

✳ *Robinson Crusoe on Mars* (1964), in which a marooned astronaut finds his Man Friday;

✳ *Mars Needs Women* (1966), in which male Martians come to Earth seeking, among other females, the go-go dancing scientist Batgirl Craig;

✳ *Total Recall* (1990), in which Arnold Schwarzenegger voluntarily has his memory altered, only to discover that he once led a complicated life as a secret agent on Mars;

✳ and most notably, *Santa Claus Conquers the Martians* (1964), a cult classic in which a very young Pia Zadora made her film debut.

♂

Asteroids:
Vermin of the Skies

he discovery of asteroids was no accident. Their existence had been suspected ever since 1781, when William Herschel discovered Uranus in the place predicted by Bode's law, a formula that dictated planetary distances. Bode's law also indicated that the gap between Mars and Jupiter should be occupied, perhaps by a planet. A German astronomer named Baron Franz Xaver von Zach was eager to find that planet, but fifteen years of searching produced nothing. So in 1800 he and several other similarly inclined astronomers formed a group known as the Celestial Police, whose purpose was to conduct an organized search for the elusive body.

They were scooped by a Sicilian monk who was director of the observatory at Palermo. On New Year's Day 1801, the first day of the nineteenth century, Father Giuseppi Piazzi was mapping the positions of the stars when he saw a very small moving body in Taurus. Because it was traveling faster than Mars but slower than Jupiter, he reasoned that it might lie between the two. He thought the unknown body might be a comet—exactly the thought Herschel had when he discovered Uranus. Nonetheless, Piazzi sent the news to Johann Bode. Within a year, the body's orbit was calculated and confirmed; it was not the elongated orbit of a comet but just what one would expect from a planet.

William Herschel suggested that the newly discovered body with a planetary orbit be called an asteroid, which means "starlike," reasoning that the object was so small that even seen through a telescope it did not show the characteristic disklike shape of a planet. Asteroids do look like dots of light—like stars. Nonetheless, these tiny bodies act like planets and are consequently often called "minor planets."

Piazzi, whose privilege it was to name the tiny body, called it Cerere di Ferdinando—a reference to Ceres, the Roman goddess of grain and the special protector of Sicily, and an acknowledgment of Piazzi's patron, King Ferdinand of Naples and the Two Sicilies (who on occasion liked to dress up as a fish merchant and wander around the public market, presumably sampling public opinion). Today, Piazzi's discovery is known as Ceres. With a diameter of 623 miles, it is the largest of the minor planets.

Ceres

By 1807, three more asteroids—Juno, Pallas, and Vesta—had been discovered, the latter two by the German astronomer Heinrich Olbers, a founding member of the Celestial Police. Thirty-eight years passed before a fifth asteroid, Astraea, was discovered. Since then, so many have been discovered that asteroids are given not only names but numbers. Initially, the names were mythological (Circe, Europa, Pandora). These soon gave way to ordinary female names (Irene, Mildred, Marilyn, Davida), and, finally, to any name at all (Cincinnati, Jack London, Dudu, Felix, Mr. Spock, Mozart, McCartney, and NORC—the first and only celestial body to be named by some romantic soul after a computer). Asteroid 1000, discovered in 1923, is called Piazzia in honor of the man who started it all. Asteroid 1003 was named Olberia. And asteroids 3350 through 3356 were named after the seven astronauts who lost their lives in the 1986 explosion of the space shuttle *Challenger*.

Juno

The most widely accepted theory about the asteroids is that they are planetesimals, ancient chunks of matter that never coalesced into a planet. They are not the remnants of an exploded planet, as once was thought; the combined mass of all the asteroids between Mars and Jupiter equals only $1/_{2,000}$ the mass of the Earth, and thus they could never have formed a truly respectable planet. Yet these misshapen minor planets are so abundant (and faint) that the astronomer Walter Baade once called them "the vermin of the skies." Around 5,000 have been tracked and given official numbers and names; an additional 13,000 have been identified; and scientists estimate the total number at around 1 million. Lumpy and dark, they are grouped into families (among them Koronos, Themis, Hungaria), the members of which share similar orbits. Most asteroids are stony, but some are made of iron and nickel (and might be profitably mined someday), and a peculiar few reveal the presence of

Pallas

Vesta

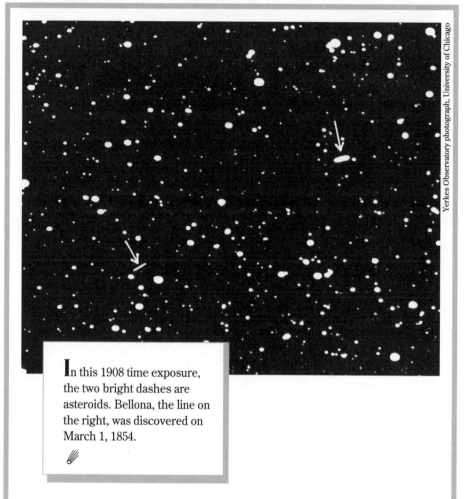

Yerkes Observatory photograph, University of Chicago

In this 1908 time exposure, the two bright dashes are asteroids. Bellona, the line on the right, was discovered on March 1, 1854.

other elements. A few, including Metis and Herculina, even have mini-moonlets. Toutatis, which flew past the Earth in December 1992 at the distance of a mere 2.25 million miles, is a binary asteroid, probably formed when two chunks of primordial rock bumped into each other and stuck. Pholus is coated with red material that may well be organic.

Spread out in a wide area between Jupiter and Mars, most asteroids are grouped into belts that are interrupted by empty lanes like the spaces between songs on phonograph records. Those spaces, called Kirkwood Gaps, were named after Daniel Kirkwood, the nineteenth-century American astronomer who figured out that any orbiting object in one of these voids would not remain in it for long. That's because orbits within the Kirkwood Gaps maintain an exact mathematical relation-ship—or resonance—to the orbit of Jupiter. For instance, an asteroid might revolve around the Sun precisely three times for every one revolution of Jupiter's. Jupiter repeatedly tugs at it, always in the same places in the circuit, until eventually the massive gravitational force sucks the planetesimal out of orbit, keeping the Kirkwood Gaps traffic free.

Not all asteroids are suspended between Mars and Jupiter. A few are farther out. Chiron was discovered in 1977 between the orbits of Saturn and Uranus; later, it developed a gaseous halo similar—indeed, identical—to a cometary coma, thereby blurring the distinction between an asteroid and a comet. In 1992, an even more distant asteroid, officially known as 1992 AD but unofficially called "the Son of Chiron," was found. It crosses the orbits of Saturn, Uranus, and Neptune. It is one of many asteroids—possibly including the ones hurled out of Kirkwood Gaps—that cross the orbits of several planets. Of these, the asteroids of greatest concern are the ones that cross the orbit of Earth. The first, discovered in 1932, was named Apollo, and since then, Earth-approaching asteroids have been known as Apollo objects. They sweep past so

THE FIRST TWENTY-FIVE ASTEROIDS

• ◞

Number & Name		Diameter in Miles
1	Ceres	623
2	Pallas	378
3	Juno	143
4	Vesta	334
5	Astraea	73
6	Hebe	121
7	Iris	130
8	Flora	94
9	Metis	94
10	Hygeia	280
11	Parthenope	93
12	Victoria	79
13	Egeria	139
14	Irene	98
15	Eunomia	169
16	Psyche	155
17	Thetis	68
18	Melpomene	93
19	Fortuna	134
20	Massalia	81
21	Lutetia	72
22	Calliope	110
23	Thalia	69
24	Themis	145
25	Phocaea	45

many planets that their orbits become jangled, making them an unstable and untrustworthy group. Icarus, for instance, crosses the orbits of Mars, Earth, Venus, and Mercury, and, like its namesake, comes perilously close to the Sun. One of these Earth-approaching asteroids could one day slam into the Earth with a force equal to that of all of the nuclear weapons in existence.

Are we in any danger? In a word, yes. The odds of the Earth being hit by an asteroid—or a comet—are much, much better than the odds of winning the lottery.

In 1972, an asteroid barreled through the atmosphere above Montana and was only 35 miles above the ground when it veered off into space. That's practically a hit. An actual hit occurred on June 30, 1908, when an asteroid (or the nucleus of a comet) about a tenth of a mile in diameter crashed through the atmosphere. A pale blue fireball exploded above central Siberia near the Tunguska River. Sonic booms rang out, a mushroom cloud bloomed in the air, and trees were uprooted and scorched for dozens of miles. An entire herd of reindeer died, and 600 miles away, windows shattered. A wave of pressure circled the Earth twice over, and for several nights the sky over Europe and Asia was luminescent. This scenario could—and will—happen again. Asteroids *ten times* as large as the Tunguska fireball are estimated to hit the Earth every few hundred thousand years.

The results of such a hit would be devastating. A major impact would destroy the immediate surroundings, send a surf of tidal waves around the world, and fill the atmosphere with dust, which in turn would cause the temperature to plummet, photosynthesis to stop, and nuclear winter to cast its appalling shadow across the Earth. Nor is it reassuring that Brian G. Marsden, the director of the International Astronomical Union's Minor Planet Center, estimates that the number of near-Earth asteroids may be as high as 10,000. Any one of them could do to us what some nameless, unnumbered asteroid did to the dinosaurs. Which may be why Congress asked NASA in 1991 to come up with some recommendations for dealing with the threat. The flashiest suggestion came from Dr. Edward Teller, who has never lost his affection for serious weapons. Widely known as the father of the H-bomb and "the midwife to Star Wars," Dr. Teller suggested that atomic bombs be kept as possible protection against an asteroid that might, someday, smash into the Earth. Less than six months after that suggestion was made and mocked, scientists predicted the possible demise of civilization on a specific date in the twenty-second century when Comet Swift-Tuttle might smash into the Earth . . . unless a powerful nuclear bomb were to be exploded somewhere out past Saturn, where the comet could be jarred into another trajectory. Thus it is that even after the end of the Cold War, the "vermin of the skies"—the minor planets and the comets—may someday provide employment for the major weapons of Earth.

T. Rex and the Attack of the Killer Rox

For 140 million years, the dinosaurs, as they say, ruled the Earth. By the end of the Cretaceous period 65 million years ago, some species (stegosaurs, for instance) had died out, but many others were thriving. And then something happened. Exactly what has been debated for a long time. The death of the dinosaurs has been blamed on changing climate, new forms of poisonous plant life, competition from mammals, volcanic eruptions, and, most reassuringly, their failure to adapt, seemingly a result of the fact that the dinosaurs had brains the size of prunes. Today, those notions have largely been discarded in favor of a theory proposed in 1980 by Walter Alvarez and his father, Luis Alvarez, two University of California scientists who hypothesized that 65 million years ago an asteroid smashed into the Earth, creating an enormous crater and saturating the air with dust, rock, and extraterrestrial matter. The thick haze, which the Sun could not penetrate, produced an ecological, worldwide holocaust. Acid rain fell, fires raced across the land, and the mother of all tidal waves pummeled the shores. Among the victims of that collision were *Tyrannosaurus rex,* its enemy *Triceratops,* and all the other dinosaurs (excluding birds, which are frequently considered to be descendants of the dinosaurs.)

Proof of the theory lay in a thin coat of iridium, an element rare on Earth but common in meteorites and asteroids. In over a hundred spots on Earth, the iridium shows up in the fossil record in a layer known as the K-T boundary that neatly divides the Cretaceous period from the Tertiary period. The iridium layer was interesting evidence, but the theory foundered on a key point: no crater.

For ten years, scientists pondered the problem. They knew that an impact powerful enough to kill the dinosaurs would have produced an impressive crater. The problem was that on Earth, craters are hard to find because they quickly disappear, their distinctive round shapes disrupted by geological movement and worn away by the forces of erosion.

Still, it had to be somewhere.

It was found in 1978, when scientists working for Petróleos Mexicanos (the Mexican national oil company) detected unusual magnetic disturbances in the Caribbean near a village, on the northern coast of the Yucatán Peninsula, named Chicxulub—Mayan for "Horns of the Devil." Geophysicist Glen T. Penfield, who plotted the undersea data on a chart in his office, noticed after several weeks, "I had half of a huge bull's-eye staring me in the face." When he found a matching ring on land, "I looked at the symmetry and just knew I was seeing a buried impact crater," he said. Estimated to be about 112 miles across, the crater was large enough to explain the global disaster that followed its creation. It was an exciting find.

No one seemed to notice. Penfield and his colleague Antonio Camargo presented their discovery at a 1981 meeting for exploration geophysicists, but Petróleos Mexicanos would not allow them to publish their data, and the news did not get around for years.

A decade later, scientific magazines heralded the discovery. Although the theory that the dinosaurs were obliterated by "killer rox" (as a New York tabloid expressed it) was not universally accepted, it was considered more than likely. Its probability increased in 1993, when new measurements indicated that the crater is wider than originally thought, with a diameter of 185 miles. To carve out a crater that large, the unknown object would have had to be five to ten miles across. Something that big slamming into the Earth would trigger a blast equal to the explosion of 300 million hydrogen bombs—easily strong enough to plunge the Earth into darkness and annihilate the thunder lizards.

●

Jupiter and the Great Red Spot

Provide ship or sails adapted to the heavenly breezes, and there will be some who will not fear even that void. . . . So, for those who will come shortly to attempt this journey, let us establish the astronomy: Galileo, you of Jupiter, I of the Moon.

—Johannes Kepler, in an
open letter to Galileo, 1610

 ohannes Kepler, son of an accused witch, was not the first person to see the ever-changing oval, known as the Great Red Spot, which swirls across Jupiter. He was, however, prescient about it. In one of his many failed attempts to decode an anagram Galileo had written about the phases of Venus, he came up with a sentence translated as "A red spot shows clearly that Jupiter rotates." He was wrong about the anagram but right about the red spot. Jupiter turns on its axis so rapidly that in less than five hours, the Great Red Spot, the most prominent feature of the largest planet, crosses from one side of the globe to the other.

The first people to see the Great Red Spot were the British scientist Robert Hooke—who supposedly saw it in 1664—and Giovanni Domenico Cassini, an Italian astronomer who moved to France in 1669 at the behest of Louis XIV, the Sun King, changed his prename to Jean-Dominique, and founded a five-generation dynasty of French astronomers. Although Cassini—who could never bring himself to believe that the planets revolved around the Sun—observed the red spot several times between 1665 and 1691, it wasn't given a name until 1887, when its unusual brightness led astronomers to suggest that it might be an island or the cloud-covered peak of some unknown Olympus. They were wrong, for Jupiter lacks even a semblance of mountains, islands, craters, or any other feature common to the interior planets. A gas giant formed primarily of hydrogen and helium with a few other compounds thrown in, Jupiter has no geography.

Jupiter's complex bands of clouds and the Great Red Spot are clearly visible in this photomosaic, taken by *Voyager 1* from a distance of 4.7 million miles.

NASA

Yet it does have weather. Sizzling bolts of lightning and aurorae worthy of the landscape painter Frederick Church light the sky, and hurricanes are always in season. They are formed by parallel bands of winds that streak across the planet in alternating directions at speeds of up to 335 mph, creating eddies and cyclones that roll like ball bearings between the billowing winds. The Great Red Spot is a cyclone* over 8,500 miles wide and 16,000 miles long—large enough to engulf the Earth twice over. Bobbing beneath it are three white oval storms. Formed in 1938, they are small in comparison to the Great Red Spot (although they are about as large as Mars), and they are probably doomed; on Jupiter, small storms are soon cannibalized by larger ones, which flourish.

Beneath the storms and variegated cloud layers lies an ocean of hydrogen that is liquid in its upper layers but gradually changes to a roiling metal. Residing at the center of the planet, where the pressure is 8 billion times the pressure we experience at sea level, is a dense core of rock and ice about fifteen times the mass of the Earth squeezed into a ball roughly double the size of our puny planet.

More massive than the rest of the planets combined, Jupiter could swallow the earth 1,330 times over. It revolves around the Sun in a lit-

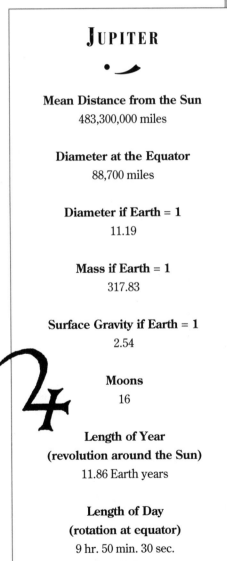

JUPITER

Mean Distance from the Sun
483,300,000 miles

Diameter at the Equator
88,700 miles

Diameter if Earth = 1
11.19

Mass if Earth = 1
317.83

Surface Gravity if Earth = 1
2.54

Moons
16

Length of Year
(revolution around the Sun)
11.86 Earth years

Length of Day
(rotation at equator)
9 hr. 50 min. 30 sec.

*Actually, astronomers refer to it as an anticyclone, not because it's made of anti-matter, but because it is in Jupiter's Southern Hemisphere and rotates in a counterclockwise direction. Only clockwise Southern Hemisphere storms technically qualify as cyclones.

THE SEARCH FOR LIFE GOES ON. . . .

To the casual observer, the possibility of life on Jupiter seems infinitely more remote than the chances for life on Mars. A planet without a surface, Jupiter has a rocky core wrapped in a wide sea of metallic hydrogen that transmutes into an ocean of liquid hydrogen, which in turn gradually becomes gas. Yet scientists believe that life could exist there because in many ways the environment on Jupiter resembles that of the Earth in its infancy. Jupiter is aswirl with methane and ammonia, gases that once blanketed the Earth and are necessary for the creation of life. On Earth, those compounds, zapped by lightning, formed amino acids, which eventually combined into DNA. So why couldn't stormy Jupiter, which also has hydrogen and water, witness a similar series of events? The answer is: It could. It might take billions of years, as it did on Earth, but it could happen.

So what would Jovians look like? We haven't a clue, but one thing is sure: Life on Jupiter would evolve very differently from life on Earth. Robert Burnham proposes that "a drifting aerial form of life" could populate the warm, colored clouds of Jupiter, while Robert Jastrow imagines that highly evolved Jovians might be "small, squat and toad-like, adapted to withstand the crushing force of Jupiter's gravity—yet, perhaps with a highly developed brain."

tle less than twelve years and turns on its axis in a shade under ten hours, making its day the shortest in the solar system. Encircled by a thin and insubstantial ring, one of the many surprising discoveries of *Voyager 2,* Jupiter is surrounded by doughnut-shaped belts of radiation that emit the strongest radio waves in the solar system. It has a magnetic field five times greater than the Sun's on an average day. And it trails behind it a stream of charged particles that stretches out for over 600 million miles. When the solar wind brushes against that invisible magneto-tail, particles are scattered all the way to Mercury.

Jupiter is noted for being as big as a planet can be. If some cosmic mason were to pack extra matter around it, the effort would only compress the planet, causing it to shrink. If, however, enough matter were added to equal seventy to eighty times Jupiter's actual mass, the rising temperature and inter-

Jupiter's Great
Red Spot.

nal pressure would spark nuclear reactions, and Jupiter would no longer
be a planet, but would become a star.

In the meantime, it is just a planet. Although it doesn't produce heat
from nuclear fusion (as would be the case if it were a star), it does have
sources of heat that pour out 70 percent more energy than it gets from
the Sun. This heat is essentially a slowly fading ember from the planet's
red-hot youth. Smaller now than in its proto-planetary past, Jupiter may
still be collapsing under the force of gravity, shrinking at the rate of perhaps a millimeter a year.

Many mysteries remain about Jupiter. Is its core solid or liquid? Is the
planet made of iron and silicates, as suspected, or might it have a different composition? How deep is the Great Red Spot? Why isn't there a similar spot on the planet's Northern Hemisphere? And what causes the
spot to be brick red? From time to time the Great Red Spot seems to pale,
partially because the turbulent winds forced to flow around it are occasionally darker than usual, and partially because the spot changes in

Hannah Berman

intensity. It may receive its characteristic color from sulfur or phosphorus, but no one is sure. Oddly, scientists who analyzed *Voyager*'s infrared spectrometer were unable to detect differences between the white ovals and the Red Spot. Clearly there is a difference, but the responsible molecule has yet to be identified.

Scientists hope to answer some of these questions with the help of the *Galileo* spacecraft. Launched on October 12, 1989, on a complicated, slingshot trajectory, *Galileo* will follow a path from the Earth to Venus, then all the way around back to the Earth, then past the orbit of Mars to the asteroid Gaspra and back to Earth *again,* at which point its second gravitational assist will fling it past the more distant asteroid Ida and on to Jupiter, where it will go into orbit in December 1995. In addition to monitoring cloud patterns and observing the Jovian moons, *Galileo* will eject a small metal probe whose kamikaze mission is to parachute into the giant's heart. The probe will send back bulletins for as long as possible, but its trip will be short-lived; sixty-two miles into its dive through the whirling clouds, it will, like other spacecraft on other planets before it, be crushed to death.

4

Galileo and
The Starry Messenger

alileo Galilei (1564–1642) was neither kind nor brave. But he was an inspired observer and a committed believer in the power of reason, and these qualities ultimately gave him the strength he needed to go into combat with one of the greatest adversaries science has ever faced: the Roman Catholic Church.

Born in Pisa into a family once considered noble, Galileo received his early education at home from his father, Vincenzo, who was a clothing merchant, professional composer, lutist, and author of books about music, as well as a man of a decidedly anti-authoritarian bent. When Galileo was ten years old, he was sent for a while to a Jesuit monastery, where he was taught Latin, Greek, philosophy, logic, and, most likely, Aristotelian science. When he toyed with the idea of becoming a monk, his father promptly took him out of school.

At age seventeen, at his father's prompting, Galileo began to study medicine. Two events deflected him from that path. The first occurred in 1583, when he heard a geometry lecture given by the mathematician Ostilio Ricci, a family friend. Galileo was so fascinated that he arranged with Ricci for a series of private lessons, which he kept secret from his father for as long as possible. When his father finally learned about them, he agreed to let Galileo pursue the subject—as long as it didn't interfere with the young man's medical education.

The other major event occurred one day at Mass, when Galileo became transfixed by a swaying lamp. Using his pulse, he timed its oscillations and noticed that no matter how large or small they were, each took the same number of pulses (an experiment anyone can do). Galileo, who liked to think in terms of commercial applicability, applied this discovery to the invention of a pendulum used to measure the pulse. The pulsiogium was an immediate success with those who had actually become doctors. Galileo wasn't one of them, for by then he had given up medicine completely.

In 1589, he was appointed lecturer in mathematics at the University of Pisa, where he was informed that members of the faculty were expected to wear academic togas. He campaigned vigorously against that requirement by means of a poetic satire in which he spoke out against the toga and in favor of nudity. His popularity with students was immense. At Pisa he also began his exploration of Aristotle's dictum, long accepted but incorrect, that heavier bodies fall faster than lighter bodies. Despite the myth promulgated by an early biographer, Galileo did not drop two cannonballs off the Leaning Tower of Pisa. He did, however, measure the acceleration of balls rolling down an inclined plane, and he discovered that an object's weight had nothing to do with its rate of fall.

When his three-year contract with the University of Pisa was over, he found a position at the University of Padua, in the Republic of Venice, at a greatly increased salary. This was fortunate, because after his father died in 1591, Galileo spent years supporting his siblings, paying his sisters' dowries and lending money to his ne'er-do-well brother, Michelangelo. At Padua, he was an entertaining teacher who lectured on the Ptolemaic view of the universe—a system in which he did not believe. He revealed this perfidy in 1597 in a letter to Johannes Kepler, who had sent him a copy of his book *Mysterium cosmographicum* (a book Galileo evidently did not read). In the letter, Galileo said that although he had believed for many years that the Earth orbited the Sun, he was afraid to state his opinions publicly. "I have not dared to publish, fearful of meeting the same fate as our master Copernicus who, although he has earned for himself immortal fame amongst a few, yet amongst the greater number appears as only worthy of hooting and derision; so great is the number of fools. I should indeed dare to bring forward my speculations if there were many like you; but since there are not, I shrink from a subject of this description."

Kepler, a completely open man, rebuked Galileo. "With your clever secretive manner you underline, by your example, the warning that one should retreat before the ignorance of the world. . . . Have faith, Galilei, and come forward!" For a dozen years after this exhortation, Galileo ignored Kepler completely. ("It has always hurt me to think that Galilei did not acknowledge the work of Kepler," Albert Einstein once commented.)

And then, after nearly seventeen years of teaching a system in which he did not believe, yet another momentous event occurred in Galileo's life. In 1608, Hans Lippershey, a Dutch maker of spectacles, invented the telescope. When Galileo heard about it, he set about making one for himself. On the third try, after "sparing neither labor nor expense," he constructed an instrument he considered excellent. (For a few hundred dollars, anyone today can buy a telescope infinitely superior to Galileo's; with a good pair of binoculars, it's possible to see almost everything he saw.) In 1610, he invited a group of clergymen and professors to look through the telescope, but they were unimpressed (and in fairness to them,

H. G. WELLS
ON THE INVENTION
OF THE TELESCOPE

•⁓

The development of the telescope marks, indeed, a new phase in human thought, a new vision of life. It is an extraordinary thing that the Greeks, with their lively and penetrating minds, never realized the possibilities of either microscope or telescope. They made no use of the lens. Yet they lived in a world in which glass had been known and had been made beautiful for hundreds of years; they had about them glass flasks and bottles, through which they must have caught glimpses of things distorted and enlarged. But science in Greece was pursued by philosophers in an aristocratic spirit, men who . . . were too proud to learn from such mere artisans as jewellers and metal- and glass-workers.

Ignorance is the first penalty of pride. The philosopher had no mechanical skill and the artisan had no philosophical education, and it was left for another age, more than a thousand years later, to bring together glass and the astronomer.

—from *The Outline of History*

Galileo's instrument had poorly ground lenses which threw rainbow-colored rims around everything). Others refused to look at all: Why argue with Aristotle? But when Galileo demonstrated the telescope to the doge and the Senate of Venice, they were struck by its commercial and military possibilities, for with it one could see incoming ships two hours earlier than was otherwise possible.* Galileo received a huge increase in salary and was granted tenure at the University of Padua.

Astounded at what he could see, Galileo was in a great rush to publish. *Sidereus Nuncius* (*The Starry Messenger*) pioneered a new style of writing—dry and scientific—and laid out some astonishing facts about the universe. For instance, the Moon was not, as the Aristotelians claimed, a perfect globe, but covered with mountains, valleys, and plains. The stars were far more numerous than had been thought. Galileo saw eighty new stars in Orion, more than forty in the Pleiades, and, in the Milky Way, "nothing but a congeries of innumerable stars grouped together in clusters. . . . Many of them are rather large and quite bright, while the number of smaller ones is quite beyond calculation."

And then there was Jupiter. Galileo saw four moons revolving around it, which proved that everything did not revolve around the Earth. These were serious blows to the Ptolemaic system.

Published in 1610, *The Starry Messenger* made Galileo famous. Many people didn't believe him, but Kepler defended him. In a response called *Conversations with the Starry Messenger,* Kepler wrote, "Who should be silent in the face of such a message? Who would not feel himself overflow with the love of the Divine which is so abundantly manifested here?" Nevertheless, when Kepler asked Galileo to send him a telescope because he could not get the lenses he needed in Prague, Galileo refused, even though he readily dispatched them to many other people (one of whom finally took pity on Kepler and lent him a telescope).

Four months later, Galileo left Padua for Florence, leaving behind his companion Marina Gamba and his four-year-old son, but bringing with him his two other children, nine-year-old Livia and ten-year-old Virginia. He soon tired of the responsibility. By then Marina had married and didn't want her daughters, so Galileo sent them to an impoverished convent, where they later became nuns. Virginia—Sister Maria Celeste—eventually forgave him. Livia—Sister Arcangela—never could.

From Florence, Galileo was granted permission to go to Rome, where he was eager to convince the church that his theories were correct. He presented his discoveries to Pope Paul V and the College of Cardinals, among whom was Maffeo Barberini, with whom Galileo later became friends. The visit was a success—or so it seemed. In fact, Galileo did not

*In the next century, George Washington had a similar thought, and borrowed Columbia University's first telescope during the Battle of Long Island.

Library of Congress

In this imaginary scene from the life of Galileo, diagrams on the prison wall show the Sun, the Earth, and the Moon. At the lower right, leg irons attached to the wall suggest a harsher imprisonment than Galileo actually experienced.

make his case effectively (reading Kepler's book might have helped him). Everyone agreed that the wonders Galileo had seen existed; they just weren't sure that these discoveries proved the Copernican system. They thought, for instance, that his observations might fit the cosmological scheme of Tycho Brahe, a compromise plan in which the Sun and the Moon orbited the Earth, while Mercury, Venus, Mars, Jupiter, and Saturn circled the Sun.

Galileo fought to get the church to see it his way. After years of hiding his true beliefs as a teacher, he stopped hanging back. When he heard that a Dominican friar had advocated that mathematicians—and especially those who believed that the Earth moved—should be expelled from all of Christendom, he wrote an argumentative letter to the man.

(An official of the Dominican order later apologized to Galileo for what he termed an idiocy.) He fully expected reason to triumph.

It did not. In 1615, a conservative Roman theologian expressed the opinion that the Copernican view should be treated as a hypothesis. Galileo insisted it was fact. In an edict of 1616, the Holy Office put Copernicus' book *De revolutionibus orbium coelestium* on the Index of Prohibited Books and instructed Galileo to stop defending Copernicus on pain of imprisonment.

Galileo realized that sooner or later the pope would die. A few years later his expectation was fulfilled, and his old friend Maffeo Barberini, who had defended Galileo many times, was elected pope. But absolute power corrupted Barberini so absolutely that when the birds in the Vatican interrupted his thoughts, he had them poisoned. Barberini—now Pope Urban VIII—affirmed the edict of 1616.

Galileo kept on. For six years, encouraged by his friendship with the pope, Galileo worked on a book entitled *Dialogue Concerning the Two Chief World Systems.* In it, he followed the letter of the law; he presented his ideas as a hypothesis explained by a character named Salviati. The church's point of view was presented by a character named Simplicio. Insult was intended and taken. In 1632, the book was banned. The following year, Galileo was brought to trial in front of members of the Inquisition. He denied belief in the Copernican system, caved in at every point, and was presented with a confession stating, "I have been pronounced by the Holy Office to be vehemently suspected of heresy; that is to say, of having held and believed that the Sun is the center of the world and immovable and that the Earth is not the center and moves." He kneeled, read it aloud, and signed it. Legend says that he then muttered, "Eppur si Muove" ("Nevertheless, it moves"). This story is not true, writes physicist George Gamow, "and it only gave ground to an old anecdote according to which Galileo was watching the wagging tail of a friendly dog which entered by mistake into the Holy Office of the Church." Still: If Galileo didn't stand up for himself in that way, he should have. Some legends are worth perpetuating.

Galileo was sentenced to imprisonment and the repetition of seven psalms once a week for three years, but the pope reduced the seventy-year-old astronomer's punishment to house arrest. Galileo spent the rest of his life confined to his villa in Florence (where he was once visited by John Milton). Until her death, his daughter Sister Maria Celeste took care of him. (A geographical feature on Venus bears her name.) During this period, Galileo went blind, possibly as a consequence of gazing at the Sun. All pleasure was not denied him, though; until his death in 1642, he played the lute, a skill he had learned from his father.

A Chronology of Observational Technology

c. 2800 B.C.: Stonehenge. Early construction includes a ditch, an earthen bank, the 35-ton heelstone, and fifty-six pits called Aubrey Holes, which may have been used to predict eclipses. Between 600 and 1,000 years later, the famous circle of stones will be added.

c. 2600 B.C.: The Great Pyramid at Giza is built, oriented toward the belt of Orion and Thuban in Draco the Dragon—the North Star of its time.

c. 440 B.C.: Moose Mountain Medicine Wheel is built in Saskatchewan, Canada, oriented toward the position of the Sun at the summer solstice.

52 B.C. to A.D. 132: Chinese astronomers devise an armillary sphere for measuring the positions of heavenly objects. Beginning with a metal ring representing the equator, it eventually includes a ring representing the path of the planets, a ring representing the meridian, and a water clock.

A.D. 150: Equipped with a plinth—a block of stone inscribed with a calibrated arc and used to plot the elevation of the Sun—and a triangular rule called a triquetrum, Ptolemy marks the position of stars.

927: A Muslim instrument maker named Nastulus makes the oldest known astrolabe, a metal map of the heavens showing the apparent movement of the stars around the Pole Star and relative to the horizon.

1000: The Mayans build an observatory at Chichén Itzá on the Yucatán Peninsula. Known as the Caracol, it is aligned with the sun at the solstices as well as with the stars Castor, Pollux, Fomalhaut, and Canopus.

1391: Geoffrey Chaucer's *A Treatise on the Astrolabe* shows how to build and use an astrolabe to compute the position of a star.

1576: Tycho Brahe begins construction of Uraniborg, his island observatory. His equipment includes a wall quadrant, a large armillary sphere, and a sextant covering 30° of the sky and equipped with fixed and movable arms to measure distances between stars.

1608: Dutch optician Hans Lippershey invents the telescope.

In 1896 William Morris's Kelmscott Press published a new edition of the works of Geoffrey Chaucer (*c.* 1342–1400). With illustrations by Sir Edward Burne-Jones and border designs by Morris, the sumptuous volume included Chaucer's *A Treatise on the Astrolabe*, written for his son Louis.

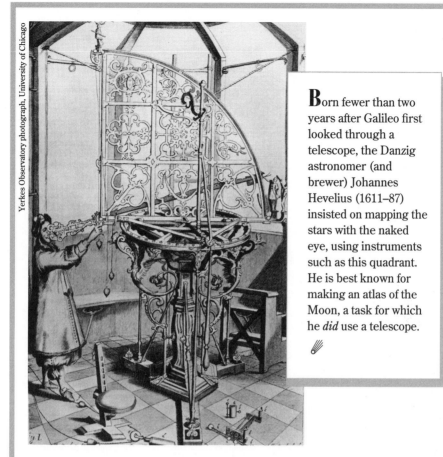

Yerkes Observatory photograph, University of Chicago

Born fewer than two years after Galileo first looked through a telescope, the Danzig astronomer (and brewer) Johannes Hevelius (1611–87) insisted on mapping the stars with the naked eye, using instruments such as this quadrant. He is best known for making an atlas of the Moon, a task for which he *did* use a telescope.

1609: Galileo builds his first telescope. A refractor with two glass lenses (a convex objective and a concave eyepiece), it magnifies about thirty times.

1611: Johannes Kepler, refining the telescope, substitutes a convex eyepiece for the concave one, thereby enlarging the field of view but turning the image upside down.

1636: French mathematician and friar Marin Mersenne suggests using mirrors to construct a reflecting telescope.

1668: Isaac Newton builds a reflector telescope using a concave mirror instead of an objective lens. Because different colors are refracted differently, the refractor telescopes in use at this time produce rainbow-colored rims around images. The reflector eliminates this chromatic aberration because colors are *reflected* evenly. Another advantage is that a mirror, unlike a lens, can be supported from the back and is thus less given to distortion.

French physician N. Cassegrain designs a telescope in which light is reflected from a convex secondary mirror through a hole in the primary mirror, an improvement over the large Newtonian reflector, in which the eyepiece is on the upper part of the telescope, requiring the observer to climb a tower or ladder to reach it. With the Cassegrain telescope, the observer remains on ground level. According to Newton, "The advantages of this device are none."

1733: Chester Moor Hall sandwiches together two kinds of glass to create an objective lens free of chromatic aberration.

1758: Using Hall's design to create a lens of flint glass and crown glass, John Dolland makes an achromatic lens, which he presents to the Royal Society.

1789: William Herschel builds a telescope with a 49-inch mirror.

William Herschel used this telescope, with its 49-inch mirror and 40-foot tube, to discover Enceladus and Mimas, two of the moons of Saturn.

1845: William Parsons, the earl of Rosse, builds a reflecting telescope with a 72-inch mirror, the largest in the world until 1917. It is known as the Leviathan of Parsonstown.

1888: A 36-inch telescope refractor is completed at Lick Observatory.

1897: The world's largest refracting telescope is built at Yerkes Observatory in Wisconsin. It has a 40-inch objective lens and a 64-inch tube.

1908: The 60-inch reflector at Mount Wilson is completed.

1917: The 100-inch Hooker reflector at Mount Wilson is completed.

1930: Bernhard Schmidt invents the Schmidt telescope, which uses a correcting lens to eliminate distortion around the edges of the mirror and to produce accurate wide-angle photographs of the sky.

1936: After designing the world's first radio telescope, Illinois engineer Grote Reber erects a 30-foot metal dish in his backyard and begins to map the Milky Way, a project he completes eight years later.

Courtesy of Caltech

This drawing, made by Russell W. Porter in 1938, presents a cross section of the 200-inch Hale telescope and dome. Note the figure at the top of the stairs.

Courtesy of Caltech

Russell W. Porter's 1940 drawing shows a nattily clad astronomer observing from the prime-focus capsule of the Hale telescope.

1948: The 200-inch reflector at Mount Palomar is completed.

1962: A small rocket detects X-rays from beyond the solar system.

1970: The first X-ray satellite is launched.

1978: The solar-powered International Ultraviolet Explorer (IUE) spacecraft is launched.

The Einstein Observatory, carrying a high-resolution X-ray telescope, is launched.

1980: The Very Large Array, a Y-shaped collection of twenty-seven radio telescopes, begins operation in New Mexico.

c. 1981: The charge-coupled device (CCD) makes the photograph obsolete. Whereas a photograph uses a fraction of the light from an object to create a chemical change on film, the far more sensitive CCD responds to almost all the light and sends electrical currents directly to a computer.

1983: The Infrared Astronomy Satellite (IRAS) is sent into orbit.

1989: NASA's Cosmic Background Explorer (COBE) satellite is launched.

1990: The Hubble Space Telescope is launched from the space shuttle *Discovery.*

1991: The fifteen-ton Compton Gamma Ray Observatory (GRO) is launched from the space shuttle with four gamma-ray detectors on board.

1992: On April 14, the Keck Telescope begins observation with all thirty-six of its hexagonal mirrors in place. On August 24, its twin, Keck II, receives the first of its thirty-six coordinated mirror segments.

1993: December. Space-walking astronauts install new solar panels, gyroscopes, a new camera, and other devices to correct the vision of the Hubble Space Telescope.

Coming attractions expected to be operational by the year 2000 include the Keck Telescope II; the orbiting Stratospheric Observatory for Far-Infrared Astronomy (SOFIA); the Advanced X-ray Astrophysics Facility (AXAF); the Space Infrared Telescope Facility (SIRTF); Princeton University's Sloan Telescope, slated to make a redshift map of a million galaxies; and the European Southern Observatory's computer-controlled multiple mirror telescope in Chile known as the VLT (Very Large Telescope).

Galileo's Starlets

These numerous love passages of Zeus (and other gods as
well), related by ancient poets, appear to us, as it is known
they appeared to the right-thinking men amongst the ancients
themselves, unbecoming of the great ruler of the universe.
The wonder is how such stories came into existence.
— Alexander S. Murray, *Manual of Mythology* (1874)

On January 7, 1610, Galileo looked through his newly constructed telescope at the planet Jupiter and saw what he called "three starlets" aligned with the giant's bulging waist. Their positions changed on a

In 1620, the Jesuits noted the changing positions of Jupiter's four Galilean moons. Galileo made this copy of their observations.

Yerkes Observatory photograph, University of Chicago

nightly basis, and less than a week later, he observed a fourth star. Eager to curry favor, he called them Medicean stars in honor of his former student, now the grand duke of Tuscany, Cosimo II de' Medici. "Your virtue alone, most worthy Sire, can confer upon these stars an immortal fame," Galileo wrote, and he sent along a telescope. Within four months Galileo was appointed court mathematician, and he moved to Florence.

Galileo proposed many possible names: Maria, Catharina, Franciscus, Ferdinandus, Cosmus Major, and Cosmus Minor. But no sooner had he announced his discovery in *The Starry Messenger* than Simon Marius, a German astronomer whom Galileo called a "poisonous reptile," claimed to have seen them first. Whether he did or not is unclear, but it is certainly true that Marius had better instincts about naming. He recommended that the newly discovered bodies be named after Jupiter's lovers, including "three virgins, whose love Jupiter secretly coveted and obtained, namely: Io, the daughter of the river god Inachus, then Callisto, daughter of Lycaon, and finally Europa, the daughter of Agenor. Yet even more ardently did he love the beautiful boy Ganymede. . . ." These names were adopted. In descending order of size, they are Ganymede, Callisto, Io, and Europa.

Their orbits were calculated and predictions made of the times when they would be eclipsed by Jupiter. In 1675, Olaus Roemer, a Danish astronomer, noticed that the predictions were a little off. When the Earth was closer to Jupiter, the eclipses came a fraction too soon, whereas when the distance between the two planets was greater, the eclipses were late. This discrepancy led him to deduce that light does not arrive instantaneously. The concept, obvious to us, was heretical at the time; René Descartes, for example, believed that light had an infinite velocity. Roemer proved that it travels at a finite speed and that the farther away an object is, the longer the light takes to reach us.

Little more was learned about the Galilean satellites until *Voyagers 1* and *2* showed us four remarkable worlds.

Ganymede, the largest moon in the solar system (bigger even than Mercury or Pluto), is half frozen water and half rock. Its dirty, icy surface is a patchwork of heavily cratered dark areas, and grooved and mountainous light areas, which were possibly flooded in the primordial past with slush from the interior.

Callisto, dark and inert, is similar to Ganymede but more uniform and heavily cratered. From pole to pole, it is a world of craters gone wild.

Europa, the smallest of the Galilean satellites, has the smoothest sur-

This collage, assembled from photographs taken by *Voyager 1*, shows Jupiter at the upper right, Io at the upper left, Europa in the center, Ganymede at the lower left, and Callisto in the lower right-hand corner.

NASA

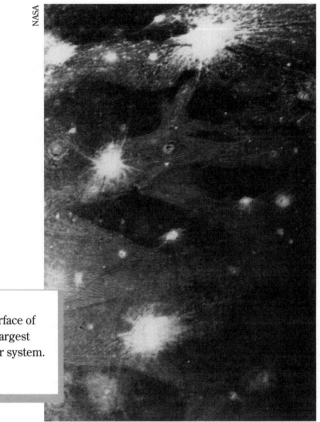

NASA

The cratered surface of Ganymede, the largest moon in the solar system.

face of all the moons in the solar system. It is almost without craters, an icy world whose highly reflective surface may hide an underground ocean of liquid water. It is also crosshatched by long, gently curving fractures and cracks so reminiscent of the nonexistent canals on Mars that they inspired one NASA scientist to ask, "Where is Percival Lowell now that we need him?"

And then there is Io, a geologically active world pulled this way and that between Jupiter's enormous gravity in one direction and its companion Galilean satellites in the other. It bulges and twists in its gravitational vise, causing the interior to expand and heat up beneath the thin crust. Hot spots dot the surface. Near the volcano Loki, named after the Norse god of fire, is a lava lake, and geysers and volcanoes erupt continually, spurting gas plumes into the thin sulfurous atmosphere and pouring molten sulfur over the Ionian landscape. Patches of sulfur-dioxide snow and blotches of red, orange, yellow, and black blanket the surface,

NASA

Io. The prominent round, ringed feature in the center is an active volcano.

which was characterized for all time by scientist Bradford A. Smith with one oft-repeated comment: "I've seen better-looking pizzas."

The Galilean satellites are not the only moons of Jupiter. In 1892, a fifth moon, named Amalthea, was discovered by E. E. Barnard. Today, there are sixteen confirmed moons. They are named after Jupiter's lovers, his children, his nurse, and even, in the case of Sinope, a platonic friend: When Jupiter promised to give her whatever she desired, Sinope told him she wanted to remain a virgin, and going against his nature, he granted the request.

The fates of Jupiter's Galilean lovers varied. Ganymede was grabbed by an eagle and taken to be Jupiter's cupbearer. Io was turned into a white heifer and, later, back into a woman. Callisto was changed into a bear either by Jupiter, Juno, or Artemis, the goddess of the Moon, and then shot down with an arrow and transformed into the constellation Ursa Major. Europa, after whom Europe is named, was abducted across the Mediterranean by Jupiter, who turned into a bull for the occasion. She bore him three sons before she married the king of Crete and had a daughter.

As for Jupiter's wife, her name appears nowhere near the planet but rather in the asteroid belt, where the 3rd asteroid to be discovered was called Juno, her Roman name, and the 103rd asteroid bears her Greek name, Hera.

The Star of Bethlehem

To figure out what the Christmas star might have been, you have to know when Jesus was born, and on this point there has been disagreement. Johannes Kepler, who dropped out of divinity school to become a teacher of mathematics, believed that Jesus Christ was born in the year 5 or 4 B.C. Other estimates range from 7 B.C. to A.D. 1. To reconstruct the date, historians have looked to such events as the death of Herod, which was known to have taken place, according to the Jewish first-century historian Josephus, after a lunar eclipse but before Passover, and the timing of a census in which citizens of the Roman Empire had to come to Bethlehem to swear allegiance to Augustus Caesar. Astronomers have tried to explain the star by looking for notable celestial events that occurred around those dates.

The trouble is that notable celestial events, especially in an era without light pollution, are not uncommon. Eclipses, comets, and conjunctions happen all the time. From an astronomical point of view, determining the thrilling circumstances that might have been interpreted as the Christmas star depends on what date you select. The date of choice right now, according to astronomer John Mosley, is 3 or 2 B.C.—which eliminates certain occurrences.

For instance, it wasn't a comet. We know this because the Chinese—who, unlike the Europeans, kept accurate astronomical records—reported no comets in 3 or 2 B.C. (Halley's Comet returned in 12 B.C.)

A supernova, which looks like a brilliant new star but is actually the explosion of a dying star, is unlikely for the same reason: The Chinese failed to report one, although they did record a supernova in the year 5 or 4 B.C.

The conjunction of two or more planets is another story. Conjunctions, which happen when celestial bodies are visually aligned and thus appear to be physically close, are ordinary events in which the planets usually remain far enough apart to be easily distinguished. Three times in the years 3 and 2 B.C., Jupiter was in conjunction with the star Regulus in the constellation Leo. In August of 3 B.C., Jupiter and Venus, the two brightest planets, were so close that, seen from Babylon, they would almost have appeared to be touching. And on June 17 of 2 B.C., the two planets would seem to have merged magnificently. They separated, but a few months later they were conjunct a third time.

We can only imagine how brilliant the conjunction of June 17 may have been; during the twentieth century, not a single conjunction has been so close and bright. The last time Jupiter and Venus were that close was 1818. The next time is the year 2065.

While many astronomers believe that the Star of Bethlehem was a conjunction, Kepler believed it was actually *created* by a conjunction. Although this is impossible (since a conjunction is strictly a visual phenomenon), Kepler remembered that in 1604, when Mars, Jupiter, and Saturn were closely conjunct, astrologers thought the event might presage the arrival of a comet. Instead, a new star appeared between Saturn and Jupiter: a supernova since known as Kepler's star. Kepler wondered if the conjunction had somehow created the star. In which case, he reasoned, perhaps a similar event had occurred during the reign of King Herod. By careful computation, he learned that Mars, Jupiter, and Saturn had been closely conjunct in 6 B.C. Perhaps that gathering of planets had also produced a new star in the sky.

Today, astronomers are leaning toward matching the Star of Bethlehem and the brilliant conjunction of June 17, 2 B.C. But they are quick to point out that this proves nothing. For the astute observer—and the three Magi may have been astrologers who examined the sky carefully—the heavens are regularly filled with surprising events.

★

Johannes Kepler

My aim is to show that the heavenly machine is not a kind of divine, live being, but a kind of clockwork.

— Johannes Kepler

opernicus was scared; Galileo was cold; Newton, according to Stephen Hawking, was gifted with "talents for deviousness and vitriol." But Johannes Kepler—now, there was a human being. Introspective, demanding, loyal, anxious, honest, both contentious and self-effacing, egotistical and easily hurt, he was a brilliant astronomer, a successful astrologer, and a vivid, confessional writer. A sort of Woody Allen of the Renaissance, he was beset by troubles all his life: physical, familial, professional, religious, financial. Yet he laid the foundation for a true scientific view of the universe.

Born in Germany in 1571, he had a miserable youth. His father, Heinrich, whom Johannes described in a revealing family horoscope as "vicious, inflexible, quarrelsome, and doomed to a bad end," worked as a mercenary and a tavern keeper, was almost hanged in 1577 (for reasons unknown to us), and deserted his family for good in 1588. His mother, an herbalist, was "gossiping, and quarrelsome, and of a bad disposition." Growing up, Johannes was cursed with bad digestion, boils, myopia, double vision, crippled hands (as a result of a near-fatal bout with smallpox), and a bizarre assortment of skin ailments including mange and "chronic putrid wounds in my feet. . . ." On New Year's Eve, when he was twenty-one, he had sex "with the greatest possible difficulty, experiencing the most acute pains of the bladder." It is probably redundant to add that he was unpopular with his classmates. Nor was his self-concept exactly stellar. In a lively third-person account, he described himself as having "in every way a dog-like nature. . . ." Fortunately, he was also brilliant.

Johannes Kepler entered the University of Tübingen as a teenager, graduated when he was twenty, and stayed on in pursuit of a degree in Protestant theology. At Tübingen, he heard a lecture supporting Ptolemy's geocentric universe. Kepler took the opposite point of view and became a strong advocate of the heliocentric Copernican system. This gained him no friends, especially among the Lutherans, at whose hands he continually suffered. Nonetheless, when he was offered a job as a professor of mathematics and astronomy in the Austrian city of Graz, he was ambivalent about taking it because it interrupted his plans to become a Lutheran pastor. He accepted the job despite his misgivings.

As a teacher, he was effusive and perhaps overenthusiastic. (His lengthy letters reveal the same qualities.) His lectures, he wrote, were "tiring, or at any rate perplexing and not very intelligible." On July 9, 1595, during just such a lecture, he experienced what he—and he alone—considered the greatest insight of his life. While drawing on the blackboard, he was pondering the fact that although there were five Platonic solids (bodies that, like the cube, are the same on every side), there were six planets. Surely planets and Platonic solids should exist in equal numbers. Then, in an exultant moment, he understood. The planets, he saw, revolved in the interstices of the Platonic solids, which nested within one another like a great cosmic toy. He described his revelation thus:

> The Earth's orbit is the measure of all things; circumscribe it around a dodecahedron, and the circle containing this will be Mars; circumscribe around Mars a tetrahedron, and the circle containing this will be Jupiter; circumscribe around Jupiter a cube, and the circle containing this will be Saturn. Now, inscribe within the Earth an icosahedron, and the circle containing it will be Venus; inscribe within Venus an octahedron, and the circle containing it will be Mercury. . . . And how intense was my pleasure from this discovery can never be expressed in words.

Although this scheme lacks all validity, he was never to give it up, both because it seemed to produce a closer approximation of the planetary orbits and because it had the Pythagorean tinge of divine geometry upon it.

In 1597, he married a widow whom he described as "simple of mind and fat of body." He also became involved in lengthy negotiations with the duke of Württemberg over the design and construction of an incredibly complicated drinking cup that was to be a model of the universe based on the Platonic solids. A sort of celestial soda fountain, it would serve, by way of hidden pipes coming from the appropriate planetary spheres, seven beverages: aqua vitae from the Sun, water from the Moon, brandy from Mercury, mead from Venus, vermouth from Mars, white wine from Jupiter, and "a bad old wine or beer" from Saturn, a planet which in astrological circles has often struggled for praise ("Grim Sir Saturne," the Elizabethan poet Edmund Spenser called him). The project was never completed. (Later projects included a weather journal, a biblical chronology, and an attempt to explain the universe by way of the Pythagorean music of the spheres. Kepler decided that the Earth's notes were "mi" and "fa," for misery and famine.)

During these final years of the sixteenth century, he was also writing. When he was ready to publish his book, *Mysterium cosmographicum,* the faculty at the University of Tübingen tried to block publication. Kepler published it with the help of his beloved professor, Michael Maestlin. He sent copies to Galileo, who evidently never read it, and to the great observer Tycho Brahe,

Kepler's nested solids from his book *Mysterium cosmographicum.* The pyramid-like tetrahedron separates the spheres of Jupiter and Mars; the large cube separates the spheres of Saturn and Jupiter; and so on.

imperial mathematician to the emperor Rudolf II in Prague. Tycho was so impressed that a few years later he hired Kepler as his assistant.

The offer came just in time, for Kepler lost his job in Graz when he refused to convert to Roman Catholicism. Kepler left for Prague on January 1, 1600. Tycho and Kepler couldn't have been more different. The red-headed Tycho was brash, self-confident, and excessive in every way; he literally threw scraps of food to a dwarf under his table, and he wore a metal nosepiece as a result of having lost most of his nose in a youthful duel. He was also the most accurate naked-eye observer in the history of astronomy. Tycho had something Kepler badly needed: reams of accurate data. Kepler had poor vision but he possessed something the aging Tycho lacked: a great theoretical mind. They were perfect for each other.

Needless to say, they didn't hit it off. Tycho's other assistants felt threatened by the young Kepler, whose reputation was already immense. Nor did Kepler improve the situation when he pledged to compute the orbit of Mars, a task that had long defeated Tycho's senior assistant, in only eight days. (In fact, it took Kepler years.)

The main problem was that Tycho withheld information. "Tycho gave me no opportunity to share in his experiences," Kepler complained. "He would only in the course of a meal, and in between conversing about other matters, mention, as if in passing, today the figure for the apogee of the planet, tomorrow the nodes of another." Finally, Kepler presented Tycho with an angry list of demands. Tycho accepted them, Kepler apologized for losing his temper, and after that Tycho willingly shared his data. Just before his death in 1601, Tycho was heard to whisper over and over again, "Let me not seem to have lived in vain," and he made Kepler his successor.

A few days after Tycho died, Kepler was officially appointed imperial mathematician. For the next eleven years, despite interference from Tycho's heirs, he worked with his predecessor's storehouse of observational data. Two tasks consumed him: the creation of a set of astronomical tables that would present Tycho's data in an organized form, and the continuing struggle with the orbit of Mars. Like every astronomer before him, Kepler assumed planetary orbits to be circular. They are not; and no matter how many circles upon circles were added, the calculated orbits still differed from the orbits observed. For over a decade, in the absence of every computational device—neither slide rules, logarithms, analytic geometry, nor calculus had been invented yet—Kepler calcu-

Johannes Kepler
(1571–1630).

lated. Adding and multiplying, he tried to come up with an orbit. Reluctantly, he abandoned the circle. Perhaps the orbit was egg-shaped. When that didn't work, he returned to the circle. He considered the oval. Again and again the idea of the ellipse flitted through his mind, but he dismissed it.

Finally, he found a formula that, calculated correctly, produces an ellipse. But Kepler did not calculate correctly. Frustrated, he set the equation aside and doggedly decided to try again, beginning with the very shape he had dismissed so many times. This, he believed, "was quite a different hypothesis." To his surprise, he discovered that the ellipse led back to the equation, and the equation produced an ellipse. "The two . . . are one and the same," he wrote. "Ah, what a foolish bird I have been."

And that is how Kepler discovered the first of his three great laws. The first two were published in 1609 in *Astronomia nova* (*The New Astronomy*). The publication of his book attracted little attention. Galileo, among others, ignored it. (Galileo's ill treatment of Kepler, his sole supporter and only astronomical equal, is a sorry chapter in the history of astronomy.) In the meantime, Kepler was also writing about optics, astrology, snowflakes, and the correct date of the birth of Christ.

In 1611, Kepler's thirty-seven-year-old wife and one of their children died, and the emperor Rudolf, Kepler's patron, abdicated. Kepler left Prague for Linz, and remarried after an exhaustive search in which he compared the merits of eleven different candidates. He also became involved with the defense of his mother, Katherine, who had been accused of witchcraft by her former best friend. At the same time, he was hard at work on another book, *Harmonice mundi* (*Harmony of the World*), which contained, in addition to astrological information and more talk of the Platonic solids, his third law of motion.

The year after this work's publication, Katherine Kepler was threatened with torture. She was released, in part because her famous son fought for her life, but she died a few months later. Kepler lived for another nine years. He completed the tables he had promised Tycho, worked on a science fiction fantasy, and cast horoscopes; although he sometimes despaired of the importance people placed upon astrology, he did not dispute its basic tenets and he made long-term predictions for his clients. In 1630, he journeyed by horseback to Leipzig in an ill-fated attempt to collect the money owed him by the emperor. He died two weeks after his arrival.

Johannes Kepler was a strange figure in many ways, mixing medieval beliefs with modern mathematics. Attached as he was to the Platonic solids, he was unaware of the importance of his contribution. But without him, Isaac Newton could never have developed the theory of gravity, and science as we know it would not exist. The astronomical advances he made were equaled by his great personal struggles. In his book *The Sleepwalkers,* a history of early cosmology, Arthur Koestler discusses Kepler's triumph: "In the Freudian universe, Kepler's youth is the story of a successful cure of neurosis by sublimation, in Adler's, of a successfully compensated inferiority complex, in Marx's, History's response to the need of improved navigational tables, in the geneticist's, of a freak combination of genes. But if that were the whole story, every stammerer would grow into a Demosthenes, and sadistic parents ought to be at a premium. Perhaps Mercury in conjunction with Mars, taken with a few cosmic grains of salt, is as good an explanation as any other." Kepler would most likely have agreed.

Kepler's Laws
of Planetary Motion

With his first law of motion, Kepler killed the circle. That law states that the planets travel on elliptical orbits with the Sun at one of the two focal points. (The ellipse in the diagram below is exaggerated. Planetary ellipses are close to circles; hence, the difficulty in discovering their true shape.)

His second law states that as a planet revolves around the Sun, its speed varies. It moves fastest when it is closest to the Sun and slowest when it is most distant. Yet during a given length of time, the area it carves out does not change: A line drawn from the Sun to the planet will sweep out an equal area of space in an equal space of time regardless of the speed at which the planet is moving. Thus, if the journey from *a* to *b* takes the same amount of time as the trip from *c* to *d,* the two dark triangles in the diagram below would be equal in area.

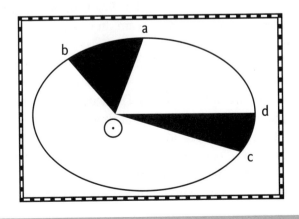

His third law, discovered some years later, establishes a relationship between a planet's distance from the Sun and the time it takes to complete one orbit. Basically, the law states that the cube of a planet's distance and the square of the time it takes to revolve around the Sun always have the same proportion to each other. Thus, Johannes Kepler, court astrologer, believer to his dying day in the importance of the five Platonic solids and the music of the spheres, quantified the laws of the universe and started the scientific revolution.

Kepler's *Dream*

Although many writers are anxious about the ways their fictional creations might affect the real-life inspirations for their works, few authors have had as much reason to worry as Johannes Kepler. His lunar fantasy, *Somnium* (*Dream*), contributed to one of the great traumas of Kepler's life: his mother's trial for witchcraft. Written at a time when Europe was embroiled in an assault against women so massive that it has been compared to the Nazi Holocaust, the book was meant to be a scientific treatise, a lunar geography written from a Copernican point of view. It had been on Kepler's mind since his student days, but because opposition to the heliocentric theory was strong, he disguised his thoughts in the form of an allegory presented as a dream. In it, a figure named Duracotus, representing science, explains that the recent death of his mother, Fiolxhilde, has freed him to tell a story. Fiolxhilde, he reports, supported herself by making charms from herbs, which she gathered and brewed "with elaborate ceremonies." One day, Duracotus cut open a charm she had been hoping to sell to a sailor, "and the herbs and patches of embroidered cloth she had put inside it scattered all about. Angry with me for cheating her out of payment, she gave me to the captain in place of the little pouch so that she might keep the money."

Having been given away by his mother (Kepler's own mother deserted him for a year when he was four years old), the fictional boy is deposited on the island of Hven with none other than Tycho Brahe, who teaches him Danish and astronomy, which "reminded me of my mother because she, too, used to commune constantly with the Moon." After several years, he returns home. He and his mother happily discuss the sky, which she tells him she has explored with the help of spirits, one of whom she agrees to summon. "My mother withdrew from me to a nearby crossroads, and after crying aloud a few words in which she set forth her desire, and then, performing some ceremonies, she returned, right hand outstretched, palm upward, and sat down beside me. Scarcely had we got our heads covered with our robes (as was the agreement) when there arose a hollow, indistinct voice, speaking in Icelandic."

The spirit, whom Kepler calls "the Daemon from Levania" (his name for the Moon, after the Hebrew), describes the physical sensations endured during a trip to the Moon (no fat people allowed, the daemon notes). After a detailed description of lunar geography, eclipses, weather, and inhabitants, the tale comes to an end familiar to every high school teacher of creative writing: "I came to my senses to find my head in fact covered by a pillow, my body wrapped in bedclothes." It was only a dream!

Although Kepler intended to circulate this book only among scholars, he sent a prepublication copy of the manuscript to the baron von Volckersdorff. Why, is not clear; Kepler was the kind of person who might have sent it on the basis of no more than a friendly chat. The baron read the manuscript and told his barber about it, and the barber told his sister Ursula about it. From then on, Kepler's mother, Katherine, was in trouble because Ursula, her former best friend, was bearing a grudge. Some time earlier, Ursula had had an abortion in order to terminate an extramarital pregnancy. She confided this to Katherine, who told her son Christoph. Soon the news was public. Ursula struck back, announcing that although she hadn't been feeling well, her distress was caused neither by pregnancy nor by abortion but by Katherine, who had cast a spell on her. It didn't help that Katherine had been raised by an aunt who was burned at the stake as a witch. And it didn't help that her famous son Johannes had written a semi-autobiographical manuscript describing "his mother's" contacts with spirits. Once Ursula pointed the finger, accusations flowed. One citizen remembered that ten years earlier, he had gotten sick after drinking from a tin cup at Katherine Kepler's home. A woman recalled a time when her husband passed Katherine on the street and experienced a pain in his thigh. A twelve-year-old girl claimed that she had once walked past Katherine on the road and felt a pain in her arm. Katherine sued Ursula for slander. But she didn't truly understand the danger she was in, for she continued to wander around town suggesting herbal remedies to everyone and offering them in the suspect tin cup.

In 1615, Katherine was charged with witchcraft. Over the next six years she was arrested, interrogated, taken to court, thrown into prison, and kept in chains. Kepler was indefatigable in her defense, writing petitions, reviewing all testimony, reminding the town council that he was court mathematician to the emperor, and loyally acting as his mother's attorney. (So dogged was he that the court record read, "The accused appeared in court, accompanied, alas, by her son, Johannes Kepler, mathematician.") Katherine was nonetheless threatened with explicit torture. According to a provost's report, "She refused to admit and confess to witchcraft as charged, indicating that even if one artery after another were to be torn from her body, she would have nothing to confess; whereafter she fell on her knees and said a pater noster." Eventually, thanks to her son, she was released, but she died shortly thereafter. Kepler, who felt guilty about whatever part he might have had in bringing about these ugly events, spent the rest of his life trying to exonerate her by annotating the *Somnium* with an extraordinary set of footnotes describing virtually every nuance of his thought processes. It was first published four years after his death, when his poverty-stricken family sold the original manuscript.

Saturn

When Galileo gazed at Saturn with his primitive telescope, he had not the slightest idea of what he was seeing. The idea of rings had never occurred to him, and so even with Saturn's flat collar of rings tilted toward the Earth, and hence most clearly visible, Galileo didn't see it for what it was. Instead, he described the blurry-looking planet as "tricorporate," or triple-bodied, as if Saturn were composed of a large planet wedged in between two small ones, and he announced his discovery of the Saturnian trinity through an anagram (printed above), which he decoded three months later. The message was "Altissimum planetam tergeminum obseruaui," or "I have observed the highest planet to be triple in form."

Two years later, Saturn reached the point in its orbit where the rings are edge on to the Earth and virtually invisible to us, something that happens twice in its three-decade crawl around the Sun. A look through his weak telescope revealed that the two smaller globes Galileo had imagined on either side of the planet seemed to have disappeared and that Saturn was no longer triple. Galileo wondered if Saturn, like the god after whom it was named, "had devoured his own children." By 1616, when the rings were once again tilted toward the Earth and visible through a telescope, Galileo described them as "half ellipses." But even though his notoriously inaccurate drawings clearly represent what we know to be

NASA

Saturn, seen from a distance of 13 million miles by *Voyager 2*. Tethys, Dione, and Thea are bright white spots visible below the planet. The dark spot on the lower part of Saturn is the shadow of Tethys.

rings, neither he nor any other astronomer of the day came up with the concept.

Then, in 1655, the Dutch astronomer Christiaan Huygens, working with his brother and with the philosopher Benedictus de Spinoza, developed a better eyepiece, a new way to grind lenses, and, hence, an improved telescope. With it he discovered Titan, Saturn's largest moon, and while observing that satellite, he solved the mystery. As Galileo had done, he published it in the form of an anagram (printed on page 163). Deciphered, it read, "Annulo cingitur, tenui plano, nusquam cohaerente, ad eclipticam inclinato," or, "Saturn is girdled by a thin flat ring, nowhere touching it, and inclined to the ecliptic."

Twenty years later, Giovanni Cassini discovered a gap in the middle of what appeared to be a single ring; three centuries later, the space program found that Cassini's Division is not empty at all but is occupied by at least a hundred tiny ringlets, and that the rings themselves are made up of countless particles, each in its own orbit.

Saturn, which could swallow ninety-five Earths, is the second largest planet. It is 886,281,264 miles from the Sun, it revolves around the Sun in 29.46 years, and it spins on its axis about every 10 hours 14 minutes, fast enough to cause a visible flattening of its polar regions. Like Jupiter, the planet it most resembles, Saturn is primarily made of gas. Its density is so low that, given a larger-than-Olympic-size swimming pool, Saturn would float. (Jupiter is slightly heavier than water and would sink, no matter how large the pool.)

Like Jupiter, Saturn is wrapped in bands of clouds. Composed of crystallized ammonia, the pale ocher clouds are less turbulent than those of Jupiter's complex paisley: On Jupiter, the winds blow the belts of clouds in alternating directions, whereas on Saturn, except around the poles, the winds all flow east.

Beneath the clouds and gaseous atmosphere is an ocean of liquid hydrogen and helium that gradually turns into metallic hydrogen. Powerful electric currents surge through it, generating Saturn's strong magnetic field.

Like Jupiter, Saturn produces more heat than it receives from the Sun. A fraction of that heat may be generated by the gravitational collapse that began in the planet's infancy and continues, albeit slowly, today, but heat is also generated within the planet by a process called helium rain, which happens when helium separates from hydrogen the way vinegar separates from oil and then tumbles inward under the force of gravity. At the heart of the planet is a relatively small core of rock and ice the size of several Earths.

Approximately every thirty years, a storm known as the Great White Spot erupts on the normally quiescent cloud cover. Seen in 1876, 1903, 1933, 1960, and 1990, the Great White Spot has a cycle close enough to Saturn's 29.46-year orbit to make astronomers take notice. They theorize that the storm is an aftereffect of

SATURN

Mean Distance from the Sun
886,281,264 miles

Diameter at the Equator
74,600 miles

Diameter if Earth = 1
9.41

Mass if Earth = 1
95.16

Surface Gravity if Earth = 1
1.08

Moons
18 named satellites

Length of Year
(revolution around the Sun)
29.46 Earth years

Length of Day
(rotation at equator)
10 hr. 14 min.

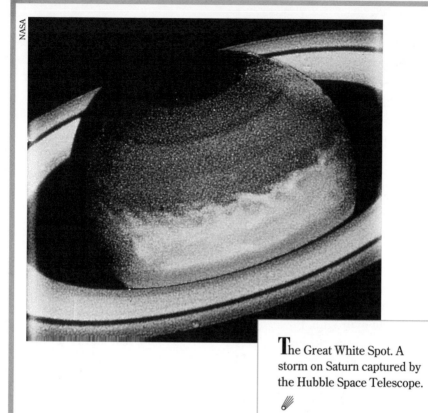

NASA

The Great White Spot. A storm on Saturn captured by the Hubble Space Telescope.

summer on Saturn. It occurs a few Earth years after the warmest season of the planet's long year, when the Saturnian Northern Hemisphere is most tilted toward the Sun. The summertime heat warms the atmosphere and causes ammonia to bubble up, crystallize, and be swept around the planet by thousand-mile-per-hour winds. During its last appearance, in 1990, the Great White Spot was initially about 15,000 miles across (twice as large as the Earth), and so luminous that an amateur astronomer described it as "like a flashlight, five or six times brighter than the rest of the planet." Within a month, it grew into a thick white stripe wrapped around the entire planet.

Yet no matter how impressive the Great White Spot might become, it will never be able to compete with the spectacle of the rings. Viewed through a small backyard telescope, Saturn glows like a magical toy. Up close, its beauty grows. *Voyager 2* captured sights Galileo couldn't have imagined: Saturn half lit, with the rings casting the narrowest blade of a shadow across the planet's face; Saturn fully lit, its amber-colored surface gleaming through Cassini's Division; Saturn viewed from its dark side, with sunlight filtering through its back-lit rings; and the rings themselves, the most glorious special effect of the solar system, light and dark and lacy, a miracle of ice.

Rings

> Saturn's rings . . . The tourism of the
> future will take place there.
>   —Tullio Regge

For centuries, Saturn was thought to be an anomaly, the only planet with rings. The picture has changed. All four giant planets—Jupiter, Saturn, Uranus, and Neptune—have rings.

Jupiter, the largest planet in the solar system, is encircled by a thin, dark, dust-laden ring too faint to be visible from the Earth. The ring was discovered on March 4, 1979, when the *Voyager 1* camera took an eleven-minute exposure of the area adjacent to the equatorial plane on the off chance that there *might* be a ring. The long exposure paid off. Jupiter is lassoed by a narrow main band and a broader area called the gossamer ring, which in turn is surrounded by a wide, diffuse halo.

Saturn, the next planet from the Sun, has the most dazzling ring system by far—a broad, dynamic, complicated wheel of snowflakes, hailstones, and balls of ice mixed with dust. Divided into three main rings (A, B, and C) and over a thousand ringlets, the rings are varied and intricate. The A ring is marked by spiral density waves and divided in two by the Encke Gap, a virtually empty slice in the ring that is scalloped on its edges and swept clean of particles by a tiny, twelve-mile moonlet. The icy B ring has spiral density waves and spokes, a result of electromagnetic forces that cause microscopic grains to hover slightly above the tumbling snowballs of the ring itself. Cassini's Division, a dark lane between

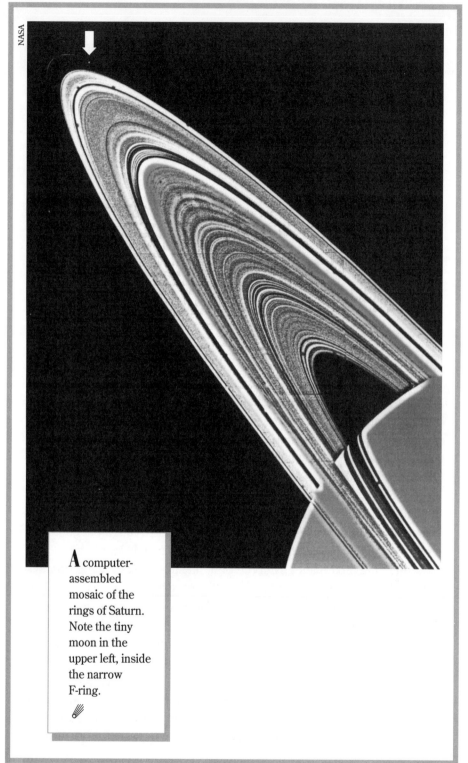

NASA

A computer-assembled mosaic of the rings of Saturn. Note the tiny moon in the upper left, inside the narrow F-ring.

the A and B rings, houses at least 100 thin rings. And Saturn's outer F ring, a narrow golden strand, is clumped, knotted, and braided. When *Voyager 1* sent back pictures of it, "Those of us watching the monitors were stunned," wrote astronomer Bradford A. Smith. "In some tabulation of ring phenomena that we least expected to see, the observed structure of the F ring would have been somewhere off the top. . . . To me, it was the most improbable picture yet sent back by either *Voyager* spacecraft."

Next comes Uranus. From the Earth, nine exceedingly dark and narrow Uranian rings had been detected prior to the *Voyager* journeys. On its approach, *Voyager* discovered two more rocky rings; and when it sped past and took a picture of the planet facing the distant Sun, it detected still more dark and dusty circles around the planet.

Finally, there is Neptune. Prior to *Voyager,* scientists thought that the rings of Neptune were incomplete, mere dashes of arc visible here and there around the planet. *Voyager* proved that Neptune's rings extend all the way around the planet but appear to be partial because some sections of the rings are thickly clumped while others are so finely spun they're hard to see.

Many questions about rings remain unanswered. For instance, why don't the icy chunks of matter within each ring spread out over a greater distance? Scientists have theorized that tiny satellites known as shepherd moons, one on each side of a ring, may gravitationally corral the material and keep it from spreading out. On Jupiter, Uranus, and Neptune, a few shepherd moons have been found that seem to do just that. But many rings seem moonless, and other rings apparently have only a single shepherd moon. It may be that some shepherd moons are too tiny to be detected by passing spacecraft—or perhaps they were just on the wrong side of the planet when *Voyager*s *1* and *2* flew by. Or maybe the theory is incomplete.

Although shepherd moons live within the ring systems, large moons never do. Rings are always closer to the planet than a major moon could possibly be. The reason is that if a moon wanders too close, it would be ripped apart by the difference between the planet's powerful gravitational pull, which creates hurricane tidal forces on the side of the moon facing the planet, and the significantly smaller gravitational pull on the other side. To preserve itself, the moon has to maintain a certain distance. This distance, known as the Roche Limit after the scientist who figured it out in 1850, is equal to 2.44 times the radius of the planet. In the entire solar system, not a single large moon lies inside the Roche Limit (although very small moons are unaffected by it). The ring systems, however, are within the Roche Limit. So perhaps they were formed when a moon or asteroid strayed within the Roche Limit and was torn to smithereens.

Or perhaps the fragments are primordial leftovers, there since the beginning yet unable to coalesce into a moon.

Or consider an alternative hypothesis. The ring systems we see may differ not so much in form as in age. It could be that the ring system of

The shadowy-looking spokes in Saturn's rings.

NASA

Saturn is new, evidence of violent collisions in the relatively recent past, while the meager rings of the other planets, perhaps equally impressive once upon a time, are the dwindled remains of gargantuan rings of an earlier era. The suggestion has been made that Saturn's rings are only about 100 million years old, and that due to the interaction of the rings with the moons, the rings will disintegrate completely 100 million years from now.

One thing is certain: Wherever we look in space, there are rings. Countless stars form rings around black holes; asteroids form bands around stars; fragments of rock and ice and great clouds of dust spread out and revolve around planets; wreaths of gas enshroud the corpses of exploded stars. The universe is bedecked with rings.

Titan and Other Saturnian Moons: The Search for Life Continues. . . .

> "There's more to Titan than just climate," said
> Rumfoord. "The women, for instance, are the most
> beautiful creatures between the Sun and Betelgeuse."
> —Kurt Vonnegut, *The Sirens of Titan*

So far, Saturn has eighteen named moons and a handful of tiny moons still unnamed. Among the distinctive satellites circling the Lord of the Rings are:

✳ Phoebe, Saturn's outermost satellite. It moves around Saturn in a retrograde direction (opposite to that of the other satellites), and may be a former asteroid captured by Saturn's gravitational force;

The identity of the poet who composed this ode to Saturn's moon Phoebe is unknown. Phoebe was discovered in 1899 by William Pickering.

Phoebe, Phoebe whirling high
In our neatly plotted sky,
Phoebe, listen to my lay,
Won't you swirl the other way?
Never mind what God has said,
We have made a law instead.
Have you never heard of this
Nebular hypothesis?
It prescribes, in terms exact,
Just how every star should act.
Tells each little satellite
Where to go and whirl at night.
(Disobedience incurs
Anger of astronomers
Who—you mustn't think it odd—
Are more finicky than God.)
And so, my dear, you'd better change;
Really we can't rearrange
All our charts from Mars to Hebe
Just to fit a chit like Phoebe.

* Iapetus, a yin-yang moon, black on one half and bright on the other, so that it's easily visible when it's traveling around one side of the planet and almost impossible to see on the other;

* Mimas, an icy world marked by a crater so huge that its creation almost broke the moon apart;

* Enceladus, a geologically active world whose clean, ice-covered surface shines so brightly that it is the most highly reflective body in the solar system;

* Janus and Epimetheus, two low-density icy worlds whose orbits are so similar that every four years they switch places when one of them threatens to overtake its companion. To avert a shattering collision, the outer moon speeds up, pulls the inner moon into its orbit, and moves into the inside lane;

* Titan, one of the few places in the solar system where life could theoretically exist. Larger than Mercury and muffled in orange smog, Titan sounds like the worst environmental disaster ever. It is a world of methane snow, methane rain, methane cliffs, and methane poles, a world whose lakes and ponds may be bordered with an organic tar typically described with words like "goo," "gunge," and "sludge." Titan also possesses a thick atmosphere that, like the Earth's, is pri-

NASA

In this photomontage of Saturn and its satellites, prepared from images taken by *Voyager 1*, Dione is in front of the planet; Tethys and Mimas, with its enormous crater, are to the right; Enceladus and Leah are to the left; and Titan is at the upper right.

NASA

Saturn's moon Enceladus.

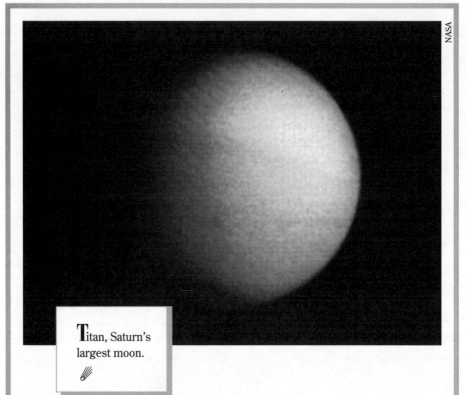

NASA

Titan, Saturn's
largest moon.

marily nitrogen, seasoned with compounds necessary to the formation of DNA. Scientists describe it as "prebiological."

Although Titan is smaller than the Earth and its gravity is
less, the thickness of its atmosphere produces a surface pressure one and a half times greater than that on Earth. Its temperature is frigid (approximately 288° F below zero), and it is
perpetually shrouded in night beneath its thick orange blanket.
That haze effectively hides the Titanic surface, so no one is certain whether it is a landscape of mountains, craters, and chasms
or a noxious-smelling prairie of goo. We will learn more in the
year 2000, when the Cassini mission, a joint project of NASA and
the European Space Agency, arrives at Saturn and ejects a probe
through the orange clouds and into the Titanic yuck.

☾

Sir Isaac Newton

Newton was not the first of the age of reason. He was the last of the magicians, the last of the Babylonians and Sumerians, the last great mind which looked out on the visible and intellectual worlds with the same eyes as those who began to build our intellectual heritage rather less than 10,000 years ago.

—John Maynard Keynes

It is fashionable today to think of Einstein as the epitome of scientific genius, and compared with us ordinary mortals, Einstein was indeed a giant. But compared with Newton, Einstein runs a very distant second.

—Subrahmanyan Chandrasekhar

saac Newton played many parts in his life. As a scientist, he is often considered the greatest genius of all time. As a public man, he was president of the Royal Society for twenty-four years, a member of Parliament, and master of the mint. As a recluse, he led a private life of the mind that to this day makes scientists uncomfortable—so much so that early biographers hushed it up. Behind closed doors, Isaac Newton was an alchemist who wrote thousands of pages detailing his thoughts about religious matters and the occult.

If his search for the philosophers' stone was unsuccessful, his search for the secret of the universe was not, for Newton did what no other astronomer had been able to do: He identified a single force as powerful in heaven as it is on Earth. Until Einstein came along, gravity explained al-

Library of Congress

Sir Isaac Newton (1642–1727) in a 1760 mezzotint.

most everything. (It doesn't explain the action inside an atom, however; the great challenge of modern physics is to connect Newtonian physics with the subatomic universe.)

Born on Christmas Day 1642 (the year Galileo died), Isaac Newton was the son of an illiterate farmer who died before his birth. When he was three, his mother remarried and left him with his grandmother. At age eleven, upon the death of his hated stepfather, he was reunited with his mother (who had given birth to three additional children). He attended school for another five years, distinguishing himself only after he sought revenge against a boy who kicked him. After fighting the boy, pulling him to the church, and rubbing his face against the wall, Newton decided to show up his adversary academically, which he did, rising from the second-to-last position to top student in the school.

Nonetheless, at age sixteen, he dropped out of school in favor of agriculture. He spent his time reading and was so preoccupied that once a horse he was leading slipped away and Newton, holding the bridle, didn't notice.

A year later, Newton returned to school, where he boarded with an apothecary and evidently formed a romantic attachment to the man's stepdaughter. The relationship withered in 1661 when Newton went to Trinity College at Cambridge. Many years later, when Newton's first love was seventy years old, she revealed the connection to a biographer who wrote that "her portion not being considerable, and he being a fellow of a college, it was incompatible with his fortunes to marry; perhaps his studies too." As far as we know, this was the only such relationship Newton ever had.

At Trinity College, Newton played cards (poorly) and studied Kepler, Descartes, Galileo, and Copernicus. After he graduated in 1665, the year the Black Death struck England, the university closed and Newton went home.

There, on his mother's farm, he passed light through a glass prism, saw the rainbow-colored band that it produced, and realized that white light is a combination of all colors. He invented calculus, as did, independently, his friend—later his foe—the German philosopher Gottfried Leibniz. And one day he watched an apple drop from a tree and wondered if the force that pulled the apple toward the Earth might also act upon the Moon.* Why didn't the Moon fall? Why should it be different from the apple? Then he realized that it wasn't different. After all, he reasoned, if nothing were attracting the Moon to the Earth, the Moon would move in a straight line past the Earth. That didn't happen because the Moon was continuously falling away from the straight line—and toward the Earth. Surely the very same force was at work. That force was gravity—a concept that Newton, it should be pointed out, did not invent. Yet prior to Newton, gravity was thought of as something that worked only on Earth. No one connected it with the force that keeps the planets revolving around the Sun; indeed, Kepler assumed that planets revolved around the Sun because of a mysterious power that was probably magnetic in nature.

> **L**ook to the heavens, and learn from them
> How one should really honor the master.
> The stars in their courses extol Newton's laws—
> In silence eternal.
>
> —Albert Einstein

Newton made the leap. He realized that gravity works in heaven and on Earth and that it diminishes as distance increases: More precisely, gravity is proportional to the inverse square of the distance from the object to the center of the Earth.† But his calculations didn't come out perfectly, possibly because he was using an incorrect figure for the diam-

---

*Newton told this story himself, but historians have persistently questioned its truth; the story appeared so many years after the event that some people suspect the great man of mythologizing himself.

†Thus, if gravity has a strength g at a certain distance, then at twice that distance its strength is $\frac{1}{4} g$; at three times that distance, its strength is $\frac{1}{9} g$; and so forth.

GEORGE BERNARD SHAW ON ISAAC NEWTON

(in an after-dinner toast to Albert Einstein)

• ⤳

Copernicus proved that Ptolemy was wrong. Kepler proved that Copernicus was wrong. Galileo proved that Aristotle was wrong. But at that point the sequence broke down, because science then came up for the first time against that incalculable phenomenon, an Englishman. As an Englishman, Newton was able to combine a prodigious mental faculty with the credulities and delusions that would disgrace a rabbit. As an Englishman, he postulated a rectilinear universe because the English always use the word "square" to denote honesty, truthfulness, in short: rectitude. Newton knew that the universe consisted of bodies in motion, and that none of them moved in straight lines, nor ever could. But an Englishman was not daunted by the facts. To explain why all the lines in his rectilinear universe were bent, he invented a force called gravitation and then erected a complex British universe and established it as a religion which was devoutly believed in for three hundred years. The book of this Newtonian religion was not that oriental magic thing, the Bible. It was that British and matter-of-fact thing, a Bradshaw [railway schedule]. It gives the stations of all the heavenly bodies, their distances, the rates at which they are travelling, and the hour at which they reach eclipsing points or crash into the earth. Every item is precise, ascertained, absolute and English.

eter of the Earth. So he put aside the universal theory of gravity and returned to Cambridge, where he was appointed Lucasian Professor of Mathematics (a post assumed in 1979 by Stephen Hawking). For almost twenty years, he made no attempt to publish his theory of gravity—possibly because the thought of publication caused him to have anxiety attacks. He might never have published it had it not been for another astronomer: Edmond Halley, after whom the great comet is named.

One day in January 1684, Halley, the architect and scientist Sir Christopher Wren, and the physicist Robert Hooke were discussing celestial motion and the nature of the force operating between the Sun and the planets. When Hooke announced that the attraction between the Sun and the planets decreases in proportion to the square of the distance (something he had discussed with Newton), Wren challenged him to prove it, offering a prize—an expensive book— as incentive. But months passed, and Hooke, who had a fairly high bluster component, failed to respond. Eventually Halley approached Newton, whom he had never met. Newton rose to the challenge. He could prove it. Indeed, he had already done so—in 1666.

But amidst his jumbled stacks of scientific, biblical, and alchemical jottings, he was unable to find his old notes. Inspired by Halley, however, he promised to redo his proof and write up his theories. *Philosophiae Naturalis Principia Mathematica,* the greatest scientific book of all time, was written in a year and a half. The Royal Society, of which Samuel Pepys was president, refused to honor a promise to publish it, partly because Robert Hooke, long an enemy of Newton's, claimed to have had the idea first. In the book, Newton acknowledged that Hooke, Wren, and Halley had had similar thoughts, but the matter didn't simmer down and the book did not get printed until Halley stepped in and, using a legacy he received after his father was murdered, paid the cost of publication himself (something Newton, who had plenty of money, could easily have done).

When *Principia Mathematica* was finally published in 1687, it made an enormous impact. Its three laws of motion and a single, universal law of gravity explained everything. ("If I have seen further," Newton wrote, "it is by standing on the shoulders of Giants.") Through rigorous mathematics and obsessional concentration, Newton united heaven and Earth. His theory, he was eager to point out, was not a hypothesis but a fact. Not since Aristotle had a worldview so all-encompassing been created.

Principia Mathematica, unlike Copernicus' *Revolutionibus,* was well received. Newton was lionized, despite having, according to the professor who occupied the Lucasian chair after him, "the most fearful, cautious, and suspicious temper that I ever knew."

Yet Newton changed. Once a virtual recluse, he became ambitious.

RALPH WALDO EMERSON ON GRAVITY

• ◞

When the fruit is ripe, it falls. When the fruit is despatched, the leaf falls. The circuit of the waters is mere falling. The walking of man and of all animals is a falling forward. All our manual labor and works of strength, as prying, splitting, digging, rowing, and so forth, are done by dint of continual falling, and the globe, earth, moon, comet, sun, star, fall for ever and ever.

When King James II attempted to make the English universities Catholic, Newton resisted, and was, as a result, elected to Parliament. In London, he asked the philosopher John Locke to help him wangle a position as comptroller of the mint. In the meantime, his alchemical experiments continued.

In 1692, he had a breakdown. Some commentators have suggested that it came about when his dog, Diamond, upset a candle, causing years of work to go up in flames. But Newton never had a dog; he wasn't that kind of man. (A story describing how he cared for a cat and her kittens is also apocryphal.) John Maynard Keynes believed the breakdown was connected with Newton's mother's death three years earlier. Or perhaps the problem was that in his assiduous attempts to make gold from lead, arsenic, antimony, and mercury, Newton had inadvertently poisoned himself.

Whatever the cause, he became insomniac, despondent, unable to concentrate, and paranoid. (Keynes used the word "gaga.") He wrote to Samuel Pepys, telling him, "I must withdraw from your acquaintance and see neither you nor the rest of my friends anymore." He accused John Locke of having "endeavored to embroil me with women." Nonetheless, his friends rallied around him, and he recovered.

Nature and Nature's laws lay hid in night:
God said, "Let Newton be!" and all was light.
　　　　　　　　　　—Alexander Pope (1688–1744)

It did not last: the Devil howling "Ho,
Let Einstein be," restored the status quo.
　　　　　　　　　　—J. C. Squire (1884–1958)

For the rest of his long life, he was busy, social, and famous. In 1696, he became warden—and, later, master—of the mint, and he was a savvy investor. In 1703, after Robert Hooke's death, the members of the Royal Society in London elected him president, a position he held for the rest of his life. In 1705, he was knighted—the first scientist to become a Sir. He moved in with his brilliant niece Catherine Barton and her husband. Catherine, who was very likely the mistress of one of Newton's friends, was a woman of her age, praised by Jonathan Swift in his *Journal to*

Stella. Thus the great scientist, no longer at the peak of his genius, hobnobbed in society. It sounds like a just and proper conclusion to a life well spent.

Except for one thing: Isaac Newton was in many respects an awful human being. Sometimes he was simply passive. For instance, when a colleague lost his job because of religious beliefs that Newton shared but had always kept secret, the great man said nothing. On other occasions he was vicious. He was so infuriated when Gottfried Leibniz published his discoveries about calculus before Newton could publish *his* that he wrote articles in his own defense under other people's names, and issued a report, through the auspices of the Royal Society, accusing Leibniz of plagiarism. John Flamsteed, the first astronomer royal, once took him to court. Nor did Newton reserve his venom for fellow scientists. During his reign over the mint, he made sure that several counterfeiters received the death penalty. "If we evolved a race of Isaac Newtons, that would not be progress," Aldous

PRINCIPIA MATHEMATICA AND THE UNIVERSAL LAW OF GRAVITY

• ➜

A mathematical equation stands forever.

—Albert Einstein

Writing in Latin, Newton laid out the three laws of motion. They are:

1. A body remains in a state of rest or a state of motion unless a force acting upon it compels it to change.

2. Change occurs in proportion to the force applied and in the same direction.

3. Every action has an equal and opposite reaction.

These laws permitted Newton to express the universal law of gravity in a single formula:

$$F = G \ \frac{m_1 m_2}{d^2}$$

F stands for the gravitational force of attraction between any two bodies. Their masses are represented by m_1 and m_2. The distance between them is represented by d, which is squared. G is the gravitational constant, determined by laboratory experiment in 1798. For those interested in such things, that number, using the International System of metric units, is 0.00000000006673.

Huxley wrote. "For the price Newton had to pay for being a supreme intellect was that he was incapable of friendship, love, fatherhood, and many other desirable things. As a man he was a failure; as a monster he was superb."

Newton saw himself differently: "I do not know what I may appear to the world, but to myself I seem to have been only like a boy playing on the sea-shore, and diverting myself in now and then finding a smoother pebble or a prettier shell than ordinary, whilst the great ocean of truth lay all undiscovered before me."

※

Uranus

Then felt I like some watcher of the skies
When a new planet swims into his ken.

—John Keats

ne night in March 1781, William Herschel, organist and choir director at the Octagon Chapel in Bath, put on a black hood that kept out extraneous light, peered through a homemade telescope, and discovered a world.

Herschel was well acquainted with the nighttime sky. A skilled telescope builder, he was in the midst of making his second survey of the sky with the help of his sister Caroline when he was surprised by an unexpected speck of light in the constellation Gemini. He knew it wasn't a star because stars look like points of light through a telescope, whereas this object at high magnification seemed to have a disklike shape. He decided it might be a comet. But as the months passed, its orbit did not become elongated, as a comet's would, but remained roughly circular—like that of a planet. Within six months of its first sighting, Herschel's discovery was accepted as a planet.

This stunning find was not a stroke of luck, a point Herschel was anxious to make. "It has generally been supposed that it was a lucky accident that brought this new star to my view," he wrote. "This is an evident mistake. In the regular manner I examined every star of the heavens, not only of that magnitude but many far inferior, it was that night *its turn* to be discovered." When the body's orbit was calculated and its distance determined, Uranus was found to be twice as far from the Sun as Saturn was. The size of the known solar system doubled.

Herschel's achievement was recognized immediately. The scientific community made him a member of the Royal Society and awarded him

Sir William Herschel (1738–1822). In this engraving, made from a drawing owned by his son, Herschel holds a scroll showing Uranus and its satellites Oberon and Titania, all of which he discovered.

the prestigious Copley Medal, and King George III gave him an annual subsidy of £200, enabling Herschel to stop giving music lessons entirely. Some scientists, however, doubted the discovery, grumbling that they hadn't seen the planet themselves (their own telescopes were not as powerful) and that, in any case, Herschel was an amateur—a music teacher. To prove his reliability, in the year following the discovery, Herschel brought his telescope to the king so that he could peer through it, and to the Royal Observatory in Greenwich so that other astronomers could verify its strength and accuracy—which they did.

The next order of business was to name the planet. The president of the Royal Society asked Herschel to come up with something fast or "our nimble neighbors, the French, will certainly save us the trouble of Baptizing it." A French astronomer did suggest a name: Herschel. In Germany, Johann Elert Bode recommended Uranus because the name fit in mythologically; Uranus was the father of Saturn, Saturn was the father of Jupiter, and Jupiter was the father of the Sun and all the other planets (except the Earth). Herschel was uninterested in the issue, although eventually, egged on perhaps by the Royal Society, he addressed a letter to King George in which he made the politic suggestion that "if in any future age it should be asked, *when* this last-found Planet was discovered? It would be a very satisfactory answer to say, 'In the Reign of King George the Third.' As a philosopher then, the name of Georgium Sidus presents itself to me."

This lack of consensus caused the planet to be known by many names.

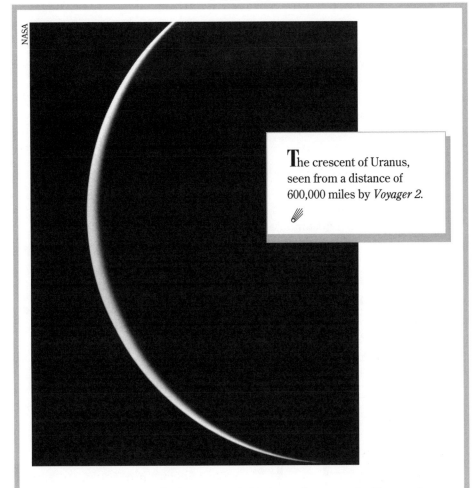

The crescent of Uranus, seen from a distance of 600,000 miles by *Voyager 2*.

In Germany it was called Uranus right from the start, in France it was called Herschel for a time, and in Great Britain it was known as the Georgian Planet. Finally, in 1850, John Couch Adams, who first calculated the position of Neptune, recommended that the planet be officially called Uranus. The name was accepted.

Like Jupiter, Saturn, and Neptune, Uranus is a gas giant. It has a diameter less than half the size of Saturn's but approximately four times the size of the Earth's. Its rocky core is surrounded by an icy slush of water, ammonia, and methane; a thick atmosphere of hydrogen, helium, and methane lends it its distinctive tint. Bands of clouds circle the planet, blown by winds that reach speeds of 360 mph. In photographs taken by NASA, Uranus looks featureless and calm, an aquamarine marble glimpsed through fog.

Encircled by eleven thin, dark rings and fifteen moons, all but two of

URANUS

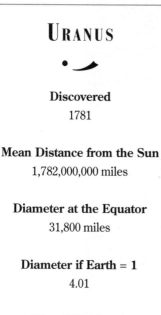

Discovered
1781

Mean Distance from the Sun
1,782,000,000 miles

Diameter at the Equator
31,800 miles

Diameter if Earth = 1
4.01

Mass if Earth = 1
14.5

Surface Gravity if Earth = 1
0.91

Moons
15

Length of Year
(revolution around the Sun)
84.01 Earth years

Length of Day
(rotation at equator)
17 hr. 14 min.

which have been given Shakespearean names, Uranus is in many ways an oddity. During its 84-year orbit around the Sun, it spins on its side, with its equator nearly perpendicular to the plane of the solar system and its poles pointing not up and down, as on Earth, but sideways. As a result of this unusual orientation, each pole experiences what is essentially a 42-year day followed by a 42-year night. Another oddity is that the Uranian magnetic field is neither parallel to the poles nor perpendicular to them but tilted at a highly unusual 60° angle to the axis. Perhaps the magnetic field was discovered in the rare act of switching directions, as happens on Earth every half million years or so. More likely, the Uranian system was once the site of a spectacular collision in which an object the size of the Earth slammed into the planet, knocking it on its side and throwing everything into disarray.

Signs of catastrophe are also apparent in the geography of the Uranian moons, the most spectacular of which is Miranda, discovered in 1948 by Gerard P. Kuiper. A tiny world with a 300-mile diameter and cliffs that measure up to 3 miles high, Miranda has a rocky landscape so varied that scientists theorize the moon was hit long ago by a marauding asteroid, which caused it to shatter. The fragments fell back together, but the fit wasn't a good one, leaving Miranda patterned with a dramatic, distinctive patchwork—evidence of the violence of the early solar system.

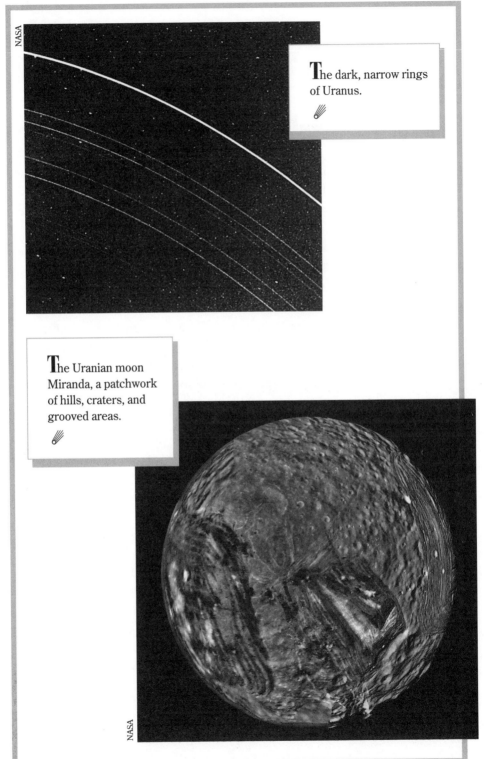

NASA

The dark, narrow rings of Uranus.

The Uranian moon Miranda, a patchwork of hills, craters, and grooved areas.

NASA

Caroline Herschel

she whom the moon ruled
like us
levitating into the night sky
riding the polished lenses
— Adrienne Rich, "Planetarium"

There is something triumphant about the life of Caroline Herschel (1750–1848). Born in Hanover, Germany, she and her brother William, twelve years her senior, were introduced to astronomy by their father, a professional oboe player. Although he told his daughter that she was neither pretty nor rich enough to marry, he also inculcated in his six children a love of music and science. "I remember his taking me on a clear frosty night into the street, to make me acquainted with several of the beautiful constellations, after we had been gazing at a comet which was then visible," she wrote. "And I well remember with what delight he used to assist my brother William in his philosophical studies. . . ." When she was seven, William joined the Hanoverian Guards as a member of the military band during the Seven Years' War with France. After watching a battle he decided that the military wasn't for him, and he left Germany to avoid further service. In England he nailed down a job as organist and choirmaster for the newly built Octagon Chapel in Bath.

When Caroline was sixteen, her father died, and her mother forced her to take responsibility for all the cooking, cleaning, and sewing. Six years later, William came to her rescue, and Caroline moved in with him. He gave her a detailed schedule of tasks. On her second day in Bath, he instructed her in English, bookkeeping, and singing, after which "by way of relaxation we talked of Astronomy and the fine constellations with whom I had made acquaintance."

Caroline hoped for a singing career, and even received some acclaim for her performances. But William's astronomical interests interfered. While still giving music lessons and working at the chapel, he was busy

Caroline Herschel
(1750–1848).

building telescopes ("It was to my sorrow that I saw almost every room turned into a workshop," Caroline noted) and spending every clear evening looking at the stars. She became his assistant, grinding lenses, making models of the large telescopes he was building, observing with him, and making certain he remembered to eat. It was a full-time job. "If it had not been sometimes for the intervention of a cloudy or moon-light night, I know not when my Brother (or I either) should have got any sleep."

After William discovered Uranus in 1781, he became famous, gave up music, and, thanks to a royal subsidy of £200 a year, devoted himself to the celestial sphere. In 1787, Caroline was awarded an annual stipend of £50, acknowledgment that she was an astronomer in her own right. The next year, William married a rich widow named Mary Pitt. The two women became friends, and Caroline also became close to her nephew, John, born in 1792.

Caroline's observations continued, with and without her brother, "every starry night on wet or hoarfrost-covered grass, without a human being within call." She worked as William's devoted assistant until shortly before his death in 1822. A French naturalist who visited the Herschels in 1784 described the scene: "The Observatory is in a garden. . . . Whenever Mr. Herschel searches for, say, a nebula, or a star of the highest magnitude, he calls from the garden to his sister, who comes to the window straightaway and, consulting one of the large handwritten tables, calls back from the window, 'Near the gamma star,' or 'Toward Orion,' or some other constellation. In truth, nothing could be more touching and agreeable than this rapport, this straightforward method."

Yet she was more than her brother's helper. She revised John Flamsteed's star catalogue, wrote *A Catalogue of the Nebulae* (she was seventy-five when she finished it), and discovered seventeen nebulae and many star clusters. She was also the first woman to discover a comet. Eventually, she found eight.

After William died, Caroline returned to Germany, where she lived for another twenty-six years, maintaining an active correspondence with her nephew, who had also become an astronomer, as well as other important scientists. She received the Gold Medal of the Royal Astronomical Society at age seventy-eight, was elected to the Royal Irish Academy at eighty-six, and, ten years later, was given a Gold Medal for Science by the king of Prussia. She died at age ninety-seven in the city of her birth. A crater on the Moon bears her name.

Bode's Law: A Formula That Shouldn't Work

It's not a law at all; it's more like a trick, and a trick that doesn't always work at that. And Bode didn't invent it. Yet Bode's law, as it is called, has played a significant role in the discovery of the asteroids and even of the planets.

It was invented by Johann Daniel Titius (1729–96), a Wittenberg professor of mathematics who translated a book by the Swiss naturalist

Charles Bonnet in which the author discussed the divinely inspired order of nature. To help illustrate Bonnet's thesis, Titius—an intrusive editor—added a paragraph about the planets in which he showed that their distances from the Sun follow a fixed formula when measured in astronomical units (one astronomical unit [AU] is equal to the distance between the Earth and the Sun). The formula worked this way: Begin a series of numbers with 0, add 3, and then double each figure thereafter. You get 0-3-6-12-24-48 and so on. Add 4 to each of those numbers, divide the results by 10, and you end up with the following progression: .4-.7-1.0-1.6-2.8-5.2-10.00-19.6-38.8.

Original Series	Add 4	Divide by 10
0	4	.4
3	7	.7
6	10	1.0
12	16	1.6
24	28	2.8
48	52	5.2
96	100	10.0
192	196	19.6
384	388	38.8

Remarkably, the first seven numbers in the last column, considered as astronomical units, roughly describe the distance between the Sun and each of the known planets, with one exception: There was an unfilled gap at 2.8 AU.

When Titius published his translation of Bonnet's book, *Contemplation de la nature,* this formula, along with the rest of the book, sank into oblivion, where it would have remained but for the attentions of Johann Elert Bode (1747–1826), a German astronomer with the popular touch. In 1772, Bode published the second edition of an introduction to astronomy he had written, and in it he included, without mentioning either Bonnet or Titius, this formula. He was also disturbed by the gap at 2.8 AU and he suggested that a search begin for a planet at that distance.

Nine years later, William Herschel discovered Uranus at 19.18 AU—a distance so close to what the formula predicts that it seemed to verify its accuracy. Writing about that discovery three years after its announcement, Bode finally credited his sources and reiterated his certainty that something had to be out there at 2.8 AU in the empty space between Mars and Jupiter.

He was right. On January 1, 1801, Giuseppi Piazzi discovered Ceres, the first and largest of the asteroids, which revolve around the Sun at about 2.77 AU—close enough to establish Bode's law.

Astronomers eagerly used Bode's law to help locate Neptune, but it proved an anomaly in the system, significantly closer than expected. However, Pluto, when it was discovered in 1930, was at a predicted distance.

It would be hard to find a scientist today who thinks Bode's formula is an immutable law of nature. And yet it hasn't been entirely discounted either. For reasons no one can explain, it hasn't been a bad predictor. If the law seems sadly mathematical (as it did to Hegel, who objected to its philosophical implications), it may be reassuring to hear that Bode, whose speculations on this point were basically Kantian, considered it predictive not only of the distances of each planet but of the spirituality of the inhabitants thereon; thus Martians were holier than Earthlings, who were closer to God than Venusians.

Planets known in 1766	Distance from ☉ predicted by Bode (in AU)	Actual distance from ☉ (in AU)
Mercury	0.4	0.39
Venus	0.7	0.72
Earth	1.0	1.0
Mars	1.6	1.52
	2.8	2.77 (asteroids)
Jupiter	5.2	5.20
Saturn	10.0	9.54
	19.6	19.18 (Uranus, discovered 1781)
		30.06 (Neptune, discovered 1846)
	38.8	39.44 (Pluto, discovered 1930)
	77.2	?

♅

Neptune

A billion miles beyond Uranus lies Neptune, the littlest giant. Striped and frosted with crystalline clouds, as beautiful as the Earth and far more blue, the planet is awash in what looks like a never-ending turquoise sea, a Caribbean without islands. It seems fitting that it should be named after the trident-bearing Greek god of the sea, Poseidon, whom the Romans called Neptune.

But it is not a sea. It is a ball of gas: a rocky, iron core hidden inside a mantle of ionized water, ammonia, and methane ice, all surrounded by a frigid (–352°F) coat of hydrogen, helium, and methane. The fastest winds in the solar system whip around Neptune, stirring up turbulence. Its surface—neither as complicated as Jupiter's or Saturn's nor as bland as that of Uranus—is marked by the Great Dark Spot, a dark oval storm topped with white clouds of methane crystals and nearly as large as the Earth; the Small Dark Spot, a storm almost the size of our Moon; and a white spot called the Scooter, which rotates so quickly that it seems to chase the other storms around the planet. Like Uranus, Neptune has a peculiar magnetic field that slices through the planet at an unexpected 47° angle. It is generated not by molten metal as on Earth but by water so hot— 4000°F—and under so much pressure that it conducts electricity and generates a magnetic field.

Neptune is encircled

NASA

Neptune. The Great Dark Spot with its methane clouds is slightly to the left of center.

NASA

Neptune's rings. *Voyager 2* was 683,000 miles past Neptune when it took this 111-second exposure.

by at least five barely visible rings, one of which features three short bright arcs named Liberté, Egalité, and Fraternité. Like all the gaseous planets, Neptune has many moons. Of the eight named satellites, six were discovered on August 25, 1989, when *Voyager 2* flew by the blue planet. The largest of those is 250 miles in diameter and the smallest is 31 miles across, which is why some astronomers prefer the term "moonlet." Recent evidence suggests that the rings may be studded with as many as a thousand of these miniature moons. Prior to *Voyager*'s historic visit only two moons were known. Nereid was discovered in 1949 by Gerard P. Kuiper (who also discovered the Uranian moon Miranda). Triton, the largest Neptunian moon by far, with a diameter of 1,681 miles, was discovered in 1846 by William Lassell, an amateur telescope maker and a professional brewer of beer.

NEPTUNE

Discovered
1846

Mean Distance from the Sun
2,792,400,000 miles

Diameter at the Equator
30,770 miles

Diameter if Earth = 1
3.89

Mass if Earth = 1
17.13

Surface Gravity if Earth = 1
1.18

Moons
8

Length of Year
(revolution around the Sun)
164.79 Earth years

Length of Day
(rotation at equator)
16 hr. 3 min.

Triton: Moon Like a Cantaloupe

The son of Neptune, Triton was a sea god, part man and part fish, who calmed the seas with a conch shell he used as a trumpet. Once, when the Argonauts' ship caught a tidal wave in the Mediterranean and was tossed about so violently that it came to rest in Africa, Triton helped the stranded sailors return to the sea and, in exchange for a golden tripod, created a beautiful new island for them. According to another story, Triton brought up Athena, the daughter of Zeus, along with his own daughter, Pallas. One day the two girls had a fight and Athena killed her playmate; grief-stricken, she took her friend's name for her own and was known forever after as Pallas Athena. Another story says that Triton attacked the women of Boeotia when they were bathing in the ocean and preparing to make sacrifices to Dionysus, the god of wine. Dionysus was incensed, and the two gods fought. Triton lost, but not as disastrously as when he butchered the cattle at Tanagra. That slaughter made people so

NASA

Triton, Neptune's largest satellite.

angry that they got him drunk and beheaded him. Like his father, Triton was not a peaceful god.

And the distant moon that bears his name is not a peaceful world. Astronomers expected the largest of Neptune's eight known satellites to resemble most of the other moons in the solar system: dull, predictable, pocked by craters, dead. Triton isn't anything like that.

To a trained observer it doesn't even look like the other moons. Between its pale, peach-colored polar caps made of nitrogen ice tinged by hydro-

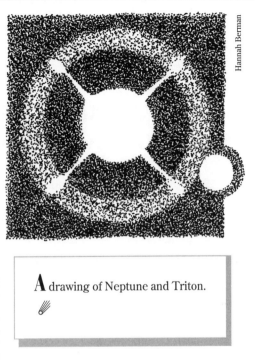

Hannah Berman

A drawing of Neptune and Triton.

carbons, its craters have been filled in and smoothed by a slush of ammonia and water that oozed up through the surface, flooded the crater basins, and froze. Its dimpled complexion is usually compared to the rind of a melon. This cantaloupean landscape exists nowhere else.

Nor does Triton rotate in the usual way. The large moons of Jupiter, Saturn, and Uranus circle the equators of their planets; Triton rotates backward at a steeply inclined angle to the equator.

Triton is also the coldest place in the solar system. The heat of the far-away Sun, which from that distance looks like a bright star in a midnight sky, is too feeble to melt the nitrogen ice coating Triton's surface. But beneath the frozen surface, the rocky core produces tiny amounts of radioactive heat. When that heat combines with the warmth of the summer sun, slight though it is, some of the ice beneath the surface turns to gas. The pressure builds. At last a plume of gas and debris explodes through the frozen surface, spouting five miles up into the thin, thin atmosphere like a gigantic geyser.

Triton is riddled with these geysers. If it resembles anything known to Earthlings, it isn't our Moon. Imagine instead an utterly deserted, lifeless Yellowstone National Park—on a night when the temperature is −391°F.

The Sorry Tale of the Discovery of Neptune

A discovery made with the mind's eye, in regions
where sight itself was unable to penetrate.
—Johann Heinrich von Mädler

On December 28, 1612, Galileo drew a picture of Jupiter, its moons, and a nearby eight-magnitude "fixed star" identified today as Neptune. A month later, he noticed that the "star" had moved. It didn't occur to him that the distant dot of light might be a planet, so he doesn't get credit for the observation. Nonetheless, although the planet was not officially discovered until 1846, the first person in history to see Neptune was undoubtedly Galileo.

The identity of its official discoverer is a matter of debate. Long before Neptune was found, scientists suspected its existence because Uranus was always drifting off its predicted orbit. Various hypotheses arose to account for this, the most reasonable of which supposed that an undiscovered planet was exerting a gravitational force on Uranus and disturbing its orbit. Yet although the new planet's position was predicted, not a single astronomer with access to a major observatory felt any compulsion to look through a telescope and see what he could see.

Central to the drama was John Couch Adams (1819–92), a brilliant student of mathematics at Cambridge who decided, shortly after his twenty-second birthday, to try to discover the unknown planet. (He was inspired in this task by an 1834 book written by Mary Somerville, a popularizer of science who became one of the first two women—the other was Caroline Herschel—to be inducted into the Royal Astronomical Society as honorary members—meaning that, as women, they couldn't vote.)

Upon graduation in 1843, Adams started to attack the problem, using complex mathematics beyond the ability of his peers. But Adams relegated his astronomical work to his vacation. The rest of the time, he tutored in order to repay his parents, tenant farmers with six other children, for his education. Still, he asked James Challis, director of the Cambridge Observatory, to provide him with some observational data. Challis conveyed the request to Sir George Airy (1801–92), an egotistical, meanspirited man who was also the astronomer royal.

Airy wanted to hear from Adams directly. But Adams was busy tutoring and didn't respond immediately. Besides, by September 1845, he had exact figures predicting the location of the mysterious planet. He gave these figures to Challis, who suggested that Adams communicate directly with Airy.

Adams tried. He made sure Airy had his research and he tried to see him personally. On his first attempt, the astronomer royal was in France. The second time, Adams left a message with Airy's wife, who was one week away from childbirth and evidently forgot. A few hours later, Adams returned and was told by the butler that Airy was eating and could not be disturbed. Feeling rebuffed, Adams left.

Although Airy now had Adams' research, he did nothing with it, because he didn't like the whole idea. The concept of beginning with theoretical mathematics and then seeking to confirm its predictions by observation bothered him; he thought things should be done the other way around. Too, he believed that gravity stopped working past the orbit of Saturn, whereas Adams' work, firmly rooted in Newton, assumed that gravity was ubiquitous. He was also predisposed against Adams because Adams had no social status. Nonetheless, he passed the information on to the amateur telescope maker William Lassell—who didn't look for the planet because he was in bed with a sprained ankle.*

In November 1845, Airy finally responded to Adams, but only to question him. He still hadn't bothered to scan the skies. Adams, never an assertive sort, decided to recalculate his figures and didn't reply. Airy took affront. Meanwhile, Neptune was revolving around the Sun, unseen, in almost the precise position Adams had predicted.

In France, a similar search was going on. In 1845, Urbain Jean Joseph Leverrier (1811–77) presented a series of papers to the Paris Academy of Sciences in which he announced that the perturbations in the Uranian orbit must be caused by an unknown planet, whose position he calculated; his predicted position differed from Adams' by only 1°.

But in France, as in England, no one actually looked for the planet.

*In 1846, Lassell discovered the first moon of Neptune, Triton. He also thought he saw a ring around the planet, but his observation was dismissed—until *Voyager 2* confirmed the existence of faint rings around the planet.

Yerkes Observatory photograph, University of Chicago

Urbain Jean Joseph Leverrier (1811–77).

Shortly afterward, Leverrier's work made its way to America, and an astronomer at the Naval Observatory in Washington, D.C., asked permission to search. He was turned down.

George Airy, however, was thrilled to hear about Leverrier's calculations because Leverrier, unlike Adams, was well known. Airy announced to James Challis, John Herschel, and a group of other scientists that a new planet was on the verge of discovery. He neither mentioned nor contacted Adams, and a few days later, when Airy and another astronomer ran into Adams by accident at Cambridge University, Airy made no mention to Adams about the developments in France.

Airy did, however, tell a former professor something about what had been going on. The professor, shocked to learn that no attempt had been made to find the missing planet, prodded Airy into action. Airy told Challis to start looking. Challis put it off for a week and a half, after which he decided to conduct a search over a full 30° slice of the sky—rather than by going directly to the point suggested by Adams.

While all this was going on, Adams refined his calculations yet again and sent them to Airy, who was by then out of the country. Finally, Adams decided to speak up for himself by presenting his findings to the British Association for the Advancement of Science. But by the time he arrived, the astronomy session was over. He still had not been able to convince a major observatory to conduct a search.

Leverrier had similar problems. Since no one in England, France, or the United States was interested enough to look for the distant planet, Leverrier sent his data to Johann Galle, an assistant director at the Berlin Observatory who, a year earlier, had sent Leverrier a copy of his doctoral dissertation.

The very day that Galle received the letter, he and Heinrich d'Arrest, a student working at the observatory, pointed the telescope toward Aquarius—where both Adams and Leverrier agreed the unknown body had to be—and within an hour located the planet. Its existence was confirmed the following day.

The next question concerned the planet's name. Once again, venality ruled. Galle suggested the name Janus. Leverrier initially suggested Neptune, a name he falsely claimed had been chosen by the French Bureau of Longitudes. But upon more careful consideration, he decided that the planet ought to bear his own name. He convinced astronomer François Arago to suggest the name Leverrier to the Paris Academy of Sciences. Eventually, scientists agreed to his first suggestion: Neptune.

In England, Airy and Challis were busy trying to claim a part in the discovery. Airy, whose antipathy to Adams knew no bounds, wrote to Leverrier, "You are to be recognized beyond doubt as the real predictor of the planet's place." John Herschel, however, published an article giving credit to Adams. This infuriated the French, who felt that the English were trying to diminish the magnitude of the French discovery by claiming it for themselves. In the end, the only major players who came out unscathed were Adams and Leverrier, who met at a party given by John Herschel and became fast friends. The entire episode, a frustrating case study of the human dimension of science, is immortalized in the names of the rings of Neptune, among them Leverrier, Adams, and Galle.

$$\Psi$$

Pluto

f life were fair, Percival Lowell (1855–1916) would have discovered Pluto. Even the presence of Neptune failed to fully account for the oddities in the orbit of Uranus, and Lowell suspected the existence of yet another planet. Between 1905 and 1907, he examined thousands of photographic plates, but Planet X, as he called it, eluded him.

In 1919, it also eluded Milton Humason, the junior high school dropout, mule driver, and janitor turned astronomer, who briefly searched for the planet at the request of Lowell's rival, E. C. Pickering. Humason would have found it, too, had not there been a tiny irregularity on the photographic plate at just the critical spot.

Then, in 1929, thirteen years after Lowell's death, Clyde Tombaugh entered the scene. By his own description "a farm boy amateur astronomer without a university education," Tombaugh nonetheless longed to become a college professor. A way opened when he sent some drawings of Mars and Jupiter to the Lowell Observatory, which happened to be looking for an amateur astronomer to work with a new telescope. (They couldn't afford to pay a professional.) Tombaugh was offered the job, and a miserable job it was. It involved a detailed search of hundreds of thousands of celestial objects on numerous matched photographs of the sky taken on different nights. Each photograph showed as many as 400,000 dots of light, most of them stars or galaxies so distant that they do not appear to move, sprinkled across the plate. Tombaugh's job was to find the spots that did move, subtract all comets, asteroids, and variable stars from consideration, and by the process of elimination find and identify the missing planet.

He went about the task with a blink comparator, an instrument which allowed an observer to compare two nearly identical pictures of the night sky by illuminating them in rapid alternation. To the eye, shifting between two images, the positions of stars and galaxies would appear unchanged. But anything that moved would seem to jump back and forth. The hundreds of thousands of pinpricks of light on each photograph were almost all brighter than Pluto. Tombaugh looked at every one.

On February 18, 1930, he saw a dim light jumping back and forth in the constellation Gemini, just 6° from the position Lowell had predicted. "That's it," Tombaugh thought. The discovery was announced on March 13, 1930—Percival Lowell's seventy-fifth birthday and the 149th anniversary of William Herschel's discovery of Uranus.

Why hadn't Lowell discovered the planet? One reason is that Pluto is tiny, with a mass Tombaugh once called "shockingly small," and so distant that telescopes cannot detect surface details.

Another reason is that Lowell expected the planet to be close to the ecliptic, the path followed by the other planets. But Pluto's orbit is tilted a full 17° to the ecliptic—10° more than Mercury's orbit, which has the next greatest inclination. It was Tombaugh's good fortune that when he began his search, Pluto was closer to the ecliptic than it had been in Lowell's day.

But Tombaugh's success was due largely to his tireless efforts with the blink comparator. Tombaugh did the drudge work. "Nothing short of perfection would satisfy me," he recalled. "When I planted the kafir corn

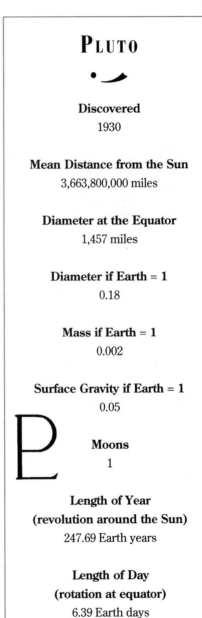

PLUTO

Discovered
1930

Mean Distance from the Sun
3,663,800,000 miles

Diameter at the Equator
1,457 miles

Diameter if Earth = 1
0.18

Mass if Earth = 1
0.002

Surface Gravity if Earth = 1
0.05

Moons
1

Length of Year
(revolution around the Sun)
247.69 Earth years

Length of Day
(rotation at equator)
6.39 Earth days

Lowell Observatory

Clyde Tombaugh (1906–   ).

and milo maize, the rows across the field had to be straight as an arrow or I was unhappy. Later, every planet suspect, no matter how faint, had to be checked out. . . . It was the most tedious work I had ever done." As a result of his discovery, he was awarded a college scholarship and became a professor of astronomy at the University of New Mexico.

Ironically, although the discovery of Pluto was based on Lowell's predictions, Pluto is so tiny that it is incapable of affecting the orbit of Uranus in the way he had thought. Lowell's calculations, in other words, were incorrect. Tombaugh may have been right when he wrote, years later, that the discovery of Pluto "was due to a remarkable chain of accidental events spanning several decades, decreed by fate."

At the suggestion of Venetia Burney, an eleven-year-old English girl who was studying mythology, the new planet was named Pluto after the Greek god of the Underworld and given the astronomical symbol ♇ as a nod toward Percival Lowell (another oft-used glyph is ♀). Pluto is a shiny world, made of rock and various kinds of reflective ice. Icy methane, nitrogen, and carbon monoxide coat its surface beneath a rarefied atmosphere of nitrogen and carbon monoxide. Slightly smaller than the Moon, with a mass approximately $1/450$ of the Earth's (so small that the suggestion has been made that Pluto is not actually a planet), Pluto takes 6.39 days to rotate on its axis; like Uranus, it does so on its side. It takes the planet 247 years 8 months to revolve around the Sun. Its orbit is wildly eccentric, so seriously tilted and so much more elliptical than other planetary orbits that at one point in its journey Pluto crosses over the orbit of Neptune, which then becomes the most distant planet. (It might sound as though Neptune and Pluto could one day collide when their orbits cross, but they are never closer to each other than 16.7 astronomical units.) Since 1979, Neptune has been the most distant planet. That will change on March 14, 1999, when Pluto crosses the Neptunian orbit and once again marks the end of the known planetary system and the beginning of the frontier.

Hale Observatories

Mars, Jupiter, Saturn, and Pluto. Pluto was photographed with the 200-inch Hale telescope; the other planets were photographed with the 100-inch telescope.

Pluto's Moon

In June 1978, one week before astronomer James W. Christy was scheduled to move his family into a new house, he was examining a photographic plate when he noticed that unlike other celestial objects on the glass plate, the hazy spot of light that was Pluto appeared strangely elongated. He considered the possibility that the planet might be emitting a flare. And then, he later recalled, "the concept—moon—jumped into my thoughts." It was right there on the photographic plates: a blurry bump on the edge of a smudge.

The next day, Christy retrieved five photographs taken eight years earlier. They too showed the elongation and enabled him to calculate the moon's orbit. "Thus, in two days, we had discovered a moon and an orbit—and I was free to start moving furniture."

As the discoverer, Christy had the privilege of naming the moon. He liked two names: Oz and Charon, after his wife Charlene. "I soon learned that the astronomical community expected the name to be taken from Greek mythology and to be associated with the god Pluto," he wrote. "But serendipity struck again, and I made my third discovery of the week as I opened my dictionary and read, 'Charon—in Greek mythology, the boatman who ferried dead souls across the river Styx to Hades, the domain of Pluto.' So be it—ours is not to reason why, at least not immediately."

With a diameter of 740 miles, Charon is slightly over half the size of Pluto, which makes it the largest moon in the solar system relative to its planet. (In terms of size alone, Jupiter's Ganymede is the largest moon.) The two bodies are locked together gravitationally and tidally, rotating at the same rate and facing each other in the same way. On Pluto, Charon is visible from only one side of the planet, where it appears fixed in the sky. On this side of Pluto, the stars appear to revolve around the planet, but Charon just hangs there. On the other side of Pluto, Charon is never visible.

In 1985, Pluto and Charon began a series of eclipses relative to the

Earth. As the moon passed in front of its planet and then behind it (revolving from the top to the bottom), astronomers were able to chart slight variations in the amount of light emitted. They learned that Pluto is brighter at its poles, which are frosted with crystals of methane ice, and darker at the equator, where the underlying rock is visible. They detected one bright spot and one large dark spot—possibly a crater—on the equatorial belt. Considering the enormous distance of the minuscule planet, it's remarkable that even this much is known about Pluto.

Although Pluto and Charon are often considered a double planet, some scientists are not convinced that Pluto is a planet at all. They think that it may be a renegade moon of Neptune, ejected from that system after a collision with Neptune's moon Triton, which was knocked into a retrograde orbit as a result of the shock. Other astronomers have suggested that Pluto, Charon, and Triton are ancient planetesimals that never formed a larger planet.

In any case, Pluto is too small to be Planet X, the body Lowell was looking for. The possibility remains that Planet X is still out there, perhaps relatively close to Pluto and too small to be seen, perhaps relatively large—even as big as Jupiter—but so distant that our telescopes cannot detect it. Possibly, as happened with Pluto itself, it is revolving around the Sun at such a steep angle to the ecliptic that no one has looked in the right place yet. Planet X could even be in an orbit perpendicular to that of the other planets. In any case, anomalies in the orbits of the planets might still be explained by the existence of Planet X—a new, unseen, Lowellian world.

♇

MANY MOONS

Planet	No. of Moons	Name	Date of Discovery	Name	Date of Discovery
Earth	(1)	Moon			
Mars	(2)	Phobos	1877		
		Deimos	1877		
Jupiter	(16)	Callisto	1610	Sinope	1914
		Europa	1610	Carme	1938
		Ganymede	1610	Lysithea	1938
		Io	1610	Ananke	1951
		Amalthea	1892	Leda	1974
		Himalia	1904	Adrastea	1979
		Elara	1905	Thebe	1979
		Pasiphae	1908	Metis	1980
Saturn	(18*)	Titan	1655	Janus	1966
		Iapetus	1671	Epimetheus	1978
		Rhea	1672	Atlas	1980
		Dione	1684	Calypso	1980
		Tethys	1684	Helene	1980
		Enceladus	1789	Pandora	1980
		Mimas	1789	Prometheus	1980
		Hyperion	1848	Telesto	1980
		Phoebe	1898	Pan	1990
Uranus	(15)	Oberon	1787	Cordelia	1986
		Titania	1787	Cressida	1986
		Ariel	1851	Desdemona	1986
		Umbriel	1851	Juliet	1986
		Miranda	1948	Ophelia	1986
		Puck	1985	Portia	1986
		Belinda	1986	Rosalind	1986
		Bianca	1986		
Neptune	(8)	Triton	1846	Larissa	1989
		Nereid	1949	Naiad	1989
		Despoina	1989	Proteus	1989
		Galatea	1989	Thalassa	1989
Pluto	(1)	Charon	1978		

*Additional moons, significantly smaller and unnamed as of this writing, also exist.

NASA

The near full Moon, photographed from *Apollo 16*. The round dark area on the horizon is Mare Crisium, the Sea of Crises, but most of the picture shows the far side of the Moon, which is not visible from Earth.

NASA

Saturn's moon Tethys.

Edmond Halley's Comet

*Nothing happened except that I was left
with a curious twitching of my left ear
after sundown and a tendency to break
into a dog trot at the striking of a match
or the flashing of a lantern.*

—James Thurber, after the return
of Halley's Comet in 1910

 dmond Halley (1656–1742) was the grandson of a haber-
dasher (named Humphrey), the son of a soapmaker, and
a man on the move. When he was eighteen, he impressed John Flamsteed, the first astronomer royal, by
announcing that planetary positions published for
Jupiter and Saturn were incorrect. Less than two years later, Halley
dropped out of Oxford, sailed across the equator to the island of St. He-
lena (where Napoleon later died), established the first observatory in
the Southern Hemisphere, and produced the first catalogue of the southern skies. Later on, he crossed the Atlantic twice in his ship, the
Paramore Pink, to measure magnetic variations at different longitudes;
discovered the globular cluster Omega Centaurus; compared the positions of the stars with those recorded by Ptolemy, and in so doing discovered proper motion—the slow movement of the so-called fixed stars
across the sky; convinced Isaac Newton to publish his great work, the
Principia Mathematica, for which Halley provided financial backing; and
became the astronomer royal. He wrote Latin poetry and was the first
person to chart life expectancy, as a result of which he could reasonably
be considered the Father of Life Insurance.

He is famous not for these accomplishments but for predicting an event that occurred after his death: the return of the comet that now bears his name. Halley was convinced that cometary orbits were neither parabolas, as was sometimes thought, nor straight lines, as Kepler believed, but ellipses. If that was true, Halley said, then comets should return on a periodic basis. He discussed this possibility with Newton, and painstakingly calculated the paths traveled by two dozen comets, among them three comets with a remarkably similar pattern: the comet of 1531, a comet Kepler saw in 1607, and a comet Halley had observed himself in 1682. In 1695, in a letter to Newton, he connected them: "I am more and more confirmed that we have seen that Comett now three times, since ye Yeare 1531."

Unfortunately, the intervals between comets were not identical. Seventy-six years passed between the first two, but it was only seventy-five years later that the third comet appeared. Halley blamed this varia-

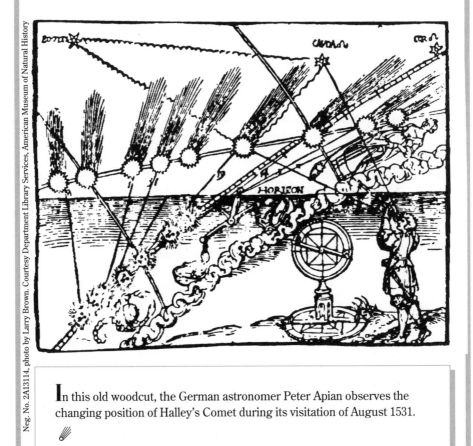

Neg. No. 2A13114, photo by Larry Brown. Courtesy Department Library Services, American Museum of Natural History

In this old woodcut, the German astronomer Peter Apian observes the changing position of Halley's Comet during its visitation of August 1531.

tion on Jupiter, whose gravitational force could perturb anything that ventured into its neighborhood. After refining his calculations to take Jupiter into consideration, he predicted that the comet of 1531, 1607, and 1682 would loop around the Sun in late 1758 or early 1759. "Wherefore if according to what we have already said it should return again about the year 1758, candid posterity will not refuse to acknowledge that this was first discovered by an Englishman," he wrote.

Halley, who died in the Greenwich Observatory at age eighty-five, a glass of wine in hand, didn't live to see that glorious event. But it did occur. By 1758, the astronomical community was agog with comet fever. Yet the first person to see the comet in the skies—to recover it—was not a professional astronomer but a German farmer, Johann Georg Palitzsch. On Christmas Eve 1758, he looked through his telescope and saw a dim, fuzzy spot near the constellation Pisces. He watched it on a daily basis until Valentine's Day 1759. The comet reached perihelion—the point in its orbit nearest the Sun—in mid-March, and afterward, when it headed back toward the outer solar system, its tail brushed across a fourth of the sky. In 1765, the phrase "Halley's Comet" entered the language. It has been used ever since to refer to all historical instances of the comet's return.

Those visitations extend as far back as 240 B.C., when Halley's Comet was seen in China and blamed for the death of the empress dowager. Since then, not a visitation has passed without its being recorded somewhere in the world. The Babylonians saw it in 164 B.C. and 87 B.C. In 12 B.C., the Romans thought it symbolized the death of the soldier and statesman Marcus Vipsanius Agrippa. In A.D. 295, the Chinese realized that the comet seen in the east (nearing the Sun) and the comet seen later that month in the west (heading away from the Sun) were the same. In 451, the defeat of Attila the Hun by the Romans was attributed to the comet, and in 684, it was blamed for a plague. (During the plague year almost a thousand years later, another bright comet was accused.) Halley's closest appearance to the Earth occurred in 837, when its tail stretched across half the sky. In 1066, an Italian cleric wrote that "its tail rose like smoke halfway to the zenith"; in England that year, the Norman Conquest and the death of King Harold II at the Battle of Hastings were attributed to its influence. In 1222, Genghis Khan, who felt the comet was his own personal starry messenger, and his Mongol troops massacred millions, and sky watchers in Korea reported that the tail was red. Johannes Kepler saw it in 1607. In September 1682, Halley himself observed the comet.

In 1910, the much-heralded reappearance of Halley's Comet caused widespread panic, especially when newspapers reported the detection of poisonous gas in the tail. In a front-page article, *The New York Times* announced that according to the French astronomer Camille Flammarion, if the Earth passed through the comet's tail, "cyanogen gas would impregnate the atmosphere and possibly snuff out all life on the planet." Other scientists pointed out that the tail of a comet was so inexpressibly thin that even if it were entirely poisonous, no one would notice. But the damage had been done. When an Eighth Avenue streetcar in New York City was hit by a fan blade that fell out of an automobile-factory window,

Hale Observatories

Halley's Comet as photographed on May 12 and 14, 1910.

a passenger screamed, "It's the comet!" Comet pills and comet insurance were sold to protect the populace. In Chicago, housewives stopped up windows and doors to keep the noxious fumes out of their apartments, and one elderly woman, terrified about what the comet might bring, committed suicide.

As everyone who remembers Bill Haley and the Comets knows, the name Halley used to be pronounced so that it rhymed with "daily" or "gaily." By 1986, the long vowel had become a short one, and the name was more often pronounced to rhyme with "Sally." Surveys of English people with the same surname indicated that the new pronunciation is the preferred one; today, some dictionaries list both pronunciations while others give the short *a* as the only pronunciation. Edmond's opinion about his own name is unknown. As was common in his era, he spelled it a number of ways, including Hawley, Hayley, and Hally.

The comet that year was a grand sight all over the world, exceptionally bright with a tail that stretched over half the sky. And although the Earth did pass through the comet's tail, the effect was minimal. "We have passed through the comet's tail and we are no wiser than we were before," said E. B. Frost, the head of the Yerkes Observatory.

The comet's visit in 1986 was a whole other story. Due to the position of the Earth, this return offered the worst apparition in two thousand years, especially in the Northern Hemisphere. But great scientific strides were made. In 1986, an international armada of spacecraft visited Halley. For the first time, scientists got a close look.

Like other comets, Halley originated in a spherical swarm of trillions of comets that envelop the solar system. Called the Oort cloud (after Dutch astronomer Jan Hendrik Oort), it is located about a light-year from the Sun and may consist of a relatively dense inner cloud and a more sparsely populated outer cloud. Occasionally, a passing gas cloud jolts a comet out of the Oort cloud and into a long slow fall to the Sun. Comets approach the Sun from all directions, some on elliptical paths so long it takes them millions of years to complete a single orbit. Others, like Halley's, go into relatively short orbit.

Halley's Comet travels in a retrograde direction, orbiting clockwise around the Sun while the planets move counterclockwise. As Fred Whipple had predicted, its rotating nucleus is a dirty snowball about 9.3 miles long and 6 miles wide—larger, actually, than had been expected—and shaped like a peanut. (The poet Diane Ackerman calls it "a waltzing iceberg.") It is marked by hills, valleys, and a crust so dark that it is one of the blackest bodies in the solar system, reflecting only about 4 percent of the available light. As it nears the Sun, pockets of ice within the nucleus turn to gas and are expelled along with a cloud of dust. The gas and dust cling to the nucleus like a mist, forming a coma that extends over 62,000 miles into space. The coma looks as thick as smoke but is so diffuse that air is 300 billion times denser. Thinner yet is the halo of hydrogen that surrounds the coma for millions of miles.

As the comet makes its celestial rounds, it pushes against the solar wind, which sweeps gas and dust into a long tail stretching behind the comet. Regardless of whether the comet is coming or going, comet tails always point away from the Sun. They are divided into straight strands of pale blue gas and curved strands of dust with a yellow tinge, and they change shape over time. In March 1986, part of the tail of Halley's Comet broke off and disintegrated; this was labeled "a spectacular disconnection event."

Nor was that Halley's last hurrah in the twentieth century. In February 1991, astronomers were surprised to discover that somewhere between the orbits of Saturn and Uranus, Halley had inexplicably brightened. The show may have been caused by a late-breaking eruption of gas or an unanticipated hit from an interplanetary marauder—although the latter is less likely because Halley was in a sparsely populated area of the solar system. Halley is still speeding away from the Sun, and further events are not expected for the twentieth century. The most famous comet in human history will make its next visit in the year 2061.

It Came from Outer Space

Nine in the fifth place means:
Flying dragon in the heavens.
—I Ching

he similarity between the words "meteor" and "meteorite" obscures the difference between them. Meteors are shooting stars that streak across the sky on their way to incineration. They come from comets and, despite appearances, are usually smaller than a sesame seed. Meteorites are large chunks of stone and metal torn from asteroids or from the exhausted hulls of comets. They blaze across the sky like neon-green fireballs, casting an oddly fluorescent glow before they crash to Earth. A few meteorites, called bolides, cut a glowing slash across the sky and explode. Perhaps because this can be a frightening sight, meteorites and meteors are often held in awe, sometimes as ill omens and sometimes as sacred objects.

Ancient people, suspecting that meteorites had an extraterrestrial origin, often revered them. A 3,047-pound meteorite was found wrapped like a mummy in ruins of the Montezuma Indians in Mexico. The Temple at Ephesus housed a stone that the ancient Greeks believed had fallen from Jupiter in the image of Artemis, the goddess of the moon. In the mythology of the Skidi Pawnee, meteorites were thought of as lesser gods created by the supreme god Tirawahat. And in German folklore, it was said that the cores of these heavenly interlopers were made of silver and gold, and that they brought with them messages from beyond and a collection of material objects including excrement, a frequently mentioned "gelatinous mass," delicatessen items both good and bad—rotten cheese, ham, a side of bacon—and the most precious substance of all: money.

Early scientists were reluctant to believe that meteorites were anything other than an atmospheric phenomenon, like rain. But by the latter half of the eighteenth century, evidence to the contrary began to accumulate in an alarming fashion. In France, a priest appeared at the Paris Museum in 1751 with a peculiar stone he claimed had fallen from the sky. Two years later, an astronomer reported another meteor fall. Both accounts were dismissed out of hand. Only when a third report, from another priest, came to its attention in 1769 did the Royal Academy of Science decide to investigate. A committee was appointed, headed by the chemist Antoine Lavoisier. It concluded that the unusual stones were simply ordinary rocks whose structure had been altered by lightning. To the enlightened citizens of the Age of Reason, the idea that rocks might have fallen from the sky was just so much folklore, too implausible to believe.

Twenty-five years later, in 1794, the German physicist Ernst Chladni published a book stating that meteorites fell from the sky and were the remnants of an exploded planet. French scientists rejected this view completely and aligned themselves with Aristotle and Newton, both of whom believed that meteorites were created when smoky vapors from the Earth rose into the air and burst into flame. Another view, mentioned in a letter Benjamin Franklin received from a friend, was that meteorites were somehow created by electricity.

Then, in 1803, three thousand stones showered down on a small town near Paris. The physicist Jean Baptiste Biot authenticated the fall and noted that many of the meteorites contained nickel, a substance rare on Earth. The reports were so compelling that French scientists reversed their previous opinion and decided that stones did drop from the sky. Their origin was unknown, but a common opinion held that they had been ejected from volcanoes on the Moon.

Nonetheless, there was still resistance to the idea. In the United States, Thomas Jefferson pointed out that these absurd reports were, after all, French. "I find nothing surprising in the raining of stones in France, nor yet had they been millstones," he wrote. "The reason is that the exuberant imagination of a Frenchman . . . creates facts for him which never happened, and he tells them with good faith."

Four years later, two Yale University professors investigated a meteorite fall in Weston, Connecticut, and concluded that the impossible had happened. Jefferson held firm. "I would rather believe that those two

A meteor streaks across the sky near the Orion Nebula.

Yerkes Observatory photograph, University of Chicago

Yankee professors would lie than believe that stones would fall from heaven," he said.

Today, the reality of meteorites has been fully accepted. On average, they are about 4.5 billion years old. Stone meteorites—including a 27-pound meteorite that crashed into a suburban high school girl's Chevy Malibu in 1992—are the most common, making up as much as 96 percent of all meteorites that fall to Earth. Known as chondrites, these meteorites are dotted with small round particles called chondrules, some of which contain carbon and possibly even organic compounds. Metal meteorites, made of iron and nickel, are distinguished by stylish geometric designs called Widmanstätten figures, which appear when the meteorite's cut surface is etched and polished. Stony irons, which make up only 1 percent of the total, combine the materials in approximately equal measure. As to where meteorites come from, their composition reflects their origin, with stony meteorites coming from stony asteroids, and so on.

A million meteorites a year fall to Earth, and though seldom seen, once in a while they cause damage. Large ones have gouged out craters, of which the 20,000-year-old Meteor Crater in Arizona, three quarters of a mile across and 191 yards deep, is the most vivid example. Small ones have hit living beings. Once, it is rumored, a meteorite hit a cat. A horse in New Concord, Ohio, was struck in 1860. And although most meteorites come from asteroids, in 1911 a dog in Egypt was killed by a meteorite from Mars. Human beings have also been struck. A man in Mhow, India, was hit in 1827, and in 1954 an Alabama housewife was asleep on her living room couch when a rock from outer space crashed through her roof and hit her on the hip, leaving an impressive bruise. It was a rude awakening.

Shooting Stars, Meteorites, and the Tears of Saint Lawrence

When the Moon is on the wave,
  And the glow-worm in the grass,
And the meteor on the grave,
  And the wisp on the morass;
When the falling stars are shooting,
And the answered owls are hooting,
And the silent leaves are still
In the shadow of the hill,
Shall my soul be upon thine,
With a power and with a sign.

—Lord Byron

On August 10, A.D. 258, four days after the execution of Pope Sixtus II, a Roman prefect told a Christian deacon named Lawrence to hand over the church valuables. When Lawrence showed up at the prefect's office with a group of poor, sick people and announced that they were the treasure of the church, the hard-nosed official ordered the deacon put to death. According to Christian tradition, this was accomplished in a horrifying way: Saint Lawrence was tied to a metal grid above a fire and roasted.

Panels of changing stars, sashes of vapor,
  silver tails of meteor streams,
  washes and rockets of fire—
It was only a dream, oh hoh, yah yah,
  loo loo, only a dream, five, six,
  seven, five, six, seven.

—Carl Sandburg

That night, the sky lit up with shooting stars that seemed to stream out of the constellation Perseus. Those shooting stars came to be known as "the Tears of Saint Lawrence." They appear regularly, always around the same date. In the 1860s, John Couch Adams, Daniel Kirkwood, and Giovanni Schiaparelli came up with a mechanism to explain the phenomenon. They suggested that as a comet orbits the Sun, it leaves in its wake a trail of debris. Every year when the Earth crosses that trail, a swarm of particles—some as large as a raspberry, others as small as a grain of sand—collides with the Earth's atmosphere. The friction of the encounter produces enough heat to vaporize the particles. A meteor shower is nothing more than comet dust, flung into the Earth's atmosphere and burning up.

That theory was confirmed in 1872, as astronomers awaited the return of Biela's Comet, which had been discovered in 1826 by an Australian amateur and which had an orbital period of 6.6 years. It returned in 1832, was not seen in 1839 because of the position of the Earth, and appeared in 1845 right on schedule. But it had by then split in two. In 1852, when the twin comets returned, they had drifted apart a little. And that was essentially the last of Biela's Comet. In 1859, the comet was not visible

BURIED TREASURE FROM OUTER SPACE

Say you're digging around in your yard one day searching for dinosaur bones or Indian arrowheads when you unearth a stone that strikes you as peculiar. How can you tell if it's a meteorite? Often, you can't; eventually, many meteorites come to resemble Earth rocks. But if it is dark, coated in part with a black crust, and embedded with glassy round or oval globules, you can be pretty sure it's a stone meteorite—specifically, a chondrite. Iron meteorites are even easier to recognize. According to Harvey H. Nininger, whose life was devoted to collecting meteorites (for which he paid a dollar a pound), an iron meteorite resembles "an old battered, rusty tin can, but of course [it] would be very heavy." Another clue is that iron meteorites are covered with indentations that look like thumbprints in soft clay; those impressions are caused when the meteorite crashes through the atmosphere. The surface of the meteorite is melted by the heat of the friction and pushed back by the force of its motion through the air.

I *would rather be a superb meteor, every atom of me in magnificent glow, than a sleepy, and permanent planet. The proper function of man is to live, not to exist. I shall not waste my days in trying to prolong them. I shall use my time.*

—Jack London

from the Earth. In 1865, it failed to appear. And in 1872, optimistic astronomers who insisted on looking for it saw instead a shower of shooting stars. Every 6.6 years thereafter, the meteor shower reappeared, the brilliant and disintegrating corpse of Biela's Comet. The orbits of Biela's Comet and of the shooting stars that appeared in its stead were identical.

That particular meteor shower is no longer visible, but others are. One of the most unusual showers occurs in mid-November. It appears annually but is especially impressive about every 33 years. Denison Olmstead, a professor at Yale University, described the shower of November 13, 1833, when the sky over the eastern United States was illuminated for over six hours by a fireworks display of shooting stars and fireballs "probably more extensive and magnificent than any similar one hitherto recorded," he wrote. "The firmament was unclouded; the air was still and mild; the stars seemed to shine with more than their wonted brilliancy. . . . For some time after the occurrence, the 'Meteoric Phenomenon' was the principal topic of conversation in every circle." Professor Olmstead noticed that the streaks of light seemed to radiate from within the constellation Leo the Lion, and that this was true even as

METEOR SHOWERS WORTH WATCHING

• ⌒

Quadrantids	January 3*
Eta Aquarids†	May 6
Perseids	August 12‡
Orionids†	October 21
Leonids	November 16
Geminids	December 13

*This is when the meteors are usually the most numerous and impressive; however, they may also be visible before and after the date given here.

†The comet responsible for these meteor showers is Halley's.

‡The best meteor shower of all.

The Leonid Meteor Shower of November 12–13, 1833, as seen at Niagara Falls.

the constellation gradually moved from east to west. As a result, the shooting stars of mid-November are called the Leonids, a word formed by combining the name of the constellation with the Greek suffix "-id," meaning children.

The source of these meteor showers—visible but insignificant in 1832 and 1834, spectacular when seen by Alexander von Humboldt in 1799— was a mystery. Heinrich Olbers suggested that they would appear in their full regalia in 1867. It happened as predicted. But 33 and 66 years later, in 1900 and 1933, the meteor showers weren't visible. Then, in 1966, the Leonids rained over Mexico and the southwestern United States, falling for over an hour at a rate estimated to be as high as 100 per second. It was one of the great celestial events of the century.

> The shooting stars attend thee;
> And the elves also,
> Whose little eyes glow
> Like the sparks of fire, befriend thee.
>
> —Robert Herrick

The Leonids are dazzling every 33.3 years because the dust and debris from the Tempel-Tuttle Comet are clumped unequally over the length of the orbit. So while every November a few shooting stars arc across the sky, every 33 years—in theory—when the Earth plows through the point of greatest distribution, the skies light up with what Robert Frost called "a fusillade of blanks and empty flashes." The Leonids are expected to return in November 1998 or November 1999, and the show should be a good one. Unfortunately, it is difficult to predict exactly where in the world it might be visible.

Fortunately, most showers are more predictable. The best way to see them is to find a dark spot in an open area (tall trees obscure the view), lie down (early morning hours are generally the best time), and wait. After no more than a few minutes, shooting stars are likely to appear near—not necessarily in—the appropriate constellation. Of course, shooting stars can slash across the sky at any time. The appropriate thing to do when you see one, in some cultures, is to wish. (In Chile, you have to pick up a stone at the same time.) In Europe, it was commonly believed that shooting stars meant someone had died. In Russia, they are devils tossed out of Heaven, and in northern California the Wintu thought of them as the souls of shamans on their way to the afterlife. A southern California tribe considered shooting stars to be "the feces of the stars," while a single fireball was deemed a cannibal spirit who roamed the night looking for lost souls to devour. And in the Philippines, meteors are the souls of alcoholics who sing a cautionary song, the lyrics of which are "Do not drink, do not drink." Their attempts to reach Heaven are doomed, and at night we can see them stumbling to Earth yet again.

PART III

THE MILKY WAY

Bloom was pointing out all the stars and the comets in the heavens to Chris Callinan and the jarvey: the great bear and Hercules and the dragon, and the whole jingbang lot. But, by God, I was lost, so to speak, in the milky way.

—James Joyce, *Ulysses*

AND BEYOND

A Star Is Born

ike Socrates, stars are mortal. They only look eternal.

A star is spawned in a cold dusty cloud so diffuse that it is emptier than a vacuum on Earth. Within that nebula, a few particles might accidentally cluster together. That clot is where it begins. The accumulation attracts particles from elsewhere, and eventually the community of particles becomes so massive that gravity does its inevitable work and the cloud collapses. Matter tumbles inward. The cloud becomes denser and more compact, the temperature rises, and the amorphous fog contracts into a protostar. It is possible to see protostars upon occasion. Although they are hidden behind dusty cocoons, which block visible light, they show up at infrared wavelengths as isolated spots silhouetted against glowing nebulosity and known as Bok Globules (after the Dutch astronomer Bart J. Bok).

Within the protostar, the collapse continues. Atoms career about crazily, and gases begin to glow. When the temperature hits about 18,000,000° F, nuclear fusion begins. Hydrogen atoms, stripped of their electrons, crash into each other so forcefully that they form helium. The time this takes varies. Millions of years must pass before a star the size of the Sun is ready to ignite, but a protostar with fifteen times the mass of the Sun requires only 10,000 years—the blink of a cosmic eye—before the nuclear furnace is switched on. At that moment, a star is born.

Afterward, gravity and the explosive nuclear energy balance each other, and the star reaches a state of equilibrium. Astronomers, referring to the Hertzsprung-Russell diagram, which charts the demographics of the stars, call this long, dull adulthood of a star's life the main sequence.

But stars don't stay on the main sequence forever. The length of a star's life, like the manner of its death, depends upon one thing: its mass. The great plebeian mass of stars—more than 88 percent of the stars in the Milky Way—are cool, dim, and smaller than the Sun. Unlike bigger, flashier stars, small stars use their fuel sparingly and live for a long time—perhaps as long as 500 billion years. If the universe is as young as most astronomers believe—somewhere in the area of 10 billion to 20 billion years old—then not one of these little stars has died.

Large stars, on the other hand, live fast, die young, and go out with a bang.

The Eagle Nebula, in Serpens (M16). Bright blue and white giants, variable stars, small dark protostars, and dramatic intersecting columns and cones of gas form this dazzling picture.

National Optical Astronomy Observatories

STELLAR NURSERIES

• ↗

At this very moment, in nebulae scattered across the firmament, stars are being born. Among the busiest stellar nurseries in the Milky Way Galaxy are:

✳ the Orion Nebula, a greenish mist billowing around a blazing clump of stars called the Trapezium. Located in Orion's sword, it is 30 light-years across—the largest nebula in the Milky Way;

✳ the Horsehead Nebula, a cold, interstellar cloud which blots out the light of the stars behind it. Located next to the star Alnitak in Orion's belt, it has a profile remarkably like that of a horse;

✳ the Rosette Nebula in Monoceros, a wreath of light dotted with stars, protostars, and mysterious dark rifts, all surrounding a cluster of bright new stars;

✳ the spectacular Eagle Nebula in Serpens;

✳ the Cone Nebula in Monoceros, a dark, star-topped tower silhouetted against churning nebulosity;

✳ the Lagoon Nebula in Sagittarius, an incandescent cloud nearly bisected by a river of dust;

✳ the Coalsack in the Southern Cross, a dark nebula that looks like a hole in the Milky Way.

Finally, the most impressive stellar nursery of all is the Tarantula Nebula in the Large Magellanic Cloud, a small companion galaxy to the Milky Way. So large and luminous that it outclasses every nebula in the sky, it is the home of some of the biggest, brightest, most massive stars in the universe.

National Optical Astronomy Observatories

The Horsehead Nebula, in Orion, is formed by a giant plume of opaque dust which absorbs the light of the stars behind it.

The Harvard College Observatory

The Coalsack Nebula is the dark smudge in the center of this picture. Directly to its right is Acrux, the bright star at the base of the Southern Cross. The bright star immediately above the nebula marks one end of the arm of the cross. Alpha and Beta Centauri are the two bright stars at the left.

William Henry Smyth
and the Colors of Stars

The way astronomers describe stars, you'd think the sky was spangled from horizon to horizon with stars the colors of flags. Those colors are correlated with temperature. Blue stars are the hottest, followed by white, yellow, orange, and red, the coolest color. As a rule, the hotter a star is, the brighter it appears, but size makes a difference: although most red stars are small and dim, a red giant or supergiant, despite its relatively low temperature, is so enormous that it can easily outshine a warm yellow star like the Sun or a hot white dwarf.

Yet those crayon-colored stars are strangely hard to find in the actual sky, which to the casual observer appears to be a panorama of black and white. Look more carefully, and tints emerge. Even with the naked eye, anyone can see that Antares, the brightest star in Scorpius, really is reddish. With an ordinary telescope, the double star Albireo in Cygnus the Swan is strikingly blue and gold, and Mira, the variable star in Cetus the Whale, is vivid red. But often the colors of the stars are so watercolor-pale that the human eye cannot discern the subtleties.

Some human eyes, however, are better than others. Consider, for example, William Henry Smyth (1788–1865). Smyth was descended from a long line of Americans that began with Captain John Smith, who arrived in Virginia in 1607 and befriended Pocahontas, and ended with Smyth's father, whose Royalist sympathies in the American Revolution forced him to leave New Jersey and move to England, where his son was born. Smyth became a vice admiral in the British Navy as well as an astronomer and a writer. His color descriptions (which Richard Hinckley Allen relied upon in his 1899 volume *Star Names: Their Lore and Meaning*) are evocative and precise.

Smyth was aware that not everyone saw colors in the same way. "The ancients recognised no blue stars; they only spoke of white or red ones," he wrote in his two-volume *Cycle of Celestial Objects*. Individuals also registered colors differently. He observed that Sir William Herschel (whom he referred to throughout his book simply as ♅, the symbol for Uranus) "saw most objects with a redder tinge than they have since proved to bear. This may be owing to the effect of his metallic mir-

ror, or to some peculiarity of vision, or perhaps both." Indeed, the inability of others to distinguish contrasting hues was, to Smyth, distressingly widespread. He noted that the chemist John Dalton was color-blind and admitted that "I was greatly surprised on finding that an intimate friend of my own could not perceive the strong colours of γ Andromedæ." Smyth described that particular multiple star as orange and emerald green, adding that, "of these colours I feel pretty positive, although the high authority of ♅ and Σ [Herschel and F. G.W. Struve] has pronounced them to be yellow and blue."

Smyth tended to shun such blatant, primary-color pronouncements. He saw double stars as light apple green and cherry red, or pale orange and sea green; he described stars like the Sun not simply as yellow but as golden yellow, topaz yellow, flushed yellow, bright yellow, deep yellow, crocus yellow, greenish yellow, pale yellow, fine yellow, and straw-colored; he distinguished among pearly white, brilliant white, pale white, clear white, flake white, creamy white, lucid white, pure white, and silvery white, not to mention ash-colored, pale gray, and dusky gray; and he populated his skies with stars colored grape red, pale rose, violet, purplish, bright lilac, plum color, ultramarine, sapphire blue, cerulean blue, indigo, and sky blue: a splendid, iridescent cosmos.

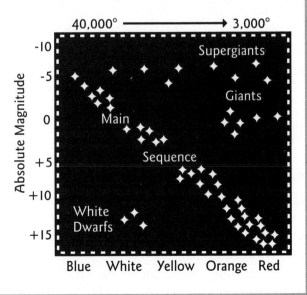

The Hertzsprung-Russell Diagram

> There is a sun, a light that for want of a better
> word I can only call yellow, pale sulphur-yellow,
> pale golden yellow. How lovely yellow is!
>
> —Vincent van Gogh

The most famous formulation in astronomy shares the names of Danish astronomer Ejnar Hertzsprung (1873–1967) and Princeton astronomer Henry Norris Russell (1877–1957), who individually discovered before World War I that when stars were charted according to intrinsic brightness and color or temperature, clear patterns arose. Rather than being scattered all over the chart, the majority of stars fell on a diagonal line that ran from the cool, dim red stars in the lower right to the hot, bright blue and white stars in the upper left. That diagonal line, where stars live out the best years of their lives, is known as the main sequence.

But not all stars are on the main sequence. A few cool red stars are also extremely bright, which suggests that they must be large. These stars festoon the upper-right-hand portion of the chart. Similarly, the white stars huddled in the lower left have to be tiny; nothing else could account for why a white—and therefore hot—star should be so dim.

This famous diagram looks like it ought to chart the life history of a star, and for a while, astronomers thought it did. It seemed as though a star should be born at the top, as a giant, and slowly work its way down the ladder to the bottom.

But that's not what what happens. At various times in its life, a star does find itself on different parts of the Hertzsprung-Russell diagram, but it does not travel up or down the main sequence. Instead, during most of its life, it inhabits one particular spot on the main sequence. In old age, it becomes a red giant and moves to the upper-right-hand corner. A period of instability, during which it varies in brightness, may follow. But in the end, the star must inevitably descend for all eternity into the stellar graveyard in the lower left.

A more detailed life history goes something like this:

After nuclear fusion begins, the newborn star stabilizes. It slides onto the main sequence and begins its long residence there. The specific point of entry is determined by its mass. A star like the Sun enters in the middle. Small stars like Barnard's Star or Proxima Centauri enter in the lower right; a big, hot star like Spica enters in the upper-left-hand corner. Once a star arrives, it neither climbs nor sinks. It just shines on and on, burning hydrogen for a period of time that depends on its mass. Small stars remain on the main sequence for a long time. Big stars stay for a relatively brief time because they burn through their fuel quickly.

When the star has turned the hydrogen in its core into helium, it turns off the main sequence and enters the final 10 percent of its life. But the action is far from over. Nuclear burning now moves to a thin shell of hydrogen surrounding the core, which has become smaller and denser, causing the hydrogen to burn much hotter and faster than it did originally. In response, the outer layers of the star puff up and the star expands dramatically. As it expands, the temperature falls; as the temperature falls, the color changes. The star becomes a red giant. It is cooler, larger, and brighter than before, and its position on the diagram moves to the upper right, where red giants (and supergiants) congregate.

> **S**tar-light, what is star-light, star-light is a little light that is not always mentioned with the sun, it is mentioned with the moon and the sun, it is mixed up with the rest of the time.
>
> —Gertrude Stein

As a red giant, a star as massive as the Sun continues to accumulate helium at the center. Not hot enough to fuse, this helium core shrinks,

The Helix Nebula, in Aquarius, is a planetary nebula, created when the star in the center blew off a shell of gas.

becoming so dense that it no longer behaves as a gas. ("Degenerate" is the word used to describe this state.) At a certain point, the density is so high that the helium atoms fuse into carbon and oxygen by pressure alone—a kind of spontaneous combustion that transforms the star, making it bluer and smaller and giving it, briefly, a new lease on life. It burns fuel much as it did when it was on the main sequence, but this time the fuel is helium, which is one tenth as efficient as hydrogen. During this phase of stellar evolution, the star may become hot enough to pulsate, changing its brightness regularly.

But soon the helium is exhausted, and now nature shows no mercy. Once more the star expands, puffing up until it has a radius 500 times that of the Sun and is 5,000 times more luminous. Bigger and brighter than ever before, the star consists of a hot core, no larger than the Earth but containing 60 percent the mass of the Sun, and a distended atmosphere so tenuous that the helium burning in the heart of the star simply blows it away. That atmosphere detaches and becomes a planetary nebula.* Within it, the spent core of the old star will ultimately become a white dwarf, cooling like an ember in the darkness of space.

*Planetary nebulae have nothing to do with planets. They're so called because when viewed through a telescope they look greenish and round, like planets. The name may seem misleading, but keep this in mind: Very little in astronomy resembles its name to the literal-minded. It takes a hyperactive imagination to look at the ragged W of Cassiopeia and see the queen of Ethiopia sitting on a chair; Pegasus resembles a winged horse far less than it does a great square; and the centaur Sagittarius, as every amateur astronomer knows, looks like a teapot. Poetic license is the rule.

Stars much more massive than the Sun have different life histories, for ultimately they explode in supernovae, turning into black holes or tiny, dense neutron stars no larger than the island of Manhattan—and nowhere to be seen on the Hertzsprung-Russell diagram, because these remnants do not shine like normal stars.

Stars slide off the main sequence in order, from the top down, with big, hot stars running through their fuel quickly, whereas small, cooler stars live longer because they burn their fuel more slowly. The higher a star is on the main sequence, the less time it spends there. As a result, an H-R diagram also indicates the age of a star cluster. With a young star cluster, such as the Pleiades, stars inhabit the main sequence right up to the top. In an older cluster, the hot blue and white stars at the top have used up their hydrogen and expanded into giants and supergiants, leaving the diagonal strip on the H-R diagram with an oddly truncated appearance, as if it had been chopped off partway down. The older the cluster, the deeper the chop.

Thus the Hertzsprung-Russell diagram, which describes stellar distribution, is a sort of heavenly census, charting the demographics of the skies.

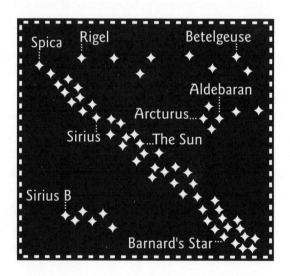

Subrahmanyan Chandrasekhar and the Death of Stars

T he Sun is halfway through its life. Five billion years from now, it will swell into a red giant and ultimately shed its outer layers by puffing out great blobs of matter in the form of a planetary nebula. At the center of that nebula is the core of the star. After consuming the remainder of its nuclear fuel, it will turn into a rigid, shrunken lump of carbon and oxygen known as a white dwarf. No larger than the Earth, a white dwarf is so dense that a teaspoonful would weigh 5.5 tons or more.

The star still shines, but now the glow derives from the leftover heat of the contraction, not nuclear fusion. Over billions of years, the star cools until at last it blinks out. It becomes a black dwarf: a cold, dark globe of crystallized carbon and oxygen—a sort of diamond. The universe is too young for any of the estimated 3 billion white dwarfs to have faded to black, so at the moment, no examples exist. Nonetheless, this is thought to be the destiny of most stars.

It is not, however, the destiny of all stars, although during the first three decades of this century, astronomers believed that it was. A white dwarf can be no larger than about 1.4 times the mass of the Sun. That figure is known as the Chandrasekhar limit in honor of the man who figured out (when he was twenty years old) that not all stars have the same fate.

His name was Subrahmanyan Chandrasekhar. As an undergraduate at Madras University in his native India, Chandrasekhar won a contest that required writing an essay on quantum physics. He requested for his prize a book he'd seen in the library: *Internal Constitution of the Stars,* by the great English astronomer Sir Arthur Eddington, who believed that stars must be fueled not by gravity or radioactivity but by nuclear fusion. ("We do not argue with the critic who urges that the stars are not hot enough for this process," Eddington wrote. "We tell him to go and find a *hotter place.*") Chandrasekhar read Eddington's book avidly. When he graduated in 1930—the same year Eddington was knighted—he was awarded a Government of India scholarship to Cambridge University. He set sail.

To entertain himself during the long days and nights at sea, Chandrasekhar thought about the death of stars. Using the theory of relativity, along with techniques he had learned from Eddington's book, he discovered that above a certain size, the equations went haywire. The star would not reach equilibrium. His figures indicated that large stars would not turn into white dwarfs. They would keep on collapsing.

This idea was not just new but heretical, as Chandrasekhar found out. He was awarded his degree and in 1934 returned to white dwarfs, refining his calculations and receiving what he took to be encouragement from none other than Eddington himself, who chatted with him several times a week about his work and obtained for him what was then a rare luxury—a hand calculator.

Eddington seemed encouraging, but unbeknownst to Chandrasekhar, the elder astronomer was convinced that his young colleague would discover the error of his ways. At a fateful January 1935 meeting of the Royal Astronomical Society, Eddington made it abundantly clear how strongly he disagreed with Chandrasekhar's idea that a sufficiently large dying star would not become a white dwarf. "The star has to go on radiating and radiating and contracting and contracting until, I suppose, it gets to a few kilometers radius, when gravity becomes strong enough to hold in the radiation, and the star can at last find peace. Dr. Chandrasekhar has got this result before, but he has rubbed it in," Eddington announced. "Various accidents may intervene to save a star, but I want more protection than that. I think there should be a law of Nature to prevent a star from behaving in this absurd way!"

The blow was devastating. Chandrasekhar felt himself to have been mocked publicly by the most highly respected astronomer of the day. Nor was it simply a onetime clash. Eddington, a Quaker who had refused to serve in the English armed forces during World War I, could be unrelentingly aggressive in scientific disputes despite his reputation for kindliness. In 1936, during a speech at Harvard University, Eddington said that Chandrasekhar "seemed to like the stars to behave that way, and believes that that is what really happens." As far as Eddington was concerned, the idea that a dying star could collapse until it was smaller than a white dwarf was strictly "stellar buffoonery."

As David to Eddington's Goliath, Chandrasekhar received no support, and the upshot of the confrontation was that he abandoned white dwarfs. He wrote in a letter to his father in 1940, "I felt that astronomers without exception thought that I was wrong. They considered me as a sort of Don Quixote trying to kill Eddington. As you can imagine, it was a very discouraging experience for me—to find myself in a controversy with the leading figure of astronomy and to have my work completely and totally discredited by the astronomical community. I had to make up my mind as to what to do. Should I go on the rest of my life fighting? After all I was in my middle twenties at that time. . . . It was much better for me to change the field of interest and go into something else. If I was right, then it would be known as right. For myself, I was positive that a fact of such clear significance for evolution of stars would in time be established or disproved. I didn't see that I had a need to stay there, so I just left it."

> **W**e forget that our sun is only a star destined to someday burn out. The time scale of its transience so far exceeds our human one that our unconditional dependence on its life-giving properties feels oddly like an indiscretion we'd rather forget.
>
> —Gretel Ehrlich,
> *The Solace of Open Spaces*

Chandrasekhar also left the British Empire behind. He emigrated to the United States in 1936, became a professor at the University of Chicago, and was naturalized in 1953. His calculations were correct, and he finally received recognition for his contribution. But it was belated. Over the years, although Chandrasekhar won many awards, not until 1974 did one of them mention white dwarfs. When he won the Nobel Prize in 1983, over half a century had passed since he had made his original shipboard calculations.

You might think, then, that Chandrasekhar and Eddington disliked each other. This was not altogether the case. In the spring of 1935, following the infamous Royal Astronomical Society meeting, the two scientists took a bicycle trip together. On another occasion, they attended a championship tennis match at Wimbledon, where they rooted for opposing players. When Chandrasekhar got married, Eddington, a bachelor who lived with his sister, invited the couple to tea. And one day Eddington showed Chandrasekhar something he said he'd never shown anyone else: a map of England marked with all the routes he had taken on his solitary bicycle rides. It was actually his second such map, he confided. His dog had destroyed the first one and he had been forced laboriously to re-create it. He told Chandrasekhar how he kept track of his bicycle mileage. "The criterion was the largest number n such that one had cycled n or more miles on n different days," Chandrasekhar later recalled. "It is perhaps touching that in every letter that he wrote to me subsequently, he included his latest value of n."

Yet at another meeting two years later, Eddington stuck with his beliefs, and the rancor resurfaced. Chandrasekhar tried to avoid the older man, but Eddington approached him when he was alone to say that he hoped Chandrasekhar wasn't angry. A curt conversation followed, after which the two men never saw each other again. Eddington died in 1944, and Chandrasekhar never forgave himself. "I regretted later, I still do," Chandrasekhar told his biographer. "I was rude, was unforgiving when he came . . . essentially to apologize."

Yet the fact is, Eddington was wrong. Stars that exceed the Chandrasekhar limit do not become white dwarfs. They have an entirely different fate.

Tycho's Star and Other Supernovae

A massive star lives and dies dramatically. Its life is short. Whereas a star the size of the Sun lives 10 billion years on the main sequence, a star with twenty-five times the mass of the Sun remains there only 7 million years. And its death is explosive. When the star has devoured the hydrogen in its core, it begins to burn helium and puffs up into a red supergiant—a cool star so bloated that the orbit of Jupiter would fit inside it. When the helium peters out, the star begins to collapse, which causes the temperature to rise. The star turns in desperation to another nuclear fuel: carbon.

In this way, the star runs through a series of elements, fusing each one into the next according to a runaway timetable in which each element burns more quickly and releases less energy than the one before it. Helium, which forms carbon and oxygen, burns for 700,000 years. Carbon burns for six centuries. Neon powers the star for a year. Oxygen burns for a mere six months. And silicon fuels the star for a single day before turning into iron. But fusing iron requires more energy than it produces. So everything stops.

In less than a second, the core disintegrates. The outer layers of the star slam into the iron core and bounce back. A tidal wave of neutrinos— subatomic particles that are either massless or virtually so—splashes out into space, and a shock wave reverberates through the star, causing it to detonate. That explosion, a conflagration so violent that the outer layers of the star are literally blown away and the star briefly outshines its entire galaxy, is called a supernova.

Within weeks, the star dims, but its luminous shards ride the shock wave into space, carrying with them the heavy elements of our world. This is the method by which stars seed the universe. Our Sun, which contains some heavy elements and therefore must be at least a second-generation star, could not have existed had it not been preceded by the death of a larger star. The same is true of ourselves. The iron in our blood, the iodine in our thyroid, the gold (and mercury) in our teeth, were manufactured by a dying star. It is not a metaphor; it is literally true: we are stardust.

Supernovae are not as rare as they seem. A star explodes about once every fifty years per galaxy. (Another type of supernova occurs when one member of a binary pair dumps material onto a white dwarf, causing it to explode.) But because so many stars are obscured by interstellar clouds and hidden by the dusty disk of the galaxy itself, supernovae are seen all too rarely. Possibly the most impressive one exploded in 1006 in Lupus. A mere forty-eight years later, in 1054, a supernova bloomed in the constellation Taurus. And that was it for over five hundred years.

Then, on November 11, 1572, a new star blazed forth in Cassiopeia. It was seen by the great Tycho Brahe, who wrote that he was "amazed," "astonished," "stupefied," and "led into such perplexity by the unbeliev-

ARCIS VRANIBVRGI. A TYCHONE BRAHE, DÑO DE KNVDSTRVP.
IN INSVLA HELLESPONTI DANICI HVENNA CONSTRVCTÆ. QVO AD TOTAM CAPACITATEM, DESIGNATIO.

Tycho Brahe's observatory, Uraniborg, built on the island of Hven.

ability of the thing that I began to doubt the faith of my own eyes." So unexpected was the sight that Tycho—who claimed to have known the positions of the stars since boyhood—stopped random passersby to ask if they, too, saw the new star. Until March 1574, when it vanished from sight, Tycho watched the star closely, noting its brightness and constructing an accurate light curve as it faded into oblivion. His measurements proved that the new object did not move across the heavens like a comet or a planet and must therefore be a star. More important, Tycho showed that the heavens were not fixed and that Aristotle was wrong. (Comets, which are far more common, might have incontrovertibly proved this point, except that they were thought to be an atmospheric phenomenon, like tornadoes or lightning, and, hence, no challenge to Aristotle.) The bright new star, proof that the great constant—change—applied even to the heavens, drew mixed reactions. A friend of John Calvin's suggested that it heralded the Second Coming. Other theologians, bound to Aristotle, concluded that the seemingly new star had simply never been noticed before. Tycho decided with some trepidation that he ought to write a book.

After being reassured by several people that this was not an inappropriate activity for a nobleman, he wrote and published *De nova stella.* (The title is misleading: what Tycho saw was not the birth trauma of a new star, as he thought, but the death throes of an old one.) Shortly afterward, Tycho taught a course at the University of Copenhagen and then took off for Germany, where he thought of settling. When news of this reached Frederick II, the king of Denmark, he contacted Tycho with the kind of offer that cannot be refused. He gave him the island of Hven, "to have, enjoy, use and hold, quit and free, without any rent, all the days of his life. . . ."

Tycho turned it into an empire. On the island's 2,000 acres, Tycho—who had also inherited wealth—built an observatory called Uraniborg, "the Castle of Heaven," which came equipped with conference rooms, bedrooms, and a large celestial globe; a smaller observatory, called Stjerneborg, "the Castle of the Stars"; workshops for the makers of the large astronomical instruments he needed; a chemical laboratory; a printing press; a paper mill; a prison; a windmill; a corn mill; fish ponds; gardens; and an arboretum with three hundred different kinds of trees.

But Tycho's abrasive personality and haughty ways cost him some friends, and after the death of Frederick II, he was forced off Hven despite the lifelong grant. (The new king, Christian IV, eventually presented the island to his mistress.) Tycho moved to Prague, where, in the last years of his life, he was appointed imperial mathematician to the emperor Rudolph II. In that capacity, he worked with a new assistant: Johannes Kepler.

In 1601, Tycho died and Kepler took over. Three years later, an official in the imperial court told Kepler he had seen a new star in the constellation Ophiuchus. After a week of overcast skies, Kepler saw the star gleaming between Jupiter and Saturn. Following Tycho's lead, he wrote a book called *De stella nova.* Just as the supernova of 1572 became known as Tycho's star, the supernova of 1604 is called Kepler's star.

Five years later, the telescope was invented. As of this writing, although exploding stars have been observed in galaxies both near and far, not one supernova has been seen in the Milky Way. In 1987, however, one came close. On February 24 of that year, the Canadian astronomer Ian Shelton of the University of Toronto and a telescope operator named

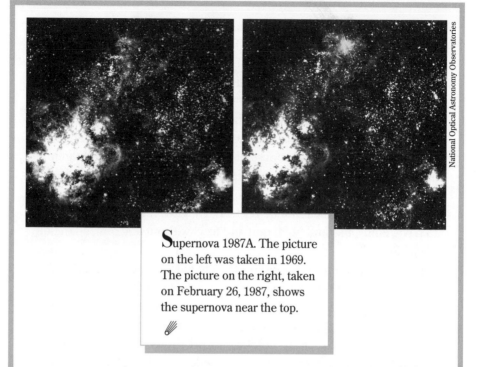

Supernova 1987A. The picture on the left was taken in 1969. The picture on the right, taken on February 26, 1987, shows the supernova near the top.

Oscar Duhalde, both working at the Las Campanas Observatory in Chile, saw an unexpected bright spot on a photographic plate of the Large Magellanic Cloud, a small sidekick galaxy to the Milky Way. On a plate taken the night before, only a faint, forgettable star appeared in that place. Had it exploded? Shelton walked outside to take a look. And there it was: the first supernova easily visible from Earth in almost four hundred years.

The scientific community was prepared. They knew, for example, something about the star that had become the supernova. Named after its discoverer, Nicholas Sanduleak of Case Western Reserve University, Sanduleak 69°202 was 163,000 light-years away. It was a blue supergiant twenty times more massive than the Sun and about 100,000 times brighter, with an iron core equal to about 1.5 solar masses: bigger than the Chandresekhar limit and enough to blow the star to smithereens.

Its behavior has reassured astronomers because, for the most part, their predictions have proven to be accurate. Various elements appeared in the predicted way, as did gamma rays. But perhaps the greatest excitement was saved for neutrinos, the subatomic particles that even now are passing through the Earth, through our bodies, and—most important—through enormous underground tanks of water (and dry-cleaning fluid), where they occasionally produce tiny blue sparks of light, which are recorded. Scientists are interested in neutrinos because, theoretically, they come surging out of a star prior to its actual explosion and thus should arrive on Earth before the supernova is visible. And that is what happened, with Sanduleak 69°202. On the day before the supernova, underground detectors in Japan and Ohio recorded nineteen neutrinos. In the world of neutrino hunting, it was a bonanza.

On the other hand, the supernova did not shine as brightly as expected, and scientists were surprised to find that the progenitor star was a blue supergiant rather than a red one. They now believe that it had been red for perhaps a million years, but that 100,000 years ago it shrank, which caused its temperature to rise and its color to change.

Supernova 1987A in the Large Magellanic Cloud. Thousands of years before it exploded, the star at the center of this picture blew off a ring of gas. Radiation from the blast raised the temperature so high that the ring began to glow.

NASA/ESA

Prior to exploding, Sanduleak 69°202 blew off a lot of material, which collected in a ring a light-year away. That ring, already in place when the star exploded, surrounds it like a racetrack. It has led scientists to predict that in the mid-1990s, when the outrushing tentacles of the explosion hit the ring, the fading ember of SN 1987A will brighten once more.

one moment your life is a stone in you, and the next, a star.

—Rainer Maria Rilke

The supernova of 1987 was a major event. But it would be nice to see one in the Milky Way, and one of these days that will happen. Likely candidates include Betelgeuse in Orion, a variable red supergiant due to self-destruct some time in the next few hundred thousand years (after which it is expected to become a black hole); Antares, the red supergiant in Scorpius; and Eta Carinae in the southern sky. A multiple, variable star hidden inside a nebula that bears its name, Eta Carinae was the second brightest star in the sky in 1843 but has not been visible to the naked eye since 1868. It emits stronger infrared radiation than anything other than the Sun and the Moon, and astronomers are confident that in the relatively near future it will explode. It remains to be seen whether that means many centuries hence—or tomorrow night.

Neutron Stars, Pulsars, and Little Green Men

Curdled stars, muddled stars, stars that had
been stirred with a spoon.

The neutron star was in the center, of course,
though I couldn't see it and hadn't expected to. It
was only eleven miles across, and cool.

—Larry Niven, *Neutron Star*

A star has three possible fates. If it's small—and a star eight times larger than the Sun can be considered small—it will sooner or later shrivel into a white dwarf, a stellar corpse the size of the Earth.

If it's large, twenty times more massive than the Sun, it will explode in a supernova and very likely end its days as a black hole.

And if it's in between, it will detonate in a supernova, blow off huge portions of its mass, and collapse into a tiny neutron star, a body perhaps ten miles across and so dense that a teaspoonful would weigh more than 100 million tons. (A teaspoonful of white dwarf, it will be recalled, weighs a mere 5.5 tons.)

It is possible to squeeze so much matter into such a small area because atoms are mostly empty space. If an atomic nucleus consisted of a few oranges on a Broadway stage, the electrons would be fruit flies buzzing around in the balconies. Ordinary matter, in other words, includes plenty of nothing. A neutron star is unimaginably dense because all that emptiness is gone, crushed out of existence by a gravitational force so great that it smashes through the sea of electrons surrounding atomic nuclei, pushes electrons and protons forcibly together, and creates what is essentially one giant nucleus—a compact star only a few miles across.

And a strange place it is, with surface gravity perhaps 100 billion times the gravity on Earth, a solid iron crust, miniature mountains less than an inch high, and a churning atmosphere of atoms and particles a few yards thick.

Neutron stars are so tiny that their existence, predicted in 1934 by Fritz Zwicky and Walter Baade, was not confirmed until 1967. The person responsible was Jocelyn Bell, a twenty-four-year-old Irish graduate student working with Antony Hewish at Cambridge University. Her duties included helping to build a giant radio antenna—a four-and-a-half-acre field of scraggly-looking antennae linked by innumerable wires strung parallel to the ground. As the Earth turned on its axis, the antennae recorded on rolls of paper almost 400 feet long radio waves streaming out of the sky. Bell's task—every bit as boring as Clyde Tombaugh's job when he discovered Pluto—was to review every marking on those rolls of paper.

In November 1967, about two months after the telescope's completion, Bell discovered on one tape "a bit of scruff." Looking back over the long rolls of paper, she found the same EKG-like squiggles repeatedly coming from the same part of the sky. She contacted Hewish. Within a month, the pattern reappeared. It had a pulse so precise that it registered every 1.3373011 seconds. No known star could transmit such a steady beat. It was so regular, so mechanical, that it didn't seem natural. Perhaps it wasn't. In recognition of that possibility, the unidentified radio sources were labeled LGM—for Little Green Men. "We did not really believe that we had picked up signals from another civilization," Bell later recalled, "but obviously, the idea had crossed our minds and we had no proof that it was entirely natural radio emission. It is an interesting problem—if one thinks one may have detected life elsewhere in the universe, how does one announce the results responsibly?"

The announcement never had to be made. Examining the miles of paper, Bell discovered similar signals elsewhere in the sky. This suggested that the phenomenon was a natural one. "It was very unlikely that two lots of little green men would choose the same improbable frequency, and at the same time to try signaling to the same planet Earth," she said. The Little Green Men theory was retired, and the objects were called "pulsars" (for "pulsating radio source").

Thomas Gold, the Austrian-born scientist associated with the Steady State hypothesis, suggested that pulsars and neutron stars might be the same animal. The idea—which has been accepted—is that as a neutron

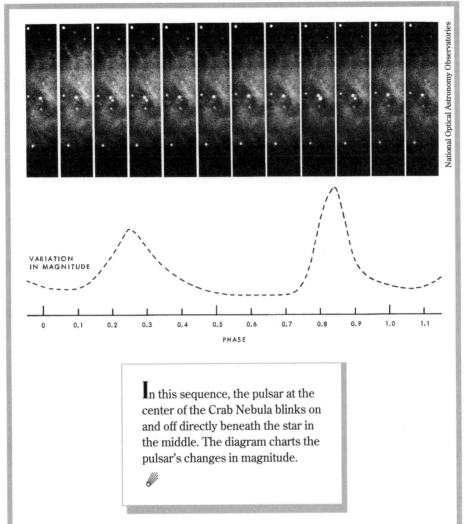

National Optical Astronomy Observatories

VARIATION
IN MAGNITUDE

0   0.1   0.2   0.3   0.4   0.5   0.6   0.7   0.8   0.9   1.0   1.1

PHASE

In this sequence, the pulsar at the center of the Crab Nebula blinks on and off directly beneath the star in the middle. The diagram charts the pulsar's changes in magnitude.

star forms, on the heels of a supernova explosion, its stellar material would be crushed into a small volume, it would spin faster and faster, and its magnetic field would intensify until it became essentially a giant magnet. Jets of electrons would spurt out of the magnetic poles, emitting various kinds of electromagnetic radiation, including visible light. As the star turns, those jets would sweep across the cosmos many times a second. If they happened to be facing the Earth, the rotating star would flash on and off, on and off, on and off.

And that is just what pulsars do. They are distinguished by sharp, precisely timed pulses so speedy—as fast as 1,000 a second—that recordings of their radio waves are as repetitive as machine-gun fire.

If pulsars are indeed rapidly rotating neutron stars, then it should be possible to find them ensconced within the remnants of a supernova. Confirmation of this possibility came from the Crab Nebula, an amorphous patch of light in Taurus named by the nineteenth-century Irish astronomer Lord Rosse. The Crab Nebula is the known remnant of a supernova; it occupies the very spot where an eleventh-century Chinese astrologer saw a "guest star." And at its center is a pulsar. It turns on its axis thirty-three times per second, and with every pirouette, a beam of radio waves and light sweeps across the universe like a giant searchlight. In addition, like other pulsars, it is slowing down. In about 10 million years, it is expected to stop broadcasting.

Since the discovery of the first pulsar, hundreds of others have been found, and the people involved in the original discovery have been rewarded for it. Antony Hewish and his codirector on the project, Martin Ryle, won the Nobel Prize in physics. Jocelyn Bell (now Burnell) did not share in the prize, although she received a great deal of publicity, much of it with an unfortunate girl-scientist cast to it. Years later, she recalled being asked "relevant questions like was I taller than or not quite as tall as Princess Margaret (we have quaint units of measurement in Britain) and how many boyfriends did I have at a time?" After her discovery, although she continued to work as an astronomer, she did not pursue the study of pulsars. But her role in the discovery is universally acknowledged, and her musical name is now and forever linked with the regular rhythm of spinning neutron stars.

Black Holes

You can't get there from here.
—Stephen Hawking

The idea has been around since the late eighteenth century, when the English geologist John Michell and the French astronomer Pierre Simon, Marquis de Laplace, recognized that if an object were sufficiently large and dense, nothing could flee from its gravitational clutches. Laplace came to this conclusion after thinking about escape velocity, the speed needed to escape from any celestial object, be it planet, moon, or star. Escape velocity varies in proportion to an object's gravity. To depart from planet Earth, a spaceship has to hit at least 7 miles per second; to escape from Jupiter, you'd have to rocket over the Great Red Spot at a minimum speed of 37 miles per second; and to flee the Sun, you'd have to reach 380 miles per second. In 1798, it occurred to Laplace that if an object were massive enough and dense enough, escape would require a velocity faster than 186,000 miles per second, the speed of light.

But nothing, including light, can outstrip the speed of light. An object such as Laplace imagined, then, would be a maximum-security prison from which nothing could escape: in short, a black hole. It didn't receive that name, however, until 1969, when physicist John A. Wheeler of Princeton University, after repeatedly using the phrase "gravitationally completely collapsed object," decided that there had to be a better name. He remembered a catchy term he'd heard: "black hole." It caught on immediately, and black holes became the celestial celebrities of our time.

A black hole is created when a large star tears itself apart in a supernova, ejecting most of its matter and leaving behind a huge, expanding nebula. Inside that glowing web is the ravenous undead corpse of the star. In some cases it will be a tiny, dense neutron star, in which the outward pressure of jammed-together neutrons balances the inward crush of gravity. But neutron stars, like white dwarfs, have upper limits. If the stellar corpse that remains behind after the supernova explosion is larger than about three solar masses, gravity becomes unstoppable, crushing the star so relentlessly that eventually, according to theory, nothing's left. Matter is not crushed into a mountain, a thimble, or a space the size of a proton. Theoretically, it is crushed into a point of zero volume and infinite density. At that point, which is known as a singularity, gravity has done its irreducible work. The singularity is the center of a black hole, which makes its presence known by devouring everything that enters its sphere of influence. It is actively destructive. But in another way, it doesn't exist. There's no there there.

Yet a black hole is entirely real. All the mass that wasn't blown away continues to exert its gravitational influence. (That's why, if the Sun turned into a black hole, the Earth would be plunged into perpetual night, but it would nevertheless keep on revolving, just as it does now. However, life would be doomed—a scenario astrophysicist George

This jet of ionized gas emanating from the giant elliptical galaxy M87 may be powered by a black hole whose mass equals that of 3 billion Suns. The ultraviolet picture was taken by the Hubble Space Telescope's Faint Object Camera.

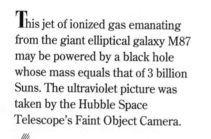

F. Duccio Macchetto/NASA/ESA

Greenstein delineates in horrifying detail in his fascinating book, *Frozen Star.*)

Surrounding the singularity like a bubble is an invisible boundary called the event horizon. It is a one-way membrane that defines the point of no return, the outer limits of the black hole. Its size depends on the mass of the original object. If the Earth were somehow to be compressed into a black hole, it would have an event horizon the size of a large marble. The Sun would have an event horizon with a radius of fewer than two miles. Inside the event horizon, matter plummets toward the singularity.

No matter how big a black hole might become, by definition you can never see it, just as you can never see the Invisible Man. But you can see his shoes, his empty-sleeved jacket, the tip of his cigarette. Something similar happens when a black hole and a star are part of a binary system, revolving around a common center of mass. When the companion star swells up at the end of its normal, main-sequence life, its expanded outer layers may come too close to the black hole. Matter from the star, unable to resist the black hole's deadly pull, spirals toward the event horizon and forms a spinning wreath of material, called the accretion disk, which shoots out jets of X-rays and other forms of radiation. The black hole itself remains invisible. But just as the dented cushions where the Invisible Man sits down reveal his presence, the accretion disk discloses the existence of the black hole and allows scientists to measure its gravitational effects.

Black holes are thought to exist in distant quasars and even in the heart of our own galaxy, where a black hole with a mass the size of a hundred thousand suns may be spewing out radio waves and infrared radiation.

A star in Monoceros, discovered in 1975 when a tidal wave of X-rays came surging out, is also a likely candidate. SS433 in Aquila the Eagle is another. It consists of a dense invisible mystery object and a big companion star. As they orbit around each other, the possible black hole (or whatever it is) pulls streams of matter from its companion onto its accretion disk and simultaneously shoots out twin jets of material five hundred light-years long.

But the first and most famous black hole—assuming it *is* a black hole—is Cygnus X-1, which came to the attention of astronomers in the late 1960s, when a satellite identified it as a powerful source of X-rays. Those X-rays turned out to be emanating from a very small area near a hot blue supergiant that revolves every 5.6 days around an unseen ob-

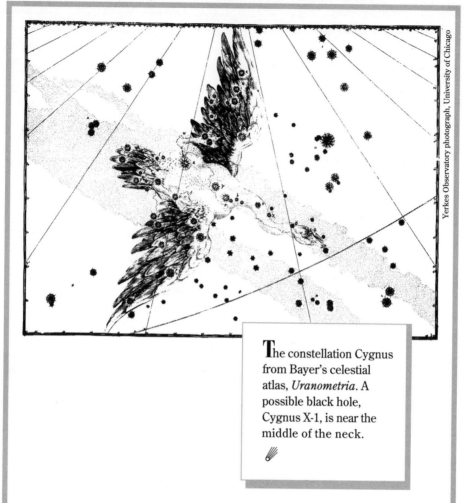

Yerkes Observatory photograph, University of Chicago

The constellation Cygnus from Bayer's celestial atlas, *Uranometria*. A possible black hole, Cygnus X-1, is near the middle of the neck.

ject. The spectrum of the blue star and the speed of the revolution allowed scientists to compute the mass of the unseen object. It is thought to be as large as six solar masses, just right for a black hole.

Cygnus X-1 is also famous as the centerpiece of a wager made in 1975 between Kip Thorne of the California Institute of Technology and the British astrophysicist Stephen Hawking, who bet against the idea that Cygnus X-1 is a black hole, even though that is by far the best explanation for its behavior. Hawking explained the bet in his best-seller, *A Brief History of Time,* as "a form of insurance policy for me. I have done a lot of work on black holes, and it would all be wasted if it turned out that black holes do not exist. But in that case, I would have the consolation of winning my bet, which would bring me four years of the magazine *Private Eye.* If black holes do exist, Kip will get one year of *Penthouse.*"

The bet was settled in 1990, when Hawking conceded that Cygnus X-1 was a black hole, just as he thought all along, and thus Thorne won his subscription. Since then, evidence in favor of the existence of black holes has continued to accumulate. One candidate may live within the hub of galaxy NGC 4261, where the Hubble Space Telescope has photographed a dusty disk of matter spinning around a hidden, high-gravity powerhouse that acts suspiciously like a black hole. Once astronomers have calculated the velocity of the spin, they'll know if the disk hides a black hole. But as for the black hole itself, the monster will never be photographed. It will remain forever invisible, its hungry presence revealed only by the maelstrom that surrounds it.

Falling into a Black Hole

The free fall toward the singularity, the trip to oblivion inside a black hole, has inspired much speculation.

There are two ways to think about falling into a black hole. The first is to imagine yourself watching some luckless astronaut spiral into eternity. Strangely, he never seems to get there. The closer he comes to the event horizon, the slower he seems to move, because the light, fighting against an increasingly powerful gravitational force, takes longer and longer to reach you. Finally, just before reaching the event horizon, where light can no longer escape, he seems to stop, as does the enormous clock strapped to his space suit. He never hurtles into the abyss at all. Instead, the light waves from his body become longer and longer, a redshift that at first turns him red and then causes him to fade away (a special effect brought to you courtesy of Albert Einstein).

That's what you see. That's not what it's like from the astronaut's point of view, though. For him, things are much, much worse. His clock keeps on ticking. He plunges through the event horizon and keeps on falling until he is first stretched and then torn apart, feetfirst (if that's the way he fell in) and atom by atom. The bigger the black hole is, the longer he falls before being stretched to death on the rack of infinity.

It's too bad anyone falling into a black hole has to meet his doom, because interesting things take place in black holes. One might simply continue to fall forever without ever arriving at the center. While in that state of permanent free fall, it is also possible, assuming one could maintain consciousness, to watch other objects, be they galaxies, stars, or competing life forms, fall into the black hole. The most exciting possibility, however, is that a black hole warps space-time the way a twenty-ton mint tossed onto a bed might warp a feather quilt: it would tunnel through it, causing the quilt to rise up around it like a pillar of down. In a black hole, something similar happens to space and time. They become a muddle.

One implication of all this is that in a black hole it might be possible to jump forward or backward in time. Here, in what we like to think of as "the real world," time flows only one way, from the past to the future. But in a black hole the arrow of time can point in any direction, thereby making it possible to hobnob with Neanderthals, stay the hand of John Wilkes Booth, or, of course, prevent your own birth (which is why science fiction writers love this stuff). It would work this way: If a black hole is rotating (as is likely), the singularity at the center would not be a dot but a ring. Enter the black hole in just the right way, and instead of being squashed until you're smaller than a quark, you could theoretically slip through the singularity, thereby arriving, by way of a space-time tunnel called a wormhole, in another universe—or another time.

This concept has aroused a great deal of enthusiasm, but Stephen Hawking is having none of it. "I'm sorry to disappoint prospective galactic tourists, but this scenario doesn't work," he has stated. "If you jump into a black hole, you will get torn apart and crushed out of existence."

Charles Messier

harles Messier (1730–1817) devoted his life to avoiding error. The mistake he wished to avoid, one he had already been stung by, was that of failing to identify a comet. It happened to him in 1758, when the astronomical community was anticipating the return of the comet that would eventually be known as Halley's. Messier, as assistant to the head of the French naval observatory, was told to find it.

Messier applied himself to the task assiduously but in vain. Hoping to become the first person in the eighteenth century to see the comet, he spent a year and a half searching in the predicted places. But when he finally did locate the comet, the head of the observatory forced him, for no good reason, to postpone the announcement for a month. By then, others, including the Saxon farmer Johann Georg Palitzsch, had already spotted the comet, and news of its reappearance had spread. When Messier finally announced his own sighting, he was mocked.

Other disappointments came his way that year, too, for although a comet is unmistakable when it is properly positioned and sports an enormous tail, distant comets look like fuzzy little blurs of light (as anyone who saw Comet Halley in 1986 knows). They resemble a lot of other celestial objects. In September 1758, Messier thought he had discovered a comet in Taurus. Subsequent observations revealed that the object—now known as the Crab Nebula—did not move and hence could not be a comet.

Despite—or because of—these letdowns, Messier became an inveterate comet hunter. But he didn't want to be fooled again. To spare himself the pain of discovering any more of these permanent objects, he decided to catalogue them so that neither he, nor anyone else, would ever mistake them for the real thing. With this in mind, he searched the heavens

SELECTIONS FROM THE *MESSIER CATALOGUE*

* M1. The Crab Nebula in Taurus. Messier described this cloud of gas as "whitish light and spreading like a flame." The remnant of a supernova that exploded in the year 1054, it holds a pulsar in its center.

* M13. When Messier saw this "round and brilliant" blur in Hercules, he thought it was "a nebula which I am sure contains no star." In fact, it is a globular cluster, a spherical bouquet of perhaps a million stars.

* M31. This is the Andromeda Galaxy, a smudge in the sky thought to be a nebula until the twentieth century, when it was proven to be another galaxy, an entirely separate wheel of stars equivalent to our own Milky Way Galaxy.

* M42. The Orion Nebula. Visible in Orion's sword (but unimpressive to the naked eye), it is a green-tinted, star-studded cloud within which stars are being born. John Herschel compared it "to a curdling liquid, or to the breaking up of a mackerel sky when the clouds of which it consists begin to assume a cirrus appearance." Some people consider it the most beautiful object in the sky.

National Optical Astronomy Observatories

The Great Nebula, in Orion.

* M45. This is the Pleiades, an open cluster so young that the largest stars are still wrapped in fuzzy wisps of nebulosity. It encompasses several hundred stars and not merely the six or seven stars visible to most people.

* M57. The Ring Nebula in Lyra. A thick circle of light surrounding a star, M57 is a planetary nebula, formed when a dying star blew off a shell of gas and illuminated by ultraviolent radiation from the star's exposed core. John Herschel thought the central area looked like "gauze stretched over a hoop."

* M81. A beautiful spiral in Ursa Major frosted with stars and bound by a sea of gas to M82, an irregular galaxy of mottled appearance.

National Optical Astronomy Observatories

The Crab Nebula, in Taurus.

and the literature and carefully sketched every object on his list. He hoped his drawings would be useful not only in identifying these non-comets but also in seeing if they changed over time.

The first entry in his catalogue was the Crab Nebula, now known as M1. "This nebula had such a resemblance to a comet, in its form and brightness," he wrote, "that I endeavored to find others, so that astronomers would not confuse these same nebulae with comets just beginning to shine."

In 1771, Messier published a catalogue of 45 comet decoys, and over the next decade he increased the list. By the time it was completed in 1784, the *Messier Catalogue* included 103 fuzzy patches of light that might waste the time of a dedicated comet hunter. The Messier numbers given to these objects are still used, although they have been supplemented, first by the 5,000 star clusters and nebulosities listed in John Herschel's 1864 *General Catalogue,* and later by the expanded *New General Catalogue,* which included nearly 15,000 objects. Among the Messier objects are nebulae, star clusters, globular clusters, and spiral and elliptical galaxies—none of which would have interested Charles Messier. He was monomaniacal in his devotion to comets; Louis XV called him "the ferret of comets." A famous story about him concerns the night his wife died, when he relinquished an opportunity to search for a comet in order to stay by her side. As a result, a comet he might ordinarily have found was discovered instead by a French amateur. Shortly thereafter, when a visitor offered condolences for the loss, a grateful Messier confided to the well-wisher how upset he was to have missed out on the discovery of what would have been his thirteenth comet. Perhaps an awkward pause in the conversation followed. In any case, Messier seems to have realized his gaffe, for he is reported to have added, "Ah! the poor woman."

Henry Draper and the
Henry Draper Catalogue

or most of his life, Henry Draper (1837–82) was a man in the wrong profession. Although he trained as a doctor and worked at Bellevue Hospital in New York City, his heart was in the stars. He built an observatory in the suburb of Hastings-on-Hudson, and on the day after his wedding went shopping with his new wife for the glass to be used for the telescope's 28-inch mirror. He organized trips to observe Venus, made expeditions to view eclipses, and photographed the spectrum of Vega and a hundred other stars. His wife, a redheaded socialite named Mary Anna, was loyal and helpful to the point of self-sacrifice. During the solar eclipse of 1878, she was asked to count the seconds as they elapsed, a task that somehow involved missing the eclipse: "Lest her vision might unnerve her, she was put within a tent and therefore saw nothing at all of the wonderful phenomenon," the astronomer Annie Jump Cannon later recalled. "Here she sat patiently and accurately calling out the seconds while the glorious and awe-inspiring spectacle was unfolded."

After the death of his father, Draper decided at long last to quit medicine. His intention was to devote himself to astronomical photography and, in particular, to the spectra of stars. (It was an interest he came by naturally; his father, John William Draper, was the president of the medical school at New York University and a well-known chemist who had taken the first daguerreotype of the Sun's spectrum in 1843.) The summer of his father's death, Henry resigned his medical position, and in October, he traveled west with friends. They hunted, went horseback riding, and camped out in the cold swirl of a Rocky Mountain blizzard.

By November, Draper was back home and ready to host a dinner

The Harvard College Observatory

party whose guests included Asaph Hall, discoverer of the moons of Mars, and Edward Charles Pickering of the Harvard Observatory. At the party, the forty-five-year-old Draper started to shiver. Five days later, he was dead of pneumonia.

Less than two months later, Pickering wrote to Mary Anna Draper, offering to help finish the work on stellar spectra. "I should be greatly pleased if I might do something in memory of a friend whose talents I always admired, and regarding whom it will always be a source of regret to me that I should not have met him more frequently," he wrote. Mrs. Draper, intent on erecting a suitable monument to her husband, entered into a lengthy, intimate correspondence with Pickering, which resulted in an ambitious plan to collect and classify the spectra of stars. She bequeathed to Harvard University all her husband's astronomical instruments and a sum of money with which Pickering purchased a glass disk ground into a prism. This item, used when observing, provided an immediate stellar spectrum for every star in the visual field and allowed Pickering and his assistants to classify close to 250,000 stars. The job, which took over forty years to complete, could not have been done without a group of women who have on occasion been called "Pickering's Harem."

As director since 1876 of the Harvard Observatory, the first observatory to include women among its staff members, E. C. Pickering championed women, whom he welcomed as volunteers and salaried assistants. Despite the low pay, which averaged $10.50 a week, applications for these jobs came flooding in from around the world, and Pickering hired as many assistants as he could. The women who worked on the catalogue were known as "computers." Bending over stellar spectra in Vic-

Members of the Harvard College Observatory around 1917. Henrietta Leavitt and Annie J. Cannon, wearing similar ties, are respectively sixth from the left and fifth from the right.

torian rooms decorated with patterned wallpaper and hung with framed portraits, these women did an enormous amount of painstaking work. Harlow Shapley, who became director of the observatory two years after Pickering's death in 1919, proudly announced in his autobiography, "I invented the term 'girl-hour' for the time spent by the assistants. Some jobs even took several kilo-girl-hours. Luckily, Harvard College was swarming with cheap assistants; that was how we got things done."

A photograph taken in 1917, thirty-five years after Henry Draper's untimely death, shows a dozen of these assistants, clad in ankle-length dresses and holding hands like a string of paper dolls. Some of these women (along with others not included in that photograph) were largely responsible for the great work known as the *Henry Draper Catalogue.* Among them were Mrs. Williamina Paton Fleming, an Irish immigrant who met Pickering when she became a maid in his household and went on, with his help, to a thirty-year career as an astronomer; Antonia Maury, Henry Draper's niece, who investigated binary stars; Henrietta Swan Leavitt, whose research into variable stars provided some of the information needed to estimate the size of the cosmos; and Annie Jump Cannon, who classified 250,000 stars in the southern sky and restructured the method of classification.

The work was largely based on photographs because, as Annie Cannon wrote, "Although the human eye is such an admirable instrument . . . it is not well adapted to observe the spectra of the stars." At night the prism purchased with Mrs. Draper's gift was fitted to the lens of the telescope, enabling the women to "behold a radiant and beautiful sight, for the twinkling starlight becomes a band showing all the rainbow colors, also crossed by the tell-tale dark lines." Pictures were taken, developed, and mounted, the lines in the spectra were analyzed, and the stars were classified.

The system of classification was a problem, though. Mrs. Fleming classified the stars according to the hydrogen lines in the spectrum and grouped them into categories labeled *A* through *O* (*J* was skipped because "in German script it is indistinguishable from *I*"). Antonia Maury, who thought that stars should be classified according to temperature, divided the stars into twenty-two classes and three divisions. Various shufflings occurred. Finally, Annie Cannon combined the two, reordering Mrs. Fleming's lettered categories on the basis of temperature and color. No longer in alphabetical order, the scheme put the hottest, bluest stars on one end and the coolest, reddest stars on the other. The categories were lettered OBAFGKM, a series taught to this day with the mnemonic "Oh, be a fine girl, kiss me." Here is a summary of those spectral classes:

Category	Color	Temperature	Example
O	blue	36,000 to 63,000°F	Zeta Puppis*
B	blue-white	27,000°F	Rigel in Orion
A	white	16,200°F	Sirius in Canis Major
F	yellow-white	12,600°F	Procyon in Canis Minor
G	yellow	9,900°F	the Sun
K	yellow-orange	7,200°F	Arcturus in Boötes
M	red	5,400°F	Betelgeuse in Orion

*O-type stars are rare. Zeta Puppis is 60,000 times more luminous than the Sun, but because it is 2,400 light-years away, it is only a second-magnitude star. The constellation Puppis represents the poop deck of the defunct constellation Argo Navis (the Argonauts' ship). Zeta Puppis is also called Naos, from the Greek word for "ship."

Annie Jump Cannon

s a child, Annie Jump Cannon (1863–1941) stargazed from the attic of her family's home and wrote down her observations by candlelight. As an undergraduate at Wellesley College, she studied astronomy. After graduation she returned to her native Delaware, where she endured a bout with scarlet fever that seriously impaired her hearing. In 1892, at age twenty-nine, she witnessed a solar eclipse in Spain, and shortly thereafter she returned to Wellesley as an assistant to her former professor Sarah Frances Whiting. Afterward she enrolled in a Radcliffe astronomy course, and by 1896 she was a permanent staff member at the Harvard Observatory.

She classified stars according to their spectra. In the beginning, she labored over 5,000 stars a month; later on, she could classify 300 an hour with barely an error, and she was noted for her remarkable memory. She ultimately wrote nine volumes of the *Henry Draper Catalogue* and classified close to 250,000 stars—"much material," she wrote, "to study the architecture of the celestial mansions and the streaming of the celestial tribes."

Her work was so extraordinary that in 1911, an appreciative E. C. Pickering suggested to Abbott Lawrence Lowell, the president of Harvard University, that Cannon's name be listed in the university catalogue and that she be given an official appointment. Lowell, whose siblings included the poet Amy Lowell and the life-on-Mars enthusiast Percival Lowell, thought it "rather better that Miss Cannon's name should not appear in the catalogue." Pickering was authorized to offer Cannon a less prestigious observatory appointment. She became curator of astronomical photographs, with a salary of $1,200 a year.

Between 1918 and 1924, the multivolume *Henry Draper Catalogue* was

The Harvard College Observatory

Annie Jump Cannon in 1925 at Oxford, where she was awarded an honorary degree.

published, "a permanent monument," according to astronomer Cecilia Payne-Gaposchkin. Pickering died in 1919, but Cannon remained with the project. She was recognized as the world's preeminent expert in the classification of stars, and in 1925, Oxford University awarded her an honorary degree; she was the first woman to be so honored. Harvard listed her name in the university catalogue, but it did not grant her an official appointment.

She never stopped working. Harlow Shapley wrote: "Miss Cannon had the disadvantage—or advantage—of having had some kind of infection while she was in college, and as a result her hearing was pretty much lost. That handicap took her out of the social life and put her into science."

Cecilia Payne-Gaposchkin—the first woman to become a full professor at Harvard, an honor she received in 1956—remembered Cannon differently. "When I think of her as a person I am at a loss for words to convey her vitality and charm. She had lost her hearing in her youth, but had had none of the suspicious pessimism so often associated with the deaf. She wore her hearing aid with an air, and made a virtue of necessity by unshipping it when she wanted to be undisturbed or to do concentrated work. She was warm, cheerful, enthusiastic, hospitable. Like many people who are hard of hearing she had a sharp metallic voice, and she often broke into a characteristic, resounding laugh."

In addition to classifying stars, Cannon discovered 300 variable stars, and in 1936, at age seventy-three, she began a study of 10,000 very faint stars. Two years later, she finally received her official Harvard appointment. She died three years later, recognized throughout the world as the most outstanding female astronomer of her time. "She was never daunted; her spirit was always gay," Payne-Gaposchkin wrote. "She died in her seventies, working to the last. But she was one of those whom the gods love, for she died young."

✳

Henrietta Leavitt's Standard Candle

H enrietta Swan Leavitt (1868–1921) graduated in 1892 from the Society for the Collegiate Instruction of Women (later known as Radcliffe College), devoted a year to postgraduate study, and in 1895 volunteered to do drudge work at the Harvard Observatory. In 1900, a family crisis forced her to move to Wisconsin. Two years later, she wrote director E. C. Pickering and told him how much she longed to return, whereupon he financed her trip back and gave her a permanent paid position. With the exception of a long convalescence in Wisconsin after an illness, she spent the rest of her life in Cambridge. Her task at the observatory was to search for variable stars on photographic plates of the southern skies. All told, she discovered about 2,400 variable stars. The most important were the Cepheid variables, yellow supergiant stars that brighten quickly and slowly fade.

Rather than studying stars scattered throughout the Milky Way, Leavitt concentrated on Cepheids in the Small Magellanic Cloud, an irregular companion galaxy to the Milky Way. It is so far away—about 200,000 light-years—that all stars therein can be considered the same distance from Earth, just as, to someone in Tacoma, everyone in Paris can be considered equally far away. With the stars at essentially the same distance, apparent differences in brightness become real; stars that look brighter *are* brighter.

When she placed each Cepheid on a chart, Leavitt discovered something astonishingly useful: The brighter the star, the slower its rate of variation. Relatively faint Cepheids—stars a couple of hundred times as bright as the Sun—pulse rapidly, contracting and expanding within a day

The Harvard College Observatory

Henrietta Swan Leavitt (1868–1921).

or two. Average Cepheids fade from maximum magnitude and return to it in about five days. Bright Cepheids, shining with the luminosity of 10,000 Suns, have a period as long as fifty-four days.

Thus, if astronomers knew the length of a Cepheid's cycle, they could figure out its intrinsic brightness; and by comparing how bright it looked with how bright it was known to be, they could estimate its distance.

For one reason or another, however, all the initial measurements utilizing this method were wrong. Ejnar Hertzsprung, famous for the Hertzsprung-Russell diagram, which charts the stars, estimated that the Small Magellanic Cloud was 30,000 light-years away—a whopping underestimate. Harlow Shapley used variable stars to calculate the distance to globular clusters, and was consequently able to estimate our position within the Milky Way, as well as the galaxy's size—but his figure was three times too big. And Edwin Hubble figured out the distance—well, half the distance—to Andromeda.

The difference between real distances (as they are now understood) and these early estimates might seem laughably immense. Yet astronomical distances are so enormous that if an estimate is off by a factor of 2—meaning the correct figure might be twice as large or half as large as

A Cepheid variable star. The arrow in the image at the left points to a Cepheid variable at the beginning of its cycle of pulsation; the brighter image below was taken five days later.

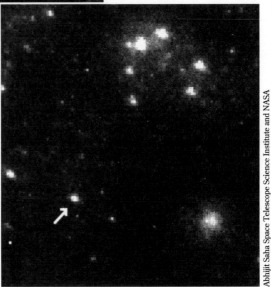

Abhijit Saha Space Telescope Science Institute and NASA

the official estimate—that's not bad, especially if you consider what a factor of 10, 100, or 1,000 would mean. Early estimates using Cepheids were wrong for several reasons, including failure to account for the effect of intervening dust and gas, and, more important, the unsuspected existence of two kinds of Cepheid variables: Type I Cepheids, found in the spiral arms of galaxies; and older, dimmer Type II Cepheids, found in elliptical galaxies, globular clusters, and the galactic halo.

Once these corrections were made, Cepheid variables allowed astronomers to calculate all kinds of distances. But Cepheids were limited. Bright as they are, they could be detected only in the nearest thirty or so galaxies. Beyond that, they could no longer be found. So astronomers sought other standard candles visible at greater distances. They used faint RR Lyrae stars, another form of variable; globular clusters, spherical swarms of as many as a million stars; planetary nebulae; and supernovae, which can be detected at distances a thousand times beyond Cepheids. Although Cepheids are more reliable indicators than any of these other, brighter objects, their territory was so limited that, in a certain way, they came to seem almost quaint.

But that was before the Hubble Space Telescope. Flawed as it was, it nonetheless detected twenty-seven Cepheid variables in a galaxy estimated to be 16 million light-years away. Their intrinsic brightness is known, and thus not only can their correct distance be estimated, but scientists can also compare them to other stars, including supernovae. Those measurements will help astronomers establish supernovae as standard candles. In the meantime, the stars decoded by Henrietta Swan Leavitt continue to be mileposts of the universe and retain their position as the best distance indicators in all the starry skies.

✳

Albert Einstein

The shortest line, Einstein replied,
Is not the one that's straight;
It curves around upon itself
Much like a figure eight,
And if you go too rapidly
You will arrive too late.

—W. H. Williams

Famous people glow, it's often said, and it's
a glow that comes from the number of times
we have seen the images of their faces.

—Leo Braudy

espite the aura of sanctity that surrounds him, Albert Einstein (1879–1955) was not a saint but a celebrity. Renowned for his genius, he was a scientific revolutionary in his youth and a stubborn conservative later on. Beloved for his humanity, he was a philandering husband and an unsentimental father who allowed his first child, a daughter born to his first wife a year before their marriage, to be given away. Admired as "the conscience of the world," he was an outspoken pacifist, a committed socialist, an opponent of McCarthyism, and an advocate of universal disarmament, yet he was also one of the important theorists whose work ultimately led to the creation of the atomic bomb.

World War II forced him into a dilemma. In 1939, a group of physicists, fearful that Nazi scientists might be working on atomic weapons, asked Einstein to help them convince President Franklin D. Roosevelt to develop the bomb before Germany did. Einstein agreed, and signed a letter to Roosevelt expressing concern about the construction of "extremely powerful bombs of a new type." Upon receiving the letter, Roosevelt appointed a committee on Uranium, and two years later, on the day before Pearl Harbor, gave the go-ahead to the Manhattan Project. Although Einstein was not involved in that (he was considered a security risk), he regretted the letter. "Had I known that the Germans would not succeed in producing an atomic bomb," he said, "I would not have lifted a finger."

It is ironic and sad that the letter and, more important, Einstein's deeply theoretical work helped unleash the greatest man-made violence the world has ever known. Einstein's goal was pure: He wanted to understand the universe in its totality. Even as a small boy, he was thrilled by the mystery of unseen forces.

He grew up in Germany in a close-knit Jewish family. Two incidents of his childhood reveal something of the extraordinary character of his mind, for although he didn't speak until he was three years old and was never successful academically, from the start his mind was constantly working. When he was four or five years old, his father showed him a compass and explained that it was a device for gauging the earth's magnetic field. The young boy was so enthralled by the way the needle always pointed in the same direction that he later described this incident as a "miracle."

When he was about twelve, a second miracle occurred. A young medical student whom the Einsteins invited to dinner once a week gave Albert a textbook on Euclidian geometry. "The clarity and certainty of its contents made an indescribable impression on me," Einstein later said. He called the gift "the holy geometry book." Around the same time, he renounced the Bible and threw himself into what he later called "a positive fanatic orgy of freethinking."

When he was sixteen, Einstein's family moved to Milan, leaving him behind to finish school. He soon dropped out and joined his parents. From Italy, he applied to the prestigious Federal Institute of Technology in Zurich but failed the entrance examination. After a year of additional study, he was admitted to the institute in 1896. Disturbed by his native country's militaristic atmosphere and compulsory military service, he renounced his German citizenship that same year.

He graduated in 1900, became a Swiss citizen, and was unable to find a job. Feeling like "a pariah, discounted and little loved . . . suddenly abandoned by everyone, standing at a loss on the threshold of life," he tutored and worked as a substitute teacher until 1902, when he landed a job in the Swiss Patent Office in Bern. Later that year his father died, and in 1903 Einstein married Mileva Maric, who had been a fellow student at the Federal Institute and was the mother of the baby girl born the previous year. Their first son, Hans, who later became a professor of engineering at the University of California at Berkeley, was born in 1904. (Their second son, Eduard, born in 1910, died in a psychiatric hospital in 1965.)

The civil service job freed Einstein to concentrate on science. In the productive year of 1905, he published his Ph.D. dissertation and five significant papers. The first, on the photoelectric effect of light, proved Max Planck's theory that light is emitted in bundles, or quanta, thus validating quantum physics. Two other papers discussed Brownian motion, which describes the way particles immersed in a fluid are bombarded with molecules, causing them to jitter.

The work for which Einstein is most famous, however, presented something revolutionary: the special theory of relativity, an idea even Einstein had trouble accepting. "I must confess," he later said, "that at the very beginning when the special theory of relativity began to germinate in me, I was visited by all sorts of nervous conflicts." Special relativity introduced the concept of a four-dimensional universe woven from space-time: the three usual dimensions plus time. At the trifling distances of ordinary life, this concept hardly makes a dent, but it matters with astronomical distances; when we look at the stars, we look out in space and back in time.

Special relativity overturned the Newtonian assumption that space and time were fixed. Operating from the principle that the only cosmic imperative is the speed of light, which in a vacuum is always 186,000 miles per second regardless of the location of the observer, Einstein realized that time and space were measurements that depended on how fast and in what direction the observer was moving. The consequences of his discovery were stunning. A stationary observer watching a clock whiz by would discover that as it approaches the speed of light, time slows down (and length and mass increase). At the speed of light, time stops, and the hands of the clock will not appear to move. Yet to a person traveling with the clock, it will tick in the usual manner. Time becomes a relative phenomenon—an idea completely contrary to the dicta of Sir Isaac Newton, who said, "Absolute, true, and mathematical time, of itself and by its own nature, flows uniformly, without regard to anything external."

Special relativity also stated clearly that matter and energy were equivalent. The most famous part of the theory was the formula $E = mc^2$, where E = energy, m = mass, and c = the speed of light. That elegant and grand equation—simple in structure, awesome in implication—opened the door to an unimagined and counterintuitive universe that immediately gained the attention of the physics community.

By 1909, Einstein was a wandering academic, working at the University of Zurich, the University of Prague, the Federal Institute of Technology in Zurich, and the University of Berlin, which offered him a research position in 1914. Shortly after moving to Berlin, his marriage disintegrated and his wife returned to Zurich with their two sons. When World War I broke out, Einstein joined a pacifist organization and, acting on his conviction that a supranationalism beyond boundaries might increase the likelihood of peace, signed a document calling for a "League of Europeans."

In 1916, his general theory of relativity was published. It was based on a thought he had one day at the patent office in 1907, when it occurred to him that a person falling through space would not feel his own weight. He called this "the happiest thought of my life," and added, "It impelled me toward a theory of gravitation."

The theory of general relativity expanded the special theory by considering what happens when velocity changes. (From this work comes the startling paradox that if one twin leaves the Earth, travels in a spaceship at a very high rate of speed, turns around, and returns, upon his arrival he will be younger than the twin who stayed at home.) The general theory showed that mass causes space to curve around it. Imagine a bowling ball on a water bed. The mattress would curve beneath the ball. Drop a marble on the water bed, and it will inevitably fall toward the bowling ball. Einstein realized that smaller masses fall toward larger masses not because the larger masses are "attracting" them but because the objects are traveling along the lines of curved space. This inevitable movement toward the heavier

When I was a kid, the only playthings we had in the whole universe were the hydrogen atoms, and we played with them all the time, I and another youngster my age whose name was Pfwfp.

What sort of games? That's simple enough to explain. Since space was curved, we sent the atoms rolling along its curve, like so many marbles, and the kid whose atom went farthest won the game.

—Italo Calvino, Cosmicomics

Einstein, in a partially buttoned coat, stands on board ship with Chaim Weizmann (to his immediate left), a chemist who became the first president of Israel. "During this voyage," Weizmann said, "Einstein kept explaining his relativity theory to me again and again, and now I believe that he has fully understood it."

mass, Einstein showed, accounted for the phenomenon known as gravity.

In a *gedankenexperiment* (the kind conducted in thought only), Einstein imagined a person stuck in an elevator in interstellar space, far from any star or planet. The person would feel weightless. But what if the elevator were to accelerate upward at precisely the right speed? The person inside would be pinned to the floor. If he dropped something, the floor would come up to meet it, but it would look as if the object had simply hit the floor. He would no longer feel that he was floating in space. Without gazing out the window, he would have no way of knowing that he wasn't on Earth. Viewed this way, gravity and acceleration were the same thing.

In 1919, he divorced Mileva (promising her, as alimony, interest on the cash award of the Nobel Prize, which he had yet to win), married his cousin Elsa, and became interested in Zionism. Perhaps more important, he received important confirmation of his theory. He had predicted that light passing a massive object, such as the Sun, would be bent a specific amount by the object's gravitational field. In March 1919, a solar eclipse permitted scientists to compare the usual positions of several stars with their positions during the eclipse, when the light streaming from those distant stars skimmed past the Sun before reaching the Earth. The measurements confirmed Einstein's predictions. From then on, although his theories remained opaque to most people (and his major contributions were behind him), he was an international celebrity.

He is as celebrated today as he was then. His absentmindedness and refusal to wear socks became as legendary as his famous formula. In the early 1950s, Einstein hired someone to help him with his taxes, and the event made headlines: "Universe like Open Book but Income Tax Baffles Father of Relativity." ("It isn't the mathematics that bothers me," he explained. "It's the philosophy of it.") Even now, his face is the most recognizable of any scientist's ever, and he has become a cultural icon. Almost four decades after his death, his image appears regularly on T-shirts, in advertisements, and even in such films as Nicolas Roeg's *Insignificance,* in which Einstein listens as his theory of relativity is breathlessly explained to him by Marilyn Monroe. "Humanity needs a few romantic idols as spots of light in the drab field of earthly existence," Einstein wrote as an old man. "I have been turned into such a spot of light."

Yet in 1919, when his theory received its spectacular proof, acclaim was far from the only reaction. "If the theory of relativity is proved right, the Germans will call me a German, the Swiss will call me a Swiss citizen, and the French will call me a great scientist," Einstein said. "If relativity is proved wrong, the French will call me a Swiss, the Swiss will call me a German, and the Germans will call me a Jew."

In Germany, Einstein's lectures were disrupted. In 1920, at a meeting held in Berlin, the Nobel Prize–winning physicist Philipp Lenard spoke out against Einstein's theory of relativity, calling it part of a Jewish plot to undermine science. "Science, like every other human product, is racial, and conditioned by blood," Lenard wrote in his four-volume work, *German Physics.* Although Lenard worked to deny Einstein the Nobel Prize, he finally received it in 1921—not for the theory of relativity but for his work on the photoelectric effect of light. Mileva got the money, as promised.

Throughout the 1920s, Einstein worked as an advocate for peace and for Zionism. (In 1952, he declined the presidency of Israel commenting,

"Politics is for the present, but an equation is for eternity.") In 1933, while Einstein was in the the United States for a visit, Hitler came to power; shortly thereafter, the Nazis ransacked his summer home and seized his bank account. Einstein never returned to Germany. (The physicist Werner Heisenberg, who remained in Germany, was allowed to teach relativity only if he agreed not to mention Einstein's name.) By the end of 1933, Einstein, Elsa, and his secretary Helen Dukas were permanently installed at Princeton. Except for a trip to Bermuda in 1935, he never left the United States again. He became a U.S. citizen in 1940.

By then, strange things were happening to Einstein. If the theory of relativity was mind-boggling to other people, Einstein had equal difficulty accepting the implications of quantum physics, even though he had been awarded the Nobel Prize for proving its most basic assumption. What troubled him most was Heisenberg's uncertainty principle—the idea that you could know either the position or the momentum of a particle, but you could not know both at the same time. That something might be unknowable was offensive to Einstein; in an almost literal way, indeterminacy was against his religion. "God may be subtle, but he is not malicious," he stated many times; he also said, "I do not believe that God plays dice" (to which Niels Bohr, his beloved friend and scientific adversary, replied, "Albert! Stop telling God what to do!"). Although Einstein recognized that quantum theory worked, he considered it "silly." His skepticism turned the most influential scientist of the twentieth century into something of a dinosaur, imposing but no longer in the mainstream of physics. In 1949, the physicist Max Born said, "Many of us regard this as a tragedy—for him, as he gropes his way in loneliness, and for us who miss our leader and standard-bearer."

Einstein sought a single, consistent synthesis of the universe, a "unified field" theory that would reconcile microcosm with macrocosm and be as true for subatomic particles as it would be for galaxies. Like today's generation of scientists, he never found it, in part because he resisted so strongly the implications of quantum theory. Although he collaborated with many colleagues and engaged in much discussion and argument, in the end, for Einstein, true science happened in one place only: his mind. Often, he didn't even read the scientific journals. He *thought*. He *imagined*. He once said of Max Planck, "The emotional state which enables such achievements is similar to that of the religious person or the person in love; the daily pursuit does not originate from a design or program but from a direct need." The same is true of Einstein.

The Great Debate

On April 26, 1920, the National Academy of Sciences held a symposium on the topic of the scale of the universe. The major participants, Harlow Shapley (1885–1972) and Heber D. Curtis (1872–1942), took the same train across the country to Washington, D.C., spent time en route discussing flowers, and upon their arrival attended a banquet so boring that in the midst of the speeches Albert Einstein leaned over to the man next to him and whispered, "I have just got a new theory of Eternity." At the symposium itself, Shapley and Curtis each read a paper and offered a rebuttal. It was not a dramatic affair; writing almost half a century later, Shapley recalled it as "a pleasant meeting," and noted that "nobody had mentioned it to me for many years. Then, beginning about eight or ten years ago, it was talked about again. To have it come up suddenly as an issue, and as something historic, was a surprise."

That event, now known as "the Great Debate," has gained in symbolic importance because it marked a pivotal shift in our conception of the universe. The official question of the symposium concerned the dimensions of the Milky Way; the unofficial issue concerned such disturbing objects as M31, the great spiral nebula—as it was then known—in Andromeda. Was it in fact a nebula, a cloud of dust and gas within a galaxy where stars were being formed? Or was it another galaxy, an entirely separate disk of stars, an island universe such as Kant described?

On one side was Harlow Shapley. The son of a Missouri hay farmer and the grandson of an abolitionist whose house was a stop on the Underground Railroad, Shapley had been a teenage reporter specializing in crime for the *Daily Sun* in Chanute, Kansas. The experience inspired him to attend the University of Missouri, but upon arrival he discovered that he could not enroll in the school of journalism for a year. "I opened the catalogue of courses and got a further humiliation," he wrote in his memoirs. "The very first course offered was a-r-c-h-a-e-o-l-o-g-y, and I couldn't pronounce it! (Though I did know roughly what it was about).

"I turned over a page and saw a-s-t-r-o-n-o-m-y. I could pronounce that—and here I am!"

Shapley is known for having estimated the shape and extent of our galaxy by using Mount Wilson's 60-inch telescope to map the globular clusters that surround the Milky Way. He realized that the globular clusters were not distributed evenly across the sky but concentrated in Sagittarius. By measuring RR Lyrae stars (also known as cluster variables), old variable stars that can be used like Cepheids to estimate distance, he was able to assess the distance of the clusters. In so doing he discovered that they formed a gigantic sphere centered in Sagittarius. The middle of

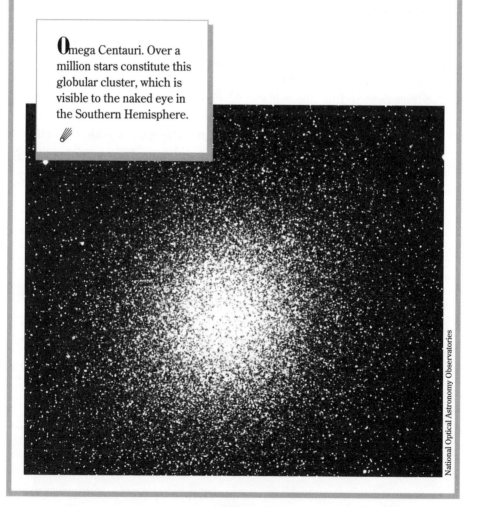

Omega Centauri. Over a million stars constitute this globular cluster, which is visible to the naked eye in the Southern Hemisphere.

National Optical Astronomy Observatories

that sphere, he theorized, was the center of the galaxy. The Sun was not the center. True, it looked central, but our view is obscured, Shapley suggested, by intervening clouds of gas and dust.

Shapley's calculations also showed that the galaxy was much larger than had been thought. Our galaxy is so big that whether we look toward Sagittarius, or in the opposite direction, toward Auriga, we see fewer than 1 percent of the stars in the Milky Way. Still, Shapley's estimate that the galaxy is 300,000 light-years across—an enormous increase over the prior estimates of 15,000 to 20,000 light-years—was about three times too large, in part because he based his calculations on erroneous assumptions about RR Lyrae stars. They are smaller and dimmer than he thought, and hence not quite as distant. Nonetheless, Shapley's use of variable stars to explore the Milky Way was considered brilliant, and his assessment of our galaxy's size, inflated though it was, revolutionary.

In the symposium that April, Shapley explained his calculations and stated that the great spiral in Andromeda was, in his opinion, entirely within the Milky Way. Unfortunately, he had miscalculated the distance of Andromeda on the basis of a luminous new star seen there in 1885. At peak luminosity, the star had reportedly been brighter than the rest of the spiral. Since it didn't seem possible that a star—any kind of a star—could outshine a galaxy, Shapley concluded that the spiral must be a nebula.

Shapley's assumption, reasonable though it seemed, was wrong. The star in Andromeda was a supernova, which had not been seen in the Milky Way since 1604. A supernova can easily outshine a galaxy, but Shapley assumed otherwise.

He had an additional reason for believing that spiral nebulae were part of our own galaxy. Adrian van Maanen, "an alert-minded person" with whom Shapley was "pals of a sort," had measured the movement of spiral nebulae and discovered that they were spinning so fast that, if they were very distant and galactic in size, their outer edges would slice through space at a rate faster than the speed of light. Since this was impossible in a non–Star Trekkian universe, the only way they could physically stay together would be if they were relatively small and close. Hence they had to be ordinary galactic objects, well within the confines of the Milky Way.

But Van Maanen's calculations were wrong. Although other scientists suspected this, Shapley "faithfully went along with my friend," an error in judgment which never ceased to rankle him. When the mistake was revealed and it became obvious that nebulae were not whirling so rapidly, Shapley felt betrayed. Describing himself in the third person, he wrote, "they wonder why Shapley made this blunder. The reason he made it was that Van Maanen was his friend and he believed in friends." (As a result of Shapley's harping on this point, Van Maanen—who also discovered a white dwarf in Pisces—has been known ever since as a man who made a mistake. Thus a shadow darkens his reputation to this day, and authors discussing his work seem compelled to comment, just in case there is doubt in anyone's mind, that he was a respectable, responsible scientist.)

Shapley's opponent in the Debate, Heber D. Curtis of the Lick Observatory, was a Latin professor turned astronomer who disagreed with Shapley on almost every point. In his presentation, he attacked Shapley's methods, suggesting that his use of variable stars was inaccurate and that Van Maanen's figures were wrong. Curtis held that the Sun was the center of our galaxy; that the universe was small, not large; and that spiral nebulae—and Andromeda in particular—were other galaxies, well beyond the Milky Way.

Thus both men were right and both were wrong. "I was right and Curtis was wrong on the main point—the scale, the size," Shapley wrote. "It is a big universe, and he viewed it as a small one." But to Shapley's irritation, Curtis was correct about the spiral nebulae. Shapley considered the point an unfair one, as if Curtis, by bringing it up, was somehow cheating. "From the beginning Curtis picked on another matter: are the spiral galaxies inside our system or outside?" Shapley wrote. "But that was not the assigned subject. Curtis, having set up this straw man, won on that."

The day after the debate, an uncredited article in the *Kansas City Star* described the event: "Dr. Harlow Shapley . . . in an exhaustive discourse, replete with terms of the higher and highest calculus, sought to demonstrate that there is but one universe, but a universe ten times larger than that hitherto conceived by the wildest astronomical calculation. . . . In the confines of his comparatively cramped universe, Dr. Curtis touched off a series of pyrotechnic displays of cosines and tangents demonstrating to the satisfaction of his partisans his conception of a league of universes, leaving the question open whether our universe belongs to a major, minor or bush league in the cosmic game. . . ." Astronomer Owen Gingerich has pointed out in *The Great Copernicus Chase* that Shapley, once a reporter for a Kansas paper, probably wrote the article himself.

The following year, Shapley left California and became director of the

Harvard Observatory. In addition to overseeing the observatory, he became an entertaining writer, one of the organizers of UNESCO, and a popular professor who invited students into his home once a month on the full moon.

But the move to Massachusetts was widely considered a mistake. Important work was still going on at Mount Wilson, where Edwin Hubble, whom Shapley disliked, found Cepheid variables in the Andromeda Nebula, which allowed Hubble to prove that Curtis was right—M31 wasn't a nebula but a separate galaxy—and that Shapley was also right: The universe was far bigger than anyone had imagined. By the end of the decade that began with the Great Debate, Hubble was able to show that, in addition, the universe was getting bigger all the time.

✳

THEODORE ROOSEVELT
AND THE
QUEEN OF THE NEBULAE

On a clear, dark autumn night, the great nebula in the constellation Andromeda looks like a faint oval smudge. But *was* it a nebula, a swirling cloud of incandescent gas? Or a "Saturniform body . . . surrounded by a series of rings"? Or was it a solar system in the making, as was suggested in 1899? By 1848, over 1,500 "minute" stars had been counted in the fuzzy patch. Did that mean it was a cluster of small, tightly packed stars, or a Kantian island universe? By 1918, Heber D. Curtis was convinced that Andromeda, the Queen of the Nebulae, was not a nebula at all but another galaxy.

Two non-astronomers who agreed were the naturalist William Beebe (1877–1962) and his friend Theodore Roosevelt. Beebe described a frequent interaction:

> After an evening of talk, perhaps about the fringes of knowledge, or some new possibility of climbing inside the minds and senses of animals, we would go out on the lawn, where we took turns at an amusing little astronomical rite. We searched until we found, with or without glasses, the faint, heavenly spot of light-mist beyond the lower left-hand corner of the Great Square of Pegasus, when one or the other of us would then recite:
>
> > That is the Spiral Galaxy in Andromeda.
> > It is as large as our Milky Way.
> > It is one of a hundred million galaxies.
> > It is 750,000 light-years away.*
> > It consists of a hundred billion suns,
> >   each larger than our sun.
>
> After an interval Colonel Roosevelt would grin at me and say: "Now I think we are small enough! Let's go to bed."
>
> We must have repeated this salutory ceremony forty or fifty times in the course of years, and it never palled.

*Current estimates place Andromeda 2.25 million light-years away.

Edwin P. Hubble

dwin Powell Hubble (1889–1953) was one imposing man. Tall, handsome, effortlessly successful in almost everything he tried, he graduated from high school with a scholarship to the University of Chicago and the strong desire to play college football—a dream he regretfully laid aside after his mother implored him to give up the game. He became a boxer instead, and was so powerful that he was asked to fight professionally against Jack Johnson, the heavyweight champion of the world. He turned the offer down but he did manage to find other opportunities for skirmish. As an undergraduate, he got into an altercation at a railroad yard, was literally stabbed in the back with a knife, and in response knocked his assailant unconscious. A few years later, while spending the summer in Germany, he saved a woman from drowning, befriended her husband, and was forced into a pistol duel when the husband heard rumors suggesting that his wife and Hubble were involved romantically. Both men shot wide of the mark.

By then, Hubble was a Rhodes scholar, studying the law at Oxford University in England. The experience influenced him so profoundly that when he returned to the United States three years later, he had an English accent he was never to lose, and he peppered his speech with Briticisms such as "by Jove" and "come a cropper." In 1913, Hubble became a member of the Kentucky bar but almost immediately gave it up in favor of astronomy. "Astronomy is like the ministry; you need a calling," he later told a reporter. "After practicing law for a year in Louisville, I got the calling." If this statement wasn't quite true—he never practiced law, although he did work as a teacher and as a basketball coach—it nevertheless captured the spirit of his conversion. Hubble returned to Chicago, got his Ph.D. in astronomy, and was offered a job at the Mount

Edwin Hubble at the 48-inch telescope.

Courtesy of Caltech

Wilson Observatory by George Ellery Hale. He accepted, but before he could head west, another opportunity arose: World War I. He enlisted immediately and was sent to France, where he served in the infantry for two years, was wounded in his right arm, and received a promotion to the rank of major—a title whose use he encouraged for the rest of his life.

He then moved to Mount Wilson to pursue faint nebulae, the topic of his dissertation. He photographed them, took their spectra (a process that sometimes required as many as ten nights), studied their shapes, and categorized them. But he still didn't know the answer to the most basic questions of all: Were these nebulae gaseous clouds trapped inside our own Milky Way galaxy? Or were they separate stellar cities floating far outside it?

When Mount Wilson's hundred-inch telescope, then the largest in the world, was installed, Hubble was able to find out. He used the new telescope to take many photographs of M31, then known as the Great Nebula in Andromeda, and M33, the Pinwheel Galaxy in Triangulum. In October 1923, on a long exposure of M31, Hubble discovered the first of many Cepheid variables. By determining their periods of pulsation (between ten and eighty days) and comparing their brightness to that of the Cepheid variables Henrietta Leavitt had discovered in the Small Magellanic Cloud, he learned their distance, and thereby discovered that Andromeda had to be at least 800,000 light-years away. Thus it could not possibly be within the Milky Way. Beginning on December 30, 1924, the American Astronomical Society held a meeting in Washington, D.C. Hubble didn't attend, but at the last moment he mailed in a paper describing his methods and results. It was read in his absence on January 1, 1925, and it changed astronomy. There was no doubt about it: Andromeda was another galaxy, and indeed the heavens were full of galaxies, all of which were incredibly more distant than had been thought.

During the next several years, Hubble and his assistant Milton

The Andromeda Galaxy (M31) and a small companion galaxy, the elliptical M32.

National Optical Astronomy Observatories

THE EXPANDING UNIVERSE: THE USUAL ANALOGIES

•⤳

It doesn't matter which galaxy you visit, the expansion of the universe would look the same. Scientists trying to explain this bizarre state of affairs generally fall back on two comparisons. Here they are:

1. The Balloon analogy. Picture a balloon covered with dots, each of which represents a galaxy. As it is blown up, the space between dots increases. Every dot appears to be moving away from every other dot.

2. The Raisin Bread analogy. Now the galaxies are raisins and the dough is space. As the bread bakes, the dough rises. Everywhere in this yeasty universe, the raisin (and walnut) galaxies rush away from one another because the dough—space—is expanding. The more dough there is between any two raisins, the greater the expansion. If the loaf doubles in size, raisins originally one inch apart will now be two inches apart; raisins that were four inches apart will be eight inches apart. The more distant the raisins, the faster and farther apart they seem to move.

Humason, an eighth-grade dropout who began his career at Mount Wilson as a mule driver and ended it as a fully recognized astronomer, took the spectra and calculated the distances of about two dozen objects. A few years earlier, Vesto M. Slipher at the Lowell Observatory in Arizona had taken the spectra of over a dozen "spiral nebulae," and had been surprised to find that almost all of them were displaced toward the red end of the spectrum, indicating that they were moving away from us. (Andromeda and the few small galaxies under the gravitational influence of the Milky Way are exceptions; they are part of our local group of galaxies and show a blue shift, indicating that they are approaching, narrowing the distance between us.) Building on Slipher's research and adding his own, Hubble placed each galaxy on a chart plotting speed against distance, which enabled him to prove something very interesting: The farther away a galaxy is, the faster it is receding. And this is true no matter in which direction you look. This is known as Hubble's law.

Hubble's law suggested either that every galaxy in the cosmos was doing its utmost to avoid the Milky Way Galaxy, or that every galaxy was moving away from every other galaxy. Thus Hubble made what could be considered the most important astronomical discovery of the twentieth century (astronomer Allan Sandage, Hubble's student, has called it "the most amazing scientific discovery ever made"): The universe is expanding—a momentous piece of knowledge so definitive that even Einstein, to whom the concept of an expanding universe was anathema, had to accept its truth.

Hubble's work underlies contemporary astronomy. A glance at the index of one typical college textbook reveals that "Hubble, Edwin," is followed by "Hubble classification," "Hubble constant," "Hubble flow," "Hubble law," and "Hubble Space Telescope." "Einstein" and "Galileo" get two listings each.

A pipe smoker and avid fly fisherman, Hubble was often—but not always—considered aloof, arrogant, and cold (whereas the jovial Milton Humason is described as "easygoing" and "lovable"). Although Hubble and his wife, Grace, had a glamorous circle of friends (including Aldous Huxley), Harlow Shapley wrote that "Hubble just didn't like people." But then, Shapley, who believed that the Milky Way Galaxy was large enough to encompass the spiral nebula in Andromeda, had reason to dislike Hubble. Hubble not only failed to acknowledge that in his own work he was following Shapley's pioneering use of variable stars; he also proved Shapley wrong.

Moreover, Hubble insulted Shapley. Here is Shapley's version of a well-known anecdote about the rivalry between the two: "I remember once somebody referred a paper of mine to Hubble for him to pass judgment on. It was a good paper; it was correct; I mean, I knew what I was talking about at that time. It was written for some journal like *Scientific American*. Hubble just wrote across it, 'of no consequence.' The editors, who told me about it, thought it was the funniest thing, because the words 'Shapley—of no consequence' got set in type."

On another occasion, Hubble took a backhanded swipe at Shapley, who was the director of the Harvard Observatory, by saying of the astronomer Cecilia Payne-Gaposchkin, "She's the best man at Harvard."

Despite these provocations, Shapley generously wrote that "Hubble went on and made himself very famous, and properly so. He was an excellent observer, better than I." Hubble died in 1953. No public funeral or memorial service was held, and the whereabouts of his ashes are unknown.

The Hubble Constant

Hubble's law states that the more distant a galaxy is, the faster it is receding. The Hubble constant, or H_0 (pronounced "H-nought"), measures the rate of that recession. Or, to phrase it differently, it measures how fast the universe is expanding. If you know the Hubble constant, you can figure out the physical scale of the universe and make a reasonable estimate of its age. So the Hubble constant is a sort of Rosetta stone. The race to figure it out has been avid and acrimonious, and it remains unresolved.

In principle, determining the Hubble constant shouldn't be too hard. All you need are recessional velocities—which can be derived from redshift—and accurate distances for a sufficiently large group of faraway galaxies. Divide the velocities by the distances and you've got it. The problem is that distances are notoriously hard to obtain. Parallax provides distances to a few nearby stars, but outside of the galaxy it's useless; Cepheid variables make it possible to determine the distances of nearby galaxies, but past a certain point Cepheids can no longer be detected; supernovae can be used as standard candles, but supernovae don't happen when and where you want them to. The brightest galaxy in a cluster is a useful measure, assuming that you measure similar kinds of galaxies and clusters—but how can you really tell? Giant elliptical galaxies may be standard candles, and the same is true of globular clusters, quasars, particular kinds of supergiant spiral galaxies, planetary nebulae, and gravitational lenses, in which the light from distant objects is split by intervening galaxies. Unfortunately, none of these methods is foolproof. Taking astronomical measurements is a complicated, error-strewn task, and the farther away the galaxy is, the wilder the estimate of its distance inevitably becomes. Different measurements of distance produce a different Hubble constant and, hence, a different rate of expansion for the universe.

Hubble himself estimated that the Hubble constant was 530 kilometers per second per megaparsec (a megaparsec is 3.26 million light-years), with a margin of error of 15 percent. That suggested a small, young universe, less than 2 billion years old—an impossibility, since Earth rocks had already been proven to be 4 billion years old. Since Hubble's time, H_0 has plummeted. In 1956, Allan Sandage, Hubble's astronomical heir, put H_0 at 180, which indicated a universe about three times as big and three times as old as Hubble's. Two decades later, Sandage and Gustav Tammann announced that the Hubble constant was 57. Then it edged downward to 50, dipped to 42, crept upward to 52, and slipped back down to 45. These figures suggest a universe between 15 billion and 20 billion years old, depending on how much mass it contains.

These figures, however, are controversial. One longtime critic is the astronomer Gérard de Vaucouleurs of the University of Texas, who has championed a Hubble constant closer to 100, which would produce a universe 10 billion or so years old.

And then there are the figures in the middle. In the early 1980s, Jeremy Mould of the California Institute of Technology (Caltech), John P. Huchra of Harvard University and the Smithsonian Astrophysical Observatory, Gregory D. Bothun of the University of Oregon, and Marc Aaronson (later accidentally killed at the age of thirty-seven when a bulkhead door slammed into him at the Kitt Peak National Observatory) came up with figures for a Hubble constant that ranged from 65 (for closer galaxies) to 90 (for more distant galaxies). In their calculations, they used a method based on the useful fact, discovered by R. Brent Tully and J. Richard Fisher, that the brighter, and hence more massive, a galaxy is, the faster it rotates; and the faster it rotates, the wider the spectral line indicating the presence of neutral hydrogen. The width of the line—a piece of data relatively easy to get—reveals the galaxy's luminosity. This technique allows scientists to probe galaxies as far away as 300 million light-years.

But nothing's foolproof. Tully-Fisher measurements begin with galaxies close enough to be calibrated by way of Cepheid variables and are then applied to more distant galaxies. So these Cepheid measurements must be reliable. But previous Cepheid measurements were made using old-fashioned photographic plates, which are inaccurate at low and high levels of light. On the other hand, wafer-thin charge-coupled devices, or CCDs, are precise. They respond to every photon. Using them, astronomers Wendy Freedman of the Carnegie Institution and (her husband) Barry F. Madore of Caltech have remeasured Cepheids, applied those figures to the Tully-Fisher relation (and other distance indicators), and come up with a Hubble constant closer to 80. That number suggests a universe younger and smaller than the Sandage universe, larger and older than the universe of de Vaucouleurs.

There's a problem, though. The globular clusters sprinkled around the galaxy are commonly considered, based on the tenets of stellar evolution, to be about 15 billion years old—a value that fits in nicely with Sandage's figures. Obviously, the universe cannot be younger than the stars. So either the quixotic Sandage is right, or the globular clusters are a few billion years younger than has been thought. This is possible because a small distance error in measuring stars in globular clusters translates into a large difference in the cluster's age. Or the physical model may be incorrect. Or maybe Einstein was right, and the universe runs on a cosmological constant. At the moment, in any case, independent measurements produce different Hubble constants, and astronomers have yet to agree on a single figure.

Even if the Hubble constant were irrefutably established, however, it would not directly predict the age of the universe. The truth is, the Hubble constant is not a constant at all. In the early years of the universe, everything was racing away from everything else and the Hubble constant was higher than it is now. Gravity has slowed things down, but the rate of that deceleration cannot be determined without a reliable estimate of the mass contained in the universe. When the mass is known, scientists will learn not only the age of the universe but its fate.

In the meantime, scientists require better distance measurements. Those may be forthcoming thanks to the final part of Hubble's legacy: the Hubble Space Telescope.

The Hubble Space Telescope

We are, by definition, in the very center of the observable region. We know our immediate neighborhood rather intimately. With increasing distance, our knowledge fades, and fades rapidly. Eventually, we reach the dim boundary—the utmost limits of our telescopes. There, we measure shadows, and we search among ghostly errors of measurement for landmarks that are scarcely more substantial.

—Edwin P. Hubble

The idea was simple. The Princeton astrophysicist Lyman Spitzer thought of it as early as 1947: Put a telescope above the atmosphere, and your view of the cosmos will be much, much clearer. This had been done in a limited way in the 1970s and 1980s with observatories such as the International Ultraviolet Explorer (IUE) satellite, the X-ray Einstein Observatory, and the Infrared Astronomy Satellite. It happened in a big way on April 24, 1990, when the Hubble Space Telescope (HST), developed over three decades at a cost of $1.5 billion, was launched from the Kennedy Space Center aboard the space shuttle *Discovery*. The telescope, which orbits the Earth every ninety-five minutes, was expected to detect objects too faint to be seen from Earth, make observations in ultraviolet light (which is absorbed by the atmosphere and thus cannot be successfully observed from Earth), and all in all provide images ten times better than those obtainable with ground-based telescopes.

Hopes were dashed when it turned out that the 94.5-inch main mirror, described in an article published shortly after launch as "the most perfect astronomical reflector ever made," was seriously—and unnecessarily—flawed. The gyroscopes were also defective, as were the solar panels, which jiggle a little when the telescope crosses the border between night and day. The argument has been made that the problem with the solar panels is worse than the problem with the doughnut-shaped mirror, but the mirror has received the bulk of the attention for one simple reason: Its flaw could easily have been caught.

The problem is spherical aberration, meaning that the mirror is unable to focus light rays at a single point because it is too shallow by about $1/50$ of a human hair—a large enough error, compared to a wavelength of light, to throw 85 percent of the light of a star into a fuzzy halo, which leaves only 15 percent in the central image. The flaw seriously incapacitates two of the telescope's instruments—the Wide Field and Planetary Camera (WF/PC), and the Faint Object Camera. The worst of it is, had tests of the primary and secondary mirrors been sufficiently heeded, the discrepancy would have been revealed.

Alas, the telescope was launched in its defective state, with the result that its vision was seriously compromised and its images were blurred and disappointing. Nonetheless, even before December 1993, when spacewalking astronauts made a series of major repairs intended to correct the telescope's myopia, computer technicians and other scientists were able to make adjustments in the data accounting for a great deal of the aberration. Thus, the uncorrected and much-maligned Hubble Space Telescope has been responsible for an impressive number of breakthroughs, including:

* the resolution of Pluto and Charon into separate objects. The existence of Pluto's moon had been deduced, but it had never actually been seen as a distinct celestial object;

* the discovery of jets, waves, a shock front, and an unknown structure that looks like a ladder in the nebula surrounding the massive star Eta Carinae, which was the brightest star in the sky in 1843;

* the discovery of an expanding elliptical ring of shining debris around supernova 1987A;

* the discovery of a dusty X across the nucleus of M51, the Whirlpool Galaxy. It is thought to mark the location of a black hole whose mass is equivalent to a million Suns. The HST has also discovered evidence of black holes in the cores of two other galaxies, M32 and M87;

* the totally unexpected discovery of young globular clusters in the peculiar galaxy NGC 1275 in Perseus—a surprise because all globular clusters were thought to be old;

* the acquisition of images of a distant radio galaxy so young that most of the stars in it are only 500 million years old—about $1/10$ the age of the Sun;

* the discovery of one of the hottest stars ever found, a 360,000° F white dwarf at the heart of the nebula NGC 2440 in the Milky Way;

* the discovery in the peculiar galaxy Arp 220 of six incredibly large, bright knots of star formation thought to have been triggered by the collision of two spiral galaxies;

* a detailed photograph of Gravitational Lens G2237 + 0305. Gravitational lenses were first suggested by Albert Einstein, who predicted that at great distances a star could act as a lens, bending and focusing the light from a more distant star. If the circumstances were right, he thought, the light could be bent into a perfect ring. Fritz Zwicky said that this wasn't likely to happen with stars but it could happen with galaxies. Zwicky was correct, but this particular gravitational lens is nonetheless known as the

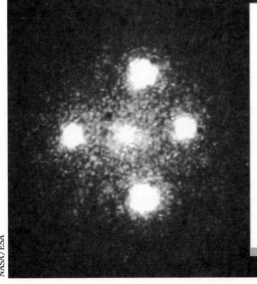

In this gravitational lens, sometimes called the Einstein Cross, the light from a quasar 8 billion light-years away has been bent by the gravity of a much closer galaxy. The light from the quasar, repeated four times, surrounds the light from the galaxy like the petals of a flower.

NASA/ESA

Einstein Cross. The picture on page 297 shows a galaxy 400 million light-years away, surrounded by four images of a more distant quasar;

* finally, the HST has enabled astronomers to zero in on a new, improved Hubble constant by pinpointing one particular type of supernova as a standard candle. These supernova explosions, which occur when one of a pair of white dwarf stars explodes, all reach the same maximum brightness. Plus, they are visible at distances literally 1,000 times farther than those at which one can detect Cepheid variables. As a distance indicator, they could be terrifically useful—but only if the absolute brightness of at least one could be accurately determined.

The best way to do that would be to discover reliable Cepheid variables in the same galaxy where this kind of supernova had been recorded. Cepheid variables are unsurpassed in usefulness because once their rate of pulsation has been determined, their absolute brightness is known. Then, by comparing their absolute brightness to their apparent brightness, scientists can figure out their distance. So the HST was directed toward IC 4182, a faint spiral galaxy 16 million light-years away that in 1937 was the site of precisely the right type of supernova. And sure enough: Cepheids were detected, enabling scientists to determine the distance and hence the absolute luminosity of the explosion.

Using this information, a group of scientists that includes Abhijit Saha, Allan Sandage, and Gustav Tammann announced that the new, improved Hubble constant is between 30 and 60 km/sec/Mpc, with the likeliest number being the one right in the middle: 45, which makes the universe about 15 billion years old.

Needless to say, that figure has not been unanimously accepted. Some scientists point out that the figures could be misleading, for IC 4182 may be filled with dust, which would make its Cepheids appear dimmer, and hence more distant, than they actually are. Consensus has not been reached, and the age of the universe is still unknown.

●

A Garden of Galaxies

Stolid indeed is the student of galaxies who has felt no sense whatever of dislocation, disenfranchisement or vertigo at the sight of them.

—Timothy Ferris

lthough the existence of galaxies other than our own was suspected for centuries, it did not become an incontrovertible fact until Edwin Hubble, using the 100-inch telescope at Mount Wilson, took a picture of the Andromeda Nebula so detailed that he was able to detect in it the pulsating Cepheid variables that allow astronomers to measure magnitude and distance. Even with the 100-inch telescope, these bright stars looked extremely dim, which indicated that the Andromeda Nebula, as it was then known, was far, far away—800,000 light-years away, in Hubble's estimation. At that distance, it had to be neither a nebula nor a cluster of stars caught in a lustrous, foggy oval, but another galaxy altogether, a brilliant wheel of stars much like the Milky Way and entirely outside it.

Today, this discovery seems almost poignant, for we now know that galaxies are as numerous as grains of sand on a beach, so commonplace that comet seekers Carolyn and Gene Shoemaker, quoted in Richard Preston's *First Light,* dismiss them with a phrase usually reserved for asteroids: "the vermin of the skies."

Hubble classified galaxies into four categories: elliptical, spiral, barred spiral, and irregular, for those asymmetrical, odd-looking galaxies, often dwarfs, whose original structure had been distorted by the gravitational force of a neighboring system or a passing galaxy.

Elliptical galaxies are made up primarily of old stars with very little

Spiral galaxy NGC 4565, edge-on to Earth.

dust or gas. Round or oval (depending upon how they're tilted relative to the Earth), they look like glowing Christmas-tree ornaments.

Spiral galaxies resemble rotating pinwheels of stars. Seen from the side, edge on, a spiral galaxy looks like a platter of dust and stars with a glowing bulge in the middle. Viewed head-on, a spiral galaxy displays spiral arms, where young stars are found, coiling around a blazing horde of old stars. On the distant outskirts of the galaxy, surrounding both the nucleus and the spiral arms, is a halo of globular clusters, a spattering of old stars, and an invisible cloud of dark matter. Small companion galaxies sometimes cling to the edges of the galaxy, while in the middle of the nucleus, it is believed, many galaxies harbor a black hole.

Barred spirals—a group that probably includes the Milky Way—resemble spirals except that the nuclei are crossed by a thick bar of stars.

Both barred and unbarred spirals rotate at an epochal pace; the Sun, spinning at the rate of 500,000 mph, will take about 200 million years to circle the center of the galaxy.

In addition, there are variable galaxies and violent, hyperactive spirals called Seyfert galaxies; luminous supergiant elliptical galaxies, found in the centers of galactic clusters and known for their tendency to cannibalize smaller galaxies; ring galaxies, which seem to lack a nucleus; polar ring galaxies, in which a captured circle of young stars orbits perpendicular to an old stellar disk; and twisted starry ribbons, the remnants of colliding galaxies.

Among the notable galaxies are:

* Peculiar galaxy NGC (for *New General Catalogue*) 5128, also known as Centaurus A, which has twin lobes of radio waves spurting out in opposite directions;

A tattered band of material crosses the luminous ellipse of the peculiar galaxy Centaurus A (NGC 5128).

National Optical Astronomy Observatories

* M87, a powerful source of X-rays and radio waves, which is ejecting a jet of hot plasma 5,000 light-years long, and probably holds in its center a black hole with the mass of 2 billion Suns;

* Peculiar galaxy Arp 220, which registers 95 percent of its energy in the infrared part of the spectrum and was possibly created when two giant spiral galaxies collided. Arp 220 is giving birth to new stars at such a prodigious rate that someday supernovae in its nucleus will explode like a string of firecrackers;

M87 (NGC 4486), a giant elliptical galaxy in Virgo. Its enormous jet is about 5,000 light-years long.

National Optical Astronomy Observatories

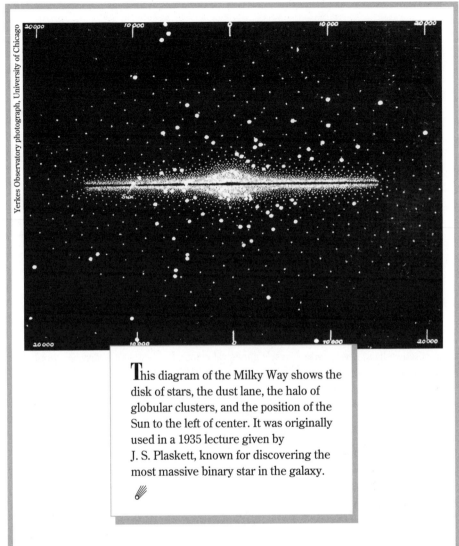

Yerkes Observatory photograph, University of Chicago

This diagram of the Milky Way shows the disk of stars, the dust lane, the halo of globular clusters, and the position of the Sun to the left of center. It was originally used in a 1935 lecture given by J. S. Plaskett, known for discovering the most massive binary star in the galaxy.

* M82, a starburst galaxy that seems to be exploding, possibly as a reaction to the passing of the bigger, brawnier spiral galaxy M81;

* Our own Milky Way Galaxy, a barred spiral approximately 80,000 light-years across. Its large, football-shaped central bulge holds billions of old stars and possibly a black hole. Astronomers have identified four spiral arms unwinding from the nucleus. Each arm is named after the dominant constellation in the part of the sky where it is most visible: Perseus, Centaurus, Sagittarius, and Orion, where our solar system is located, about 28,000 light-years from the center.

M51: THE WHIRLPOOL GALAXY

•

On October 13, 1773, Charles Messier, in the midst of compiling his catalogue of fuzzy objects that might be mistaken for comets, stumbled upon his fifty-first blur just below the tip of the handle in the Big Dipper. After he ascertained that M51 was not a comet, Messier lost interest in it.

John Herschel (1792–1871), the son of William Herschel, paid closer attention. He saw what he thought was a ring of stars with "a real physical resemblance and strong analogy of structure" to the Milky Way. But his telescope wasn't strong enough for him to see much more than that.

Lord Rosse (1800–67), an Irish nobleman who left the British Parliament in 1834 to devote himself to making telescopes, also observed M51. When he looked at it through a telescope with a 36-inch mirror, he saw nothing interesting. But in the 1840s he built a more powerful instrument, a telescope so large it was known as Leviathan. An awkward and dangerous device with a 72-inch mirror and a 50-foot wooden tube wide enough for a person to enter, it was suspended between two ivied, turreted walls at Birr Castle, Rosse's ancestral estate, by means of staircases, hooks and chains, and a large pulley. Using it required a staff of four.

Examining M51 with this reflector, Lord Rosse saw not just a ring of stars but a distinct spiral pattern—the first to be discovered. In a drawing he made of M51—or the Whirlpool Galaxy, as it is sometimes known—the spiral arms are clearly visible, as is a small, indistinct blob.

At first, astronomers thought M51 might be a solar system in formation. But in the 1920s, when Andromeda, which has a spiral structure, was shown to be a separate galaxy, M51 was also placed in that category. It is about the size of Andromeda—bigger than the Milky Way—and is located about 35 million light-years away. In 1992, pictures taken by the Hubble Space Telescope revealed a thick *X* across its nucleus. Scientists believe the *X* consists of two intersecting dust rings surrounding and feeding a black hole. A jet of bubbling hot gas and radio waves shooting out from the center suggests that the black hole at M51's heart is brighter and more energetic than the very weak black hole thought to inhabit our own galaxy.

As for the blob observed by Lord Rosse, it is a satellite galaxy, NGC 5195, whose original shape, whatever it may have been, has been distorted by its interaction with the more powerful M51. That collision took place millions of years ago; NGC 5195 is now careening away from M51.

National Optical Astronomy Observatories

The Whirlpool Galaxy (NGC 5194) and its small companion (NGC 5195).

The Mythology of the Milky Way

City dwellers on vacation are often dazzled by the Milky Way—spectacularly visible, it turns out, in Capri or Arizona. But even a few hours from the heart of New York City or Los Angeles, you can see the Milky Way lifting across the sky. Those powdery bands are nothing less than the white wheel of our galaxy, viewed from the inside. The reality of the Milky Way—a spinning pinwheel of stars at the unseen center of which may lie an omnivorous black hole—is so dramatic that mythology pales. But then, mythology isn't about the spiral arms of the galaxy or the central bulge. It's about us.

Consider the Greek story of the Milky Way's creation. It seems that Zeus and his mortal lover, Alcmene, had a child whom they named Hercules. Because the baby was mortal, Zeus placed it near his long-suffering wife, Hera, hoping that the baby might nurse from her breasts and thus become immortal. When she pushed the child away, her milk spurted across the sky and formed a milky circle—*galaxias kyklos,* from the word for milk, *gala.* Our word "galaxy" comes from the same source. The Romans called it the *via lactea*—the Milky Way. The Egyptians thought of the Milky Way as a celestial Nile streaming from the udders of the cow-headed moon god, Hathor. And although the Chinese, who saw the Milky Way as a river leading to the land of peaches, did not connect it with milk, the story they told about it did include a cow.

The cow told a cowherd named Kien-niou that if he went to a meadow where seven fairies were bathing and stole the dress of one of them, that fairy would become his wife. He went to the meadow, stole one of the dresses, and waited. When the fairies emerged from the water, six of them got into their clothes and returned to heaven. But the owner of the stolen dress, a weaver named Tchi-niu, remained behind. She and Kien-niou married, and right away, the cow grew sick. It instructed Kien-niou that after it died, he should make its hide into a sack, fill it with sand, and carry it with him, along with the cow's golden nose ring. Kien-niou obeyed.

Three years passed. A boy and a girl were born to the couple, but Tchi-niu still wanted her dress. One day, after Kien-niou inadvertently told her where it was, she took it back and ascended to her celestial home. Desolate, Kien-niou grabbed the children and, with the help of the magic cowhide, followed her.

When Tchi-niu saw him, she took a golden hairpin and drew a line in the sky. It turned into a river, which became one branch of the Milky Way. Kien-niou tossed the sand from the cowhide across it; she drew a longer line, which became the other branch of the celestial river. Kien-niou threw the ring at her; Tchi-niu fired back with her weaving shuttle. Finally the king of the gods ordered that the hostilities cease and the two remain apart.

In a friendlier version of this tale, the cowherd follows his lover to heaven, where they cavort all day. This distresses the heavenly rulers, who create the Milky Way as a means of keeping the two apart so that Tchi-niu, the weaver, can do her work.

In either case, the lovers are separated by the Milky Way. Both ver-

The Milky Way and the stars of Sagittarius.

National Optical Astronomy Observatories

sions of the story offer a small reprieve. On the seventh day of the seventh month (which, in the Chinese calendar, falls in August), if the skies are clear, the magpies, black-and-white birds streaked in a fashion faintly reminiscent of the Milky Way, fly across the river and form a bridge the lovers can cross. If it rains, their meeting has to be postponed for another year.

The cowherd in this story is the star Altair in the constellation Aquila the Eagle. The weaver is Vega in the constellation Lyra (and the children are two smaller stars). Between Altair and Vega, two bright stars in the summer triangle, runs the Milky Way.

In many cultures, the Milky Way was perceived as a road. In Scandinavia, it was the pathway to Valhalla. In eastern Europe, it was a road covered with straw that had been stolen by, depending on who's telling the story, an Armenian fire god, a Persian thief, or a band of Hungarian gypsies. In Britain, it was called Watling Street, after an early Roman road that ran south from London to Dover. To the Cherokee, it was a trail of cornmeal, and the Oglala Sioux saw it as the route followed by ghosts on their way to judgment: After passing the campfires of the dead scattered along the path, the ghosts meet an old woman who decides their fate, either offering them entrance into the other world or pushing them over a cliff and back to Earth.

To the Navajo, the Milky Way was another example of Coyote's mischief. This story begins with Black God, the fire god of the Navajo, who created the stars. As a demonstration of power, he tied the cluster of stars called Dilyehe to his ankle and stamped his foot. Each time, Dilyehe—which we know as the Pleiades—jumped to another part of his body. From his ankle, it bounced to his knee, his hip, his shoulder, and finally, the left side of his face, where it stayed.

Then Black God took a handful of crystals and set them in the sky, forming the constellations. Among them were the Man with his Feet Ajar, the Horned Rattler, the Bear, Thunder, the Revolving Male (Ursa Major and the North Star), the Revolving Female, Rabbit Tracks, and the First Big One (Scorpius). With the constellations in place, Black God added an "igniter" star so the other stars would shine. Finally, he sprinkled the leftover crystal chips across the sky, thereby creating the Milky Way. Everything had been accomplished in an orderly fashion.

The time was therefore ripe for Coyote. He blew the leftover crystal dust across the sky, obliterating Black God's neat cartography. The difference between the stars set into place by Black God and those strewn across the firmament by Coyote is that Black God's stars have names but Coyote's do not—with one exception. After adding countless, nameless stars to the cosmos, Coyote discovered that he had one crystal left. He placed it carefully. It is known as the Monthless Star or the Coyote Star, and although its identity is not certain, it is perhaps the red star Antares, located in the First Big One, opposite Dilyehe.

Clusters, Superclusters, and Voids

Stars aren't scattered uniformly throughout space, and neither are galaxies. Well before their true nature was recognized, when galaxies were still considered to be nebulae, scientists recognized their distribution into clusters and superclusters. In 1785, William Herschel commented on a "remarkable collection of many hundreds of nebulae which are to be seen in what I have called the nebulous stratum of Coma Berenices." Many of those nebulae were members of the Coma Super-

National Optical Astronomy Observatories

Some galaxies in the Virgo Cluster. Among the many spiral galaxies are two bright elliptical galaxies: M84 at the left, and M86 near the center.

cluster—a part of the sky so rich with galaxies that in 1957 Fritz Zwicky catalogued close to 30,000 galaxies in the area.

It made sense that if galaxies were collected into certain regions, then other parts of the universe must be more or less empty. Still, it was a surprise in 1981 when scientists discovered what seemed to be a void in Boötes the Herdsman. Another surprise came about when Margaret Geller and John Huchra, both of Harvard and the Smithsonian Astrophysical Observatory, and graduate student Valerie de Lapparent made a three-dimensional map of a tiny bite of the universe. The very first slice revealed astonishing structure, for it showed large voids bordered by thin sheets of galaxies running across the survey and gathered into what looks unmistakably like a stick figure. Adjacent slices confirmed the structure, showing a long row of galaxies—the Great Wall—across the figure. ("He's also called the Harvard stick man," Geller says. "That's because if you have high enough resolution you can see that he wears one of those t-shirts you can buy around the Square.")

> **W**e all learn in school that Ptolemy was this idiot who thought the Earth was the center of the universe. In my opinion that was a minor idea of Ptolemy's compared to the idea that maps are useful as scale representations of physical systems; that idea underlies much of modern science. Today, we map the human genome. We make a map of the arrangement of atoms in molecules. We learn about DNA. We make maps of the distribution of resources on Earth. We map the distribution of species in the rain forest. We map the universe. All of these things are scale representations of physical systems. The first step toward our understanding that science is making the model.
>
> —Margaret Geller

The thin arcs of galaxies and the large voids also reminded Geller of something else. "It occurred to me," she recalls, "that if you took a thin slice through bubbles in your kitchen sink and the thickness of the slice was small compared to the diameter of the bubbles, you would see the foam, the bubbles, surrounding empty regions. So we suggested that galaxies are on two-dimensional surfaces and form a bubble-like pattern. They are the largest known patterns in nature."

The sturdiest, most massive forms, it turns out, resemble the most ephemeral. How things got that way is unknown. Several scenarios have been suggested to explain the formation of clusters. According to the so-called bottom-up theory, separate clouds of gas and dust gradually gathered together and condensed into galaxies.

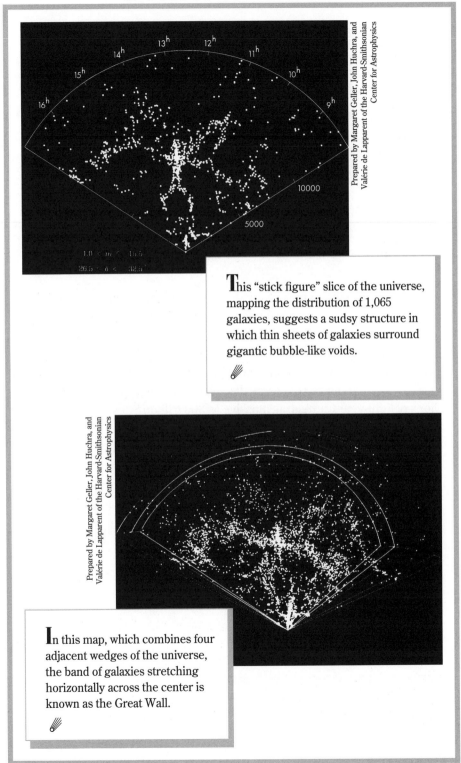

Prepared by Margaret Geller, John Huchra, and
Valérie de Lapparent of the Harvard-Smithsonian
Center for Astrophysics

This "stick figure" slice of the universe, mapping the distribution of 1,065 galaxies, suggests a sudsy structure in which thin sheets of galaxies surround gigantic bubble-like voids.

Prepared by Margaret Geller, John Huchra, and
Valérie de Lapparent of the Harvard-Smithsonian
Center for Astrophysics

In this map, which combines four adjacent wedges of the universe, the band of galaxies stretching horizontally across the center is known as the Great Wall.

Hannah Berman

Galaxy NGC 4622.

According to the opposing top-down theory, things started with a primordial gas pancake colossal enough to account for an entire cluster of galaxies. As the eons rolled by, parts of it broke away, forming separate galaxies within which still smaller condensations of matter gave birth to individual stars. Unconfirmed evidence supporting this theory appeared in 1991, when radio astronomers discovered a flat cloud of hydrogen millions of light-years across, as massive as 300 trillion Suns, and so close to the edge of the observable universe that its light has been traveling toward us for over 13 billion years (assuming the universe is roughly 15 billion years old).

Or it may be that the sudsy structure of the universe was determined in an unknown manner by the ever-mysterious dark matter that, whatever it may be, can cause the visible universe to obey its commands by the simple application of the law of gravity.

✧

THE LOCAL GROUP

• ⤳

Imagine the Milky Way at the center of a ball of space 1,000 kiloparsecs—3,260,000 light-years—in every direction. Within that sphere over two dozen galaxies constitute the Local Group, our own sparse cluster of galaxies. (The Coma Cluster, in contrast, has 1,300 major galaxies.) Among the members of the Local Group are:

✳ M31, the Andromeda Galaxy. About 130,000 light-years across, it is the largest galaxy in the Local Group and has over half a dozen satellite galaxies, including elliptical galaxy M32;

✳ the Milky Way, the second-largest galaxy in the Local Group. Its satellites include the Large and Small Magellanic Clouds, two irregular galaxies seen in the southern skies and linked to the Milky Way by a bridge of hydrogen called the Magellanic Stream;

✳ M33, the third-largest spiral in the Local Group, also known as the Pinwheel Galaxy;

✳ NGC 6822, a small irregular in Sagittarius with a diameter less than $1/10$ that of the Milky Way. Discovered in 1884 by E. E. Barnard, NGC 6822 is actively giving birth to new stars;

✳ irregular galaxies in Pegasus, Cetus, and Aquarius, and tiny dwarf ellipticals in Draco, Ursa Minor, Carina, Leo, and Sculptor, which houses an aggregate of stars so small and unimpressive that it has no nucleus, no spiral arms, and no interstellar gas clouds. The elliptical galaxy Maffei I and its spiral companion Maffei II, discovered by the Italian astronomer Paolo Maffei in 1968 and thought to be the most distant members of the Local Group, are now considered too distant to be included in the local census. But the count is not complete, and the likelihood is strong that other dwarf galaxies, sequestered within the confines of the Local Group, are our undiscovered neighbors.

The Big Bang

In the beginning the Universe was created. This has made a lot of people very angry and been widely regarded as a bad move.

—Douglas Adams,
The Restaurant at the End of the Universe

The Big Bang basically says once the universe was very hot and very dense, and now it's less so.

—Margaret Geller

lthough the general theory of relativity predicts a dynamic, changing universe, Einstein couldn't bear the thought. So strong was his belief in—or hope for—a static, eternal universe, and so disturbed was he by the idea that the universe might be expanding, that in 1917 he made what he later called "the greatest blunder of my life." He added to his equations, which predicted a changing universe, a cosmological constant whose purpose was to repel the force of gravity and stop the universe dead in its tracks.

It was too late. The concept of a dynamic universe was in the air. The Dutch astronomer Willem de Sitter came up with the dizzying possibility of an expanding universe that lacked matter entirely and consisted only of space, and the Russian meteorologist Alexander A. Friedmann (1888–1925), after recognizing an error Einstein had made, discovered several solutions to Einstein's equations. Friedmann described three possible futures, each of which included a period of expansion. In an open universe (sometimes known as "the Big Bore"), the expansion con-

Abbé Georges Lemaître (center), with Albert Einstein and the physicist Robert A. Millikan, president of the California Institute of Technology.

tinues for all eternity; in a closed universe, the expansion is followed by a turnaround, a contraction, and a horrifying final moment known as "the Big Crunch"; and in a flat universe, the expansion will slow but never cease, causing the universe to reach what is essentially a plateau.

After Friedmann wrote up his findings, he sent the paper to Einstein. Several months passed before he received a curt note from Einstein admitting that Friedmann's work was publishable. But it made Einstein uncomfortable, and after the article came out in the journal *Zeitschrift für Physick,* he wrote to the editor to point out a mistake he thought Friedmann had made. Friedmann responded: No, he had not made an error. Einstein conceded. Three years after the publication of the first of these revolutionary papers, Friedmann caught a chill while sailing above the Earth in a weather balloon and died shortly thereafter at the age of thirty-seven.

Another player came on stage. Georges Lemaître (1894–1966) was a man of God and a man of science in approximately equal measure. An engineer who had served as an artillery officer in the Belgian military during World War I, he was, according to the English astronomer Sir William McCrae, "a man of robust vigour. He appeared to be the complete extrovert; he had a stentorian laugh, which was readily provoked. To some extent, however, I think all that appearance was a protection for a sensitive personality. He was a man of courage." When the war was over, Lemaître studied mathematics and physics at the University of Louvain, got his Ph.D., and enrolled in a seminary. After being ordained as a Jesuit priest in 1923, he returned to science. He worked at the Harvard Observatory and attended a lecture by Edwin Hubble, which led him to realize that if the universe was expanding, then in the past everything must have been closer together. If you ran the film of the universe backward, he mused, galaxies would move closer and closer together until at last they would arrive at a single place. Eventually, all the matter in the universe would be packed into that spot. Lemaître thought of that spot—which he imagined to be the size of the orbit of the Earth—as the primeval atom or cosmic egg, but as his theory developed, the point shrunk until its size was essentially zero. When it exploded—an event he called "the big noise"— it disintegrated in a manner reminiscent of radioactive decay, decomposing into galaxies, stars, and cosmic rays. "The evolution of the world could be compared to a display of fireworks just ended—some few red wisps, ashes, and smoke," he wrote. "Standing on a well-cooled cinder we see the slow fading of the suns and we try to recall the vanished brilliance of the origin of the worlds."

> **E**very cubic inch of space is a miracle.
>
> —Walt Whitman

Lemaître—the son of a man who accidentally blew up his glassmaking factory—knew what a surprising notion this was. In his book *The Primeval Atom*, he wrote, "I shall certainly not pretend that this hypothesis of the primeval atom is yet proved, and I would be very happy if it has not appeared to you to be either absurd or unlikely."

Although many of the details of Lemaître's theories are incorrect— radioactive decay, for instance, is not a mechanism of expansion—his scenario explained the expansion of the universe more clearly than any other. When Hubble proved that the expansion of the universe was not a theory but a fact, Lemaître's ideas came into the spotlight, thanks to Sir Arthur Eddington, who reprinted Lemaître's paper in 1930 in the *Monthly Notices of the Royal Astronomical Society*. The next year, Einstein withdrew the cosmological constant as "theoretically unsatisfactory anyway." (He would probably have savored the fact that physicists today have become conscious of certain inconsistencies between theory and observation that can be addressed, it turns out, by none other than the cosmological constant.) Lemaître received the prize that even a Nobel cannot bequeath: He is known as the Father of the Big Bang.

The Big Bang scenario as it is understood today begins with a point of infinite density, infinitely high temperature, and a single, unified force. Something causes it to begin expanding. That expansion is a cataclysm that happens everywhere, including here. 10^{-43} second later, a moment known as Planck Time (after the physicist Max Planck), gravity "freezes out" from the original Ur-force, thereby establishing its individual presence, and particles and anti-particles tumble forth in glorious profusion. Within the next incomprehensibly tiny sliver of a second, the strong nuclear force, which binds particles together, asserts itself, and the universe inflates, doubling and redoubling until it swells from the size of a particle into a seething mass the size of a melon. The weak nuclear force, which has to do with radioactive decay, and the electromagnetic force become separate. Quarks combine to form protons and neutrons. Matter and anti-matter, in a suicidal frenzy, meet, bathing the universe in radiation. The universe is one second old.

Within three minutes, atomic nuclei begin to form.

Five hundred thousand years later, electrons swing into orbit around nuclei and form atoms.

A few hundred million years after the explosion, clouds of atoms collapse to form galaxies, and the stars light up.

And through it all, space continues to expand. It's not that galaxies are racing away from one another. Space itself is expanding, and the galaxies are moving along with it.

Although the Big Bang is now the most widely accepted theory, some scientists campaigned vigorously against it. Chief among them was the English astronomer Fred Hoyle, who threw his support behind the Steady State theory, which he developed with Hermann Bondi and Thomas Gold. "It was Gold's idea that there might be a process of continual creation that allowed a steady state to go on," Bondi recalled in 1988. "Fred and I said: 'Ach, we will disprove this before dinner.' Dinner was a little late that night, and before very long we all saw that this was a perfectly possible solution to the question." The theory posited a universe without beginning and without end. As the universe expanded—an incontrovertible fact proven by Hubble in 1929—it maintained its density by continuously creating new matter. Hoyle argued that such an event was not nearly as unlikely as it sounded. "It turns out that the required rate is very slow, amounting to about one atom per century for each unit of volume corresponding to that of the largest man-made building," he wrote. "So it is not at all difficult to understand why the process, if it really occurs, has not been detected in the terrestrial laboratory."

Ironically, it was Hoyle who, in a dismissive remark on a BBC radio program in the 1940s, referred to the theory as "the Big Bang." The name caught on, though Hoyle himself, in a popular coffee-table book published in 1962, conspicuously avoided the term he coined.

In many ways it strains credulity to believe that the entire universe, along with everything in it, was born in the explosion of a point so small as to be dimensionless. But that is the most widely accepted model of creation. And proof exists. Much of it came from one of the theory's earliest supporters, George Gamow (1904–68). The irrepressible son of Russian schoolteachers (one of his father's students was Leon Trotsky), Gamow came to the United States in 1934 and established himself as a physicist and as the author (and illustrator) of many popular science books. Gamow is associated with two pieces of evidence that support the Big Bang.

The first has to do with the abundance of the elements in the universe. Gamow and his associate Ralph Alpher showed that nuclear reactions in the primordial fireball would create hydrogen, deuterium (or heavy hydrogen), and helium—which together constitute 99 percent of everything—in proportions close to those seen in actual observations. The paper describing these chemical abundances promised to be a milestone, and Gamow, a known jokester, wanted to publicize it. Noting that Alpher's name and his own sounded like the Greek letters alpha and gamma, he appended Hans Bethe's name to the list of authors so that when the paper was published in the *Physical Review* on April 1, 1948 (a date Gamow relished), three authors were listed: Alpher, Bethe, and Gamow.

Gamow also thought that the heavy elements—not just hydrogen and helium—must have been created during the Big Bang from the substance he (and Aristotle) called "ylem." He was wrong about this—heavy elements are created in supernova explosions, as Hoyle was to show—but he was right about something else. Along with his colleagues Ralph Alpher and Robert Herman, Gamow believed that the explosion would flood the universe with heat. As the eons passed, the temperature would fall. The universe today should be awash in the remainder of that heat: low-temperature microwave background radiation.

This prediction, like Friedmann's original models, sank into obscurity until the mid-1960s, when Princeton physicist Robert Dicke and his Ph.D. student Jim Peebles began thinking along similar lines. Then, in 1965, Arno Penzias and Robert W. Wilson, two scientists working for Bell Laboratories in New Jersey, were monitoring a twenty-foot radio receiver whose function was to relay telephone calls to communication satellites. To their annoyance, they kept getting a low, persistent hum. No matter where they pointed the movable antenna, they couldn't eliminate the noise. They thought something was wrong with the antenna, and one by one they eliminated the suspected culprits. It wasn't a temperature problem. It wasn't dirt. It wasn't the two pigeons nesting in the radio dish. It wasn't the "white diaelectric" contributed by the pigeons. Mystified, Penzias called Dicke.

The hum was finally identified. It was the very radiation

BEFORE THE BIG BANG

•➔

The universe came into being in a Big Bang, before which, Einstein's theory instructs us, there was no before.

—John Wheeler

✳

There is no way that one can determine what happened before the Big Bang from a knowledge of events after the Big Bang. This means that the existence or nonexistence of events before the Big Bang is purely metaphysical.

—Stephen Hawking

✳

Naturally, we were all there,—old Qfwfq said,—where else could we have been? Nobody knew then that there could be space. Or time either: what use did we have for time, packed in there like sardines?

I say "packed like sardines," using a literary image: in reality there wasn't even space to pack us into. Every point of each of us coincided with every point of each of the others in a single point, which was where we all were. In fact, we didn't even bother one another, except for personality differences, because when space doesn't exist, having somebody unpleasant like Mr. Pbert Pberd underfoot all the time is the most irritating thing.

—Italo Calvino, Cosmicomics

✳

It may be that the universe is just one of those things that happens from time to time.

—Edward Tryon

Gamow and his colleagues had predicted—the dying light of the fireball at the beginning of time. Its temperature was everywhere the same: –454°F (3° on the Kelvin scale), close to the predicted value. Its ubiquity, the discovery of which won a Nobel Prize for Penzias and Wilson, remains the strongest single proof of the Big Bang (even though Wilson felt himself philosophically aligned with the Steady State theory). News of the discovery reached Georges Lemaître on his deathbed.

A quarter century later, in November 1989, the Cosmic Background Explorer was launched. It surveyed the skies, measuring the weak microwave radiation—the background to the universe—that is everywhere a remnant of the creation. Initial results seemed to confirm that no matter where you look, that radiation is the same: no lumps, no bumps, no quantum fluctuations. This discovery fit in nicely with the idea of the Big Bang.

And yet it also caused many people to question the Big Bang. Because although the universe might have been smooth as silk in its early micro-moments, it isn't that way now. Here, there are galaxies; here, there are none. Matter collects in streams and bubbles. The universe is a lumpy stew. Opponents of the Big Bang claimed that any theory of creation that failed to account for the existence of galaxies and stars wasn't much of a theory. It didn't sound like an unreasonable objection.

And then in April 1992, George Smoot of the University of California at Berkeley announced the discovery of subtle temperature fluctuations in the microwave radiation. It wasn't entirely even after all, although the differences were minute: six parts in a million. But over time, these tiny ripples would grow, and the differences would become magnified. According to Smoot, these ripples, which were as long as 10 billion light-years, eventually coalesced into the large-scale foam of the universe—the bubbles made up of galaxies, clusters of galaxies, clusters of clusters, and empty space.

This discovery created a flurry of headlines and *Nightline* appearances. Reporters used terms like "holy grail," and lingering doubts about the Big Bang became even flimsier, for it was now clear that the seeds of structure were embedded in the early smoothness. Thus at this moment there is much reason to believe that once upon a time, all matter, all space, all everything, was compressed into a single, mind-boggling point of pure energy.

✳

Anti-matter

There are two hotels in Djang: the Hotel Windsor and, across the street, the Hotel Anti-Windsor.

—Bruce Chatwin, *The Songlines*

he physicist Paul A. M. Dirac, famous as a man of few words, once had a conversation with a woman who was knitting. Afterward, he found himself thinking about the rhythmic pattern of her knitting needles, and it occurred to him that by varying the way she held the needles, she could produce a mirror-image sister stitch. When he announced his discovery to her, he learned that he had rediscovered purling, without which cabled sweaters and ribbed sweater cuffs would not exist.

The discovery, which Dirac boasted about, was a peculiarly appropriate one for him to make, for the relationship of knitting to purling, techniques that produce equal but opposite stitches, is similar to that of matter to anti-matter, whose existence he posited in 1928 on the basis of an equation that could be solved either positively or negatively. One solution described the electron, a very small subatomic particle with a negative charge; the other solution described a mysterious mirror-image particle—an anti-electron—with the same tiny mass but an opposite charge. It had to exist. Dirac dismissed the possibility that the proton, which has a positive charge, might be the anti-electron because the proton is 1,836 times more massive than the electron. Dirac's unknown particle would have the same mass as the electron, but it would act differently; the physicist George Gamow called these strange particles of anti-matter "donkey electrons."

Dirac's prediction was fulfilled in 1932 by Carl Anderson. Using a cloud chamber with a lead plate, Anderson tracked particles as they crashed into the plate and were deflected from their straight path onto one that curved. The electrons always curved off in the same direction. While examining a photograph of these particle paths, Anderson noticed a few lines that arched in the opposite direction. Those curls were the trails of the positive electrons. Anderson called these anti-electrons positrons.

The discovery—which netted a Nobel Prize for both Dirac and Anderson—was the first of many in the field. Dirac contended that every particle could exist as an anti-particle, and research has supported this prediction. The anti-proton was discovered in 1955, the anti-neutron in 1956. Anti-muons, anti-tauons, anti-neutrinos, and anti-quarks are known to exist, and scientists believe that all matter theoretically has its antithesis.

But if so, a surprisingly small amount of anti-matter has been seen. No one has discovered a single anti-element or anti-galaxy, or seen any sign of the powerful explosions that should occur when matter rubs up against anti-matter. Everything seems to be made of matter—a curious situation, because particles and anti-particles are created in equal measure.

Not only that, they are destroyed in equal measure. No sooner do they meet than they annihilate each other in a burst of energy and gamma rays, leading scientists to repeat the question asked in 1714 by Gottfried Leibniz, Newton's great rival: "Why is there something rather than nothing?"

PERILS OF MODERN LIVING

•‿

Well up above the tropostrata
There is a region stark and stellar
Where on a streak of antimatter,
Lived Dr. Edward Anti-Teller.

Remote from Fusion's origin,
He lived unguessed and unawares
With his antikith and kin,
And kept his macassars on his chairs.

One morning, idling by the sea,
He spied a tin of monstrous girth
That bore three letters: A. E. C.
Out stepped a visitor from Earth.

Then, shouting gladly o'er the sands,
Met two who in their alien ways
Were like as lentils. Their right hands
Clasped, and the rest was gamma rays.

—Howard Furth

One answer is that matter and anti-matter were not created equal. For every 100,000,000,000 antiprotons, there were 100,000,000,001 protons. Even mutual annihilation could not then prevent the triumph of matter.

Alternatively, perhaps during the Big Bang, particles and anti-particles were torn away from each other, preventing their mutual suicide and requiring them to exist on their own. Maybe anti-planets, anti-stars, and anti-galaxies are out there right now, living their wrong-way lives. (Richard Feynman said that anti-particles are "ordinary particles going backward in time.") You can't tell by looking because optical light is emitted by matter and anti-matter alike. But an anti-galaxy colliding with a regular galaxy would explode in an extravaganza of gamma rays.

No evidence has been found for such an explosion. Anti-matter can be created in a laboratory, but if it comes into contact with matter, both particles detonate, wiping each other out.

Yet anti-matter *does* exist. Somewhere near the center of our galaxy, an unknown object, possibly a neutron star or a black hole, is emitting positrons. And a balloon experiment conducted in 1979 detected anti-protons high above Texas. They might have been created through the smashing together of particles in space. Or they might have been created in the Big Bang. NASA intends to send an anti-matter collector into space to determine exactly how much anti-stuff is out there. They will probably find very little, for everything suggests that this is a universe in which matter predominates. But what if there are, say, anti-galaxies? Or what if scientists were to discover one day evidence of a few anti-neutrinos in a vat of dry-cleaning fluid here on planet Earth? That might imply that somewhere an anti-star had erupted, releasing into the universe rivers of anti-neutrinos and proving the truth of Thomas Hardy's observation: "While many things are too strange to be believed, nothing is too strange to have happened."

Dark Matter

The idea that the universe is filled with invisible matter is not a new one. In the late eighteenth century, John Michell and Pierre Simon de Laplace explored the possibility that some objects might be so massive that light could not escape. Their ideas, developed almost two centuries before the term "black hole" was coined, belong to a time that the astronomer Virginia Trimble calls the "Neolithic" period in the history of dark matter.

Fritz Zwicky, known for his work with supernovae and galaxies, belongs to Trimble's "Classical" period. In 1933, after studying clusters of galaxies in Coma Berenices, Zwicky calculated how much gravitational force would be needed to keep the galaxies from spinning off in all directions and how much mass was required to create that gravity. The mass required turned out to be many times more than what he could see. Something else—Zwicky called it missing mass but it is more commonly known as dark matter—had to be out there, and in enormous quantities. Zwicky calculated that a minimum of 90 percent of the universe consists of this unknown, invisible substance.

Thus, in a sense, Fritz Zwicky can be said to have deduced the existence of almost everything.

If Zwicky had been a different kind of man, or if the 1930s had been a different kind of decade, this accomplishment might have been more quickly recognized. Despite the contributions of dozens of scientists, however, the question of dark matter was more or less quiescent until 1977, when the astronomer Vera Rubin's surprising research indicated that even within their own borders, galaxies required far more mass than was visible. She discovered this by measuring the rotational speed of gas near the centers of galaxies. It was thought that galaxies would rotate like the solar system, in which the outer planets rotate more slowly than those near the center. Instead, she discovered that the rotational speed in spiral galaxies does not diminish near the edges. In the Milky Way, for instance, stars fly through space at about 150 miles per second, regardless of location. Stars near the center of the galaxy and stars in the trailing fingers of the spiral arms seem to rotate at the same speed.

The reason is that spheres of dark matter surround galaxies like ghostly halos. The galaxy we see, like the yolk in the egg or the eye in the head, is part of something greater than itself. The gas and stars in the outer arms rotate at the same speed as those in the center because, in a sense, they're all in the center. Rubin calculated that spiral galaxies held between five and ten times as much matter as they displayed. Almost everything, it turns out, is invisible and undetectable.

So what *is* dark matter? It cannot be hydrogen gas, which reveals itself by emitting radiation, nor dust, which can be seen. Black dwarfs, the stellar cinders that remain after a white dwarf star stops shining, aren't a possibility at the moment because the universe is too young for them to have been created. Among the other candidates are undetected dwarf galaxies (billions of them) and MACHOs (massive compact halo objects), which include black holes, large planets, and would-be stars called brown dwarfs. A brown dwarf—which the English astronomer Patrick Moore has called "a star which has failed its Common Entrance Examination"— is not hot enough to spark nuclear reactions in the core but may emit enough infrared energy and light to be barely detectable. On the other hand, trillions of brown dwarfs would be needed to account for the missing mass, and so far astronomers have not conclusively discovered even one.

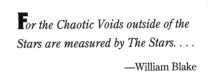

For the Chaotic Voids outside of the Stars are measured by The Stars. . . .
—William Blake

Or maybe dark matter is made of exotic particles. Neutrinos, for instance, exist in such quantity that should they have even a little mass they could account for the gravitational anomalies. But it is not clear whether they have mass, so other particles are being considered. Among them are slowly moving, weakly interacting photinos, gravitinos, and gluinos. Like neutrinos, these WIMPS (weakly interacting massive particles) can pass right through us with impunity, but unlike neutrinos, they have significant mass. Given sufficient numbers, any of these hypothetical particles could be the elusive dark matter. And the same is true for axions, squarks, selectrons, sneutrinos, cosmic strings, and a puzzling, ghostlike substance called "shadow matter" which can be neither felt nor seen but which would be detectable by its "gravitational effects—

like the invisible djinns of the Arabian Night," Nobel Prize winner Abdus Salam suggests.

Dark matter directly determines the fate of the universe. It doesn't matter whether it consists of brown dwarfs and Jupiter-size planets, immense clouds of slightly massive neutrinos arching from one galaxy to another, or any other possibility. "Astronomers are fond of saying that the dark matter could be cold planets, dead stars, bricks, or baseball bats," writes Vera Rubin. "Physicists are fond of saying that it could be billions of mini–black holes, or somewhat fewer maxi–black holes, or, indeed, any one of a number of exotic particles from the zoo of objects permitted by physical theories but never yet observed. Whatever it is—and it could be of more than one type—it must be the major constituent of our universe." The more there is of this mysterious stuff, the more slowly the universe will expand. At the moment, as pervasive as dark matter is, there doesn't seem to be enough of it to stop the expansion. But if dark matter is lurking between clusters and superclusters, if it has seeped into the great intergalactic voids, if in fact over 99 percent of everything is dark matter, the universe could well slow down, turn around, and collapse.

■

Fritz Zwicky

ven in the rarefied domain of science, individuality occasionally overwhelms accomplishment. Take the case of Fritz Zwicky (1898–1974). A polyglot with a name full of fricatives and a personality to match, he was a first-rate scientist with an original, inventive mind (and fifty patents to prove it). Nonetheless, his contributions have sometimes been dwarfed by his personality.

Born in Bulgaria to Swiss parents, Zwicky attended the Federal Institute of Technology in Zurich (the same institute where Einstein, after failing the entrance examination, became a professor), and earned his Ph.D. in 1922 with a thesis about salt crystals. Three years later, he emigrated to the United States and established a lifelong association with the California Institute of Technology. During World War II, he became director of research at the Aerojet Engineering Corporation and was awarded the Freedom Medal by President Harry Truman for his work in jet propulsion. After the war, he worked for the United States government in Germany and Japan, where he was assigned the task of assessing damage caused by the nuclear bombs dropped on Hiroshima and Nagasaki.

Among other scientific accomplishments, Zwicky

* predicted, in collaboration with Walter Baade, the existence of neutron stars. "With all reserve, we advance the view that supernovae represent the transition from ordinary stars into neutron stars," they wrote. More than three decades later, pulsars were discovered, thereby confirming the prediction;

* used photographs taken with an eighteen-inch Schmidt telescope on Palomar Mountain to conduct a highly successful search for supernovae in distant galaxies;

* recognized that galaxies have a tendency to form clusters and discovered clusters in Pegasus, Pisces, Cancer, Coma Berenices, and elsewhere;

* produced a six-volume catalogue of galaxies and clusters of galaxies;

* predicted that galaxies could be used as gravitational lenses that would enable scientists to calculate the mass of even more distant galaxies;

* measured the relationship between light and mass in the Coma Cluster. This led him to discover the mysterious and invisible dark matter, which at minimum accounts for over 90 percent of the mass of the entire universe.

Despite these accomplishments, Zwicky sometimes felt at odds with the scientific establishment, and he thought of himself as a lone wolf. He was "the last of the scientific individualists, a breed that is dying out in an age of teamwork," wrote the Harvard astronomer Cecilia Payne-Gaposchkin in an obituary published in *Sky & Telescope*. "Aggressively original, outspoken to the point of abrasion, he seemed to his contempo-

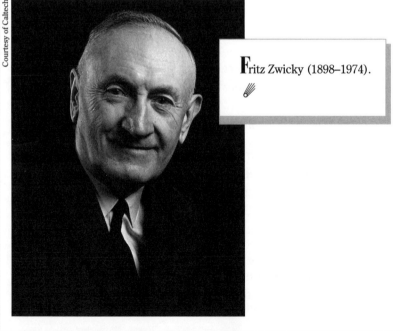

Courtesy of Caltech

Fritz Zwicky (1898–1974).

raries stubbornly opinionated. His ideas were so fertile and his projects so vast that he could have employed all the facilities of a great observatory. Looking back on his rugged determination and his slightly Renaissance flavor, one is reminded of Tycho Brahe: brilliant, opinionated, combative, a superb observer, and a very human person."

If he irritated people, he also inspired loyal friendships. "I found him extraordinarily entertaining and very stimulating and not at all difficult," reports the British-born Caltech astronomer Wallace Sargent. "He had a sense of humor which was a little different from other people's. He deliberately exaggerated statements in order to provoke a reply and often said things which were far more exaggerated than he believed." This was not always obvious to people.

Nor was it widely known that Zwicky was, in Payne-Gaposchkin's words, "one of the kindest of men, with a deep concern for humanity." After World War II, he became actively involved with the Pestalozzi Foundation, which helped to support orphanages. He also organized the Committee for Aid to War-Stricken Scientific Libraries, which distributed astronomical books and journals to decimated libraries and observatories.

Although problems of the most difficult kind still confront us, no determined morphologist will give up hope that eventually we shall achieve the impossible through the morphological approach. And in the meantime, during some leisure hours of solitary contemplation, we shall allow ourselves the luxury of dreaming about a clean world, which does not disgust us, and in which every individual has realized his genius and is happily prepared to recognize the fact that in some way or another everyone of us fellow men is unique, incomparable, and irreplaceable.

—Fritz Zwicky, *Discovery, Invention, Research: Through the Morphological Approach*

But beyond all that, Zwicky was a man with an idea. Just as Johannes Kepler considered his theory of the Platonic solids to be the apex of his work, Zwicky put great faith in a complex, overarching theory called the morphological method. As he made clear in *Morphological Astronomy,* published in 1957, his system was applicable to all fields of human endeavor. Its first goal was "the evolution and completion of a mental *World Image*—which will allow us to visualize and comprehend all of the essential interrelations among physical objects, phenomena, concepts and ideas, as well as to evaluate the human capabilities needed for all future constructive activities." Its principles were universal but required "absolute detachment from all prejudice" and "a deep understanding of the psychological and spiritual traits of individuals and peoples, paying particular attention to the aberrations of the human mind." These Zwicky found especially vexatious. "The correct World Image probably would have been recognized long ago were it not for aberrations of the human mind, inertia of thought and action, and the paralyzing influence of sterile dogmas," he wrote.

On at least one occasion, however, Zwicky admitted that the morphological method might have limitations. Zwicky, who had three daughters, once confided to Wallace Sargent in his lilting Swiss-German accent that the girls were having boyfriend problems.

"Surely the morphological method could do something about that," Sargent suggested.

"There are some things that even the morphological method cannot solve," Zwicky replied.

Fritz Zwicky died in 1974 and is buried in Switzerland. He was provocative, bitter, thoughtful, funny, and so brilliant that, according to one prominent astronomer, "Had he not been such an acerbic character, he would have gone down as the greatest astronomer in the twentieth century."

★

Quasars

he story of quasars is the story of a perceptual shift. Discovered in the early 1960s, they were enigmatic objects that looked like ordinary, faint blue stars but poured forth radio waves. This combination of a starlike appearance with a cascade of radio waves caused them to be labeled "quasi-stellar radio sources," or "quasars" for short. But what was really unusual about a quasar was its spectrum. When Allan Sandage photographed one of them, a sixteenth-magnitude dot of light in Triangulum known as 3C 48 (because it was the forty-eighth object in the *Third Cambridge Catalogue of Radio Sources*), he discovered that it had "the weirdest spectrum I'd ever seen."

Explanations were bandied about. Perhaps quasars were dense stars made up of exotic heavy elements. Maybe they were the radiating corpses of recent supernovae in which electrons had been stripped away, turning familiar elements into something strange. Or maybe quasars weren't stars at all. Sandage noted "the remote possibility that [3C 48] may be a very distant galaxy of stars."

The necessary leap was made by Maarten Schmidt, a Dutch astronomer working at the California Institute of Technology in Pasadena. Schmidt's interest in astronomy began in Groningen, Holland, during the blacked-out nights of World War II. "In the summer of '42—which is a strange thing to say, but that's the summer it was—I visited an uncle who was an amateur astronomer. He showed me his telescope and indicated to me that you could make one," Schmidt remembered. "I used a toilet-paper roll as a tube, and I managed to get an image. And there I had my first telescope."

Twenty years later, Schmidt was using the largest telescope in the world: the 200-inch Hale Telescope at California's Mount Palomar. On December 27, 1962, the eve of his thirty-third birthday, he recorded a spectrum from one of these puzzling objects: 3C 273, a radio source seemingly made up of an ordinary thirteenth-magnitude star in Virgo and a nearby wispy jet shooting away from the star.

"My initial suspicion was that the jet was a peculiar galaxy associated with the radio source and that the star of magnitude thirteen was a foreground object," Schmidt recalled in 1988. He fully expected the star's spectrum to reveal that it was indeed a relatively close, faint, ordinary object.

The spectrum, however, was puzzling. On February 5, 1963, while writing an article about it, Schmidt realized why. The standard emission lines indicating the presence of hydrogen were there all right, but not in their usual positions. Instead, they were shifted way over toward the red end of the spectrum. The implications were stunning because the higher an object's redshift, the faster it is receding from us; and the faster it is receding, the farther away it is.

3C 273's redshift implied that it was speeding away from us at the startling rate of almost 16 percent of the speed of light, which meant that rather than being a foreground object, 3C 273 was incredibly distant—2 billion light-years away. And although it appeared faint, its great distance meant that 3C 273 was actually incredibly luminous, 100 times brighter than an entire galaxy. Stunned, Schmidt spoke with his colleague Jesse Greenstein, who immediately reexamined the spectrum of 3C 48. Greenstein discovered an even larger redshift, indicating that 3C 48 was rocketing toward eternity at 37 percent of the speed of light. It was as if what looked like a nearby speck of light—the glowing tip of a cigarette or a sputtering birthday candle—turned out to be a bonfire blazing on the horizon. The discovery galvanized astronomers. Robert Burnham of *Astronomy* magazine describes it as "the moment when astronomy broke out of its rut and got interesting."

Quasars are the brightest and the most distant objects in the universe by far.* It is an astronomical truism that the more distant something is, the farther back in time and the closer to the Big Bang it is, which means that quasars are also the earliest objects ever found. Since 3C 273 was the brightest quasar, it was probably also the nearest. In which case less luminous quasars might turn out to be even more distant and remote in time—astronomical fossils from a still earlier age. "It was immediately

---

*The brightest quasar on record, as of this writing, is HS 1946 + 7658 in Draco: It shines with the incomprehensible light of 1.5 quadrillion Suns.

clear," Schmidt states, "that you could start studying the history of the universe."

Despite appearances, quasars are not stars, and only about 10 percent are powerful radio sources. Quasars are found in violent young galaxies that harbor in their centers a black hole with the mass of perhaps 100 million Suns. The rampaging black hole pulls gas, dust, and stars into a

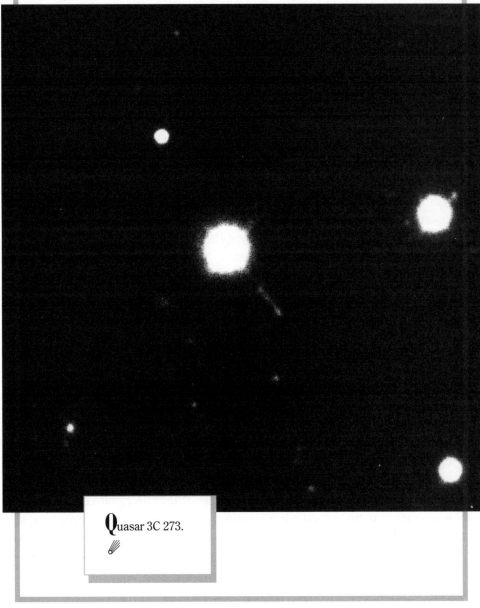

National Optical Astronomy Observatories

Quasar 3C 273.

whirling accretion disk that surrounds the black hole like a gigantic doughnut. As the objects sucked into the black hole spiral to their doom, the temperature rises to hundreds of millions of degrees, releasing a flood of energy that includes light, X-rays, ultraviolet and infrared radiation, possibly gamma rays, and sometimes—but not always—radio waves.

It is possible that quasars are born when two gas-rich galaxies collide. Data from the Infrared Astronomical Satellite (IRAS) suggests that behind the gigantic dust storms whipped up by the collision, some interacting galaxies may be concealing quasars.

Several thousand quasars have been located since the 1960s, each one farther away than the next. As with other explosively violent objects, we don't see them in our own part of the universe because we don't see them in the present. They are denizens of the primeval universe. When astronomers look at a quasar that is, say, 7 billion light-years away, they're seeing light that began its journey to planet Earth long before there was a planet Earth. Quasars have been detected at distances so great, so close to the Big Bang, that they are virtually at the edge of time. Quasars illuminate the space between them and us—between then and now—like cosmic flashlights. When their light finally reaches us, it reveals huge primordial clouds of neutral hydrogen. But beyond the quasars, space is opaque. They represent, Schmidt states, "the edge of discrete objects."

To look at a quasar is to see the universe as it existed billions of years ago—or so goes the standard theory. A contrary view, held most notably by astronomer Halton C. Arp, hypothesizes that quasars may not be as distant as is commonly thought, and that large redshifts may not indicate large distances. In support of this heretical opinion, he has pointed out that many quasars with large redshifts appear to be tied by filaments of light to nearby galaxies with smaller redshifts. A luminous bridge, for instance, connects the seemingly distant quasar Markarian 205 with the relatively nearby galaxy NGC 4319. Even 3C 273, the brightest and most famous quasar, seems to be interacting with a giant cloud of hydrogen—even though the redshift of 3C 273 indicates that it is 2 billion light-years away, while light from the cloud started traveling toward the Earth around the time when the dinosaurs breathed their last, a mere 65 million years ago. So something is wrong. One suggestion is that quasars display a huge redshift because they have been ejected with preternatural speed from their home galaxies. The problem with that presumption is that if quasars were rocketing out of galaxies, a few should have been ejected in our direction. But no quasars have been found with blue shifts, suggesting that this model is inaccurate.

Or maybe 3C 273 and the cloud, like other quasars and galaxies, only appear to be connected, much as a star seen through a window and a fleck of paint on the glass might be accidentally aligned. Since many instances are known in which quasars are grouped with galaxies whose redshift they share, the orthodox view continues to be that quasars really are small, blindingly shiny relics from when the universe was very young. As Maarten Schmidt points out, "Mature, nearby galaxies, whose light takes less time to reach us, don't have quasars. But the ancient skies glittered with quasars still visible in galaxies far, far away."

❋

The Life and Work
of Stephen Hawking

n ebullient media star grappling with a deadly disease, Stephen Hawking seems to have inherited Einstein's mantle of fame and reputation for genius. The man is lionized—perhaps, some scientists suggest, out of all proportion. His book *A Brief History of Time,* a major best-seller probably more purchased than read, became, against all reason, a movie, and it would be pleasant to think that cosmology has become required reading for cocktail-party conversation. The truth is, Hawking inspires not simply because of the intellectual breakthroughs he has made but because he has made them in the absence of any support from his body, a vessel so frail that Hawking might seem to be a unique form of disembodied intelligence. Yet that image doesn't fit the man, whose magnetism derives in part from his brilliance, courage, and vulnerability—and in part from his quick wit, his penchant for Marilyn Monroe posters, and his irksome humanity.

Born on January 8, 1942, the 300th anniversary of the death of Galileo (a fact he often cites), he grew up in the suburbs of London in the kind of eccentric household that seems to pass for ordinary in England. It was, according to his younger brother Edward, "a little bit like the Munsters"; in the basement, the family kept bees. His father, who spent a good deal of time in Africa, was a doctor specializing in research, but Hawking rejected biology and by age fourteen had decided to devote himself to mathematics and physics. Three years later he enrolled at Oxford, grew his hair, and proceeded to give academic work short shrift. A popular student with a reputation for being smart enough not to study, he played bridge in the evening and by day acted as coxswain for his college crew;

a picture taken in 1961 shows him sitting at the front of a boat, natty in a white suit and straw boater, while eight larger men in striped jerseys row. "Steve and I had to be on the river every afternoon, six days a week," physicist Gordon Berry later recalled. "Something had to give, and it was definitely the experimental labs."

So when Hawking took his examinations at the end of his undergraduate career, after several years of dozing in class, his scores straddled the borderline between First and Second. Admission to Cambridge, the school of his choice, required a First. Called up in front of the examiners, he explained the situation forthrightly. "If I get a First, I shall go to Cambridge," he told them. "If I get a Second, I will remain at Oxford. So I expect that you will give me a First." They did.

At Cambridge, his occasional clumsiness and tendency to slur his words, which had first appeared at Oxford, got worse. Tying his shoes became difficult. Over Christmas vacation, his father noticed these problems. Hawking, not yet twenty-one, went to a specialist and within a few weeks was diagnosed with amyotrophic lateral sclerosis (ALS), also known as Lou Gehrig's disease. It is a progressive degenerative disease that causes the muscles—but not the mind—to atrophy. The disease, which usually strikes older people, progressed quickly at first. Given about two years to live, Hawking plunged into depression. "I felt somewhat of a tragic character," he told an interviewer. "I took to listening to Wagner."

Two years later, things started to improve. He married Jane Wilde, a high school student he had met before his diagnosis, and began to apply himself. His thesis adviser, Dennis Sciama, recommended that Hawking meet the mathematician Roger Penrose, who was studying what happens when a star uses up its fuel and collapses. Penrose showed that in an expanding universe governed by Einstein's general theory of relativity, once a star collapses past a certain point, it must inevitably become a singularity, the hypothetical point inside a black hole where matter is crushed to infinite density and where space, time, and the laws of physics become undone. Excited by this idea, Hawking threw himself into the exploration of completely collapsed stars and found his lifework. As the writer Dennis Overbye has observed, "It was hard not to think of Hawking as his own metaphor."

It occurred to Hawking that if a star could collapse into a singularity, then the process could also go the other way. A singularity could be a beginning as well as an end. In which case the universe, which was known to be expanding, might have started as a singularity. Hawking was able to prove more than that: An infinitely expanding universe, he showed, *must* have begun in a singularity.

But what if the universe doesn't expand infinitely? What if it has enough matter so that the expansion slows down and reverses itself, ending in the fatal implosion known as the Big Crunch? Would that universe also have had to begin as a singularity? The answer, Hawking said, was yes. In 1970, he and Penrose published a paper proving that the universe must have begun as a Big Bang singularity.

That November, while preparing for bed ("My disability makes that rather a slow process, so I had plenty of time"), Hawking had another realization: Since nothing can escape from a black hole, it can never get smaller. It can only stay the same or get larger; it cannot be divided, it cannot shrink, it cannot be blown to bits. With every new piece of matter it ingests, its mass increases and the event horizon bulges out a little more.

A research student at Princeton, Jacob Bekenstein, picked up this idea. Bekenstein saw a parallel between a black hole and the concept of entropy, the measure of random chaos in a system. According to the second law of thermodynamics, the amount of disorder in a closed system can only grow with time; entropy, like a black hole, always increases. Since every system has entropy, every time a black hole swallows another piece of matter, its entropy must increase along with its event horizon. The size of a black hole and the amount of its entropy might be equivalent.

Hawking refuted the analogy. His objection was that any system with a certain amount of disorder, or entropy, would also have to have a temperature, and anything with a temperature, no matter how low, emits radiation. "But by their very definition, black holes are objects that are not supposed to emit anything," he wrote. Hence, he decided, the comparison had to be wrong. Besides, Bekenstein irritated him.

Two Soviet physicists convinced Hawking to consider the possibility that black holes might, nonetheless, emit particles. When Hawking recalculated, he found, "to my surprise and annoyance, that even nonrotating black holes should apparently create and emit particles at a steady rate." At lectures, Hawking projected a transparency on the wall displaying the simple statement "I was wrong."

He reached this conclusion by looking at black holes from the point of view of quantum mechanics and the uncertainty principle, according to which space is never completely empty. Rather, it is populated by vagabond pairs of "virtual" particles—matter and anti-matter twins—that blink into existence and annihilate each other, all in a fraction of a fraction of a nanosecond, too fast to be observed. Hawking suggested that if such a pair were to pop up near the event horizon, the anti-matter particle could be sucked into the black hole, while the other, a tad farther away, might slip past the monster and into the quotidian universe. The particle would appear to be streaming out of the black hole. In which case, as Hawking puts it, "Black holes ain't so black."

The radiation from the black hole isn't actually coming from the black hole itself but from the layer of space that surrounds it. Nonetheless, so-called Hawking radiation takes a toll on the black hole, for as one particle swirls into eternity like water down a drain, never to return, its widowed mate, which cannot annihilate itself in the absence of its partner, has no choice but to become matter. Which requires energy. That energy has to come from the black hole.

But energy, Einstein taught, is just another form of mass, and vice versa. So when the black hole gives the virtual particle a bit of energy, it also loses a minuscule amount of mass—which isn't supposed to happen. The black hole shrinks a little and radiates more rapidly.

Eventually, the black hole evaporates in a mighty explosion equivalent

STEPHEN HAWKING, CASTAWAY

• ⤳

On December 25, 1992, Hawking was interviewed for a BBC radio program called *Desert Island Discs*, which asks its guests to pick eight pieces of music they would like to have with them in the unlikely event that they are marooned on a desert island with a good sound system. Hawking's Choices:

Poulenc's *Gloria*
The Brahms Violin Concerto
Beethoven's String Quartet, Opus 132
Wagner's *The Valkyrie*, Act One
The Beatles' "Please Please Me"
Mozart's *Requiem*
Puccini's *Turandot*
Edith Piaf's "Je Ne Regrette Rien"

to a billion one-megaton hydrogen bombs. This won't happen anytime soon; the average black hole will be around for 10^{67} years before it vanishes.

Hawking has described this process in an even more exotic way, based on the idea that the uncertainty principle makes it theoretically possible for a particle to travel faster than light. "The probability is low for it to move a long distance at more than the speed of light, but it can go faster than light for just far enough to get out of the black hole, and then go slower than light," Hawking said in a 1991 lecture. He warns, however, that this is unlikely to happen with large black holes. Even black holes whose mass equals that of the Sun are too big, because particles would have to exceed the speed of light for as long as several miles before they could reenter the ordinary universe.

But what about exceptionally small black holes? That's a different story. Hawking suggests that when the universe was young and much denser than it is now, primordial black holes, morsels with the mass of a mountain, might have been created. Those mini–black holes, artifacts of creation, wouldn't take long to evaporate. Hawking figures they ought to be evaporating just about now, disappearing in bursts of gamma rays. Scientists, many of whom doubt the reality of these miniature monsters, have yet to detect the telltale signs of such an event. That doesn't mean it isn't going to happen.

Thus Hawking's idea that black holes cannot get smaller was disproved by the discovery of the concept of Hawking radiation, which shows that black holes can disappear entirely. Something similar happened with his ideas about the singularity at the beginning of time. He began to reconsider. General relativity, it was true, demanded the existence of singularities; but at the point of singularity, where matter was compressed to infinite density, general relativity broke down. Maybe quantum mechanics, which traffics in uncertainty, could keep the Big Bang singularity from ever having existed.

Hawking decided that his original belief that the universe began with a singularity was wrong. Maybe the space-time universe didn't begin at all. The argument goes something like this: If you get close enough to the beginning of the universe, time doesn't exist; if time doesn't exist, there is no moment of creation, no moment of genesis, no moment at all. Without time, there is no time.

Unfortunately, ordinary mortals have trouble with this line of thought. Hawking points out that in any case the universe would *appear* to begin and end at a singularity. ("Thus, in a sense, we are still all doomed," he writes.) But in another sense—a highly conceptual sense involving many possible universes as well as a mathematical concept called "imaginary time"—time is like a circle, having neither beginning nor end. In this "no-boundary" proposal, Hawking typically compares the universe to the Earth. No matter where you start, it never ends. It never begins. In a similar way, "Asking what happens before the Big Bang is like asking for a point one mile north of the North Pole," Hawking writes. "The quantity that we measure as time had a beginning, but that does not mean spacetime has an edge, just as the surface of the Earth does not have an edge at the North Pole, or at least, so I am told; I have not been there myself."

Hawking has also speculated about baby universes, by-products of Alan Guth's inflationary model of the universe, according to which, for a volatile fraction of an instant, the early universe inflated wildly. If that process created tiny bumps in the fabric of space-time, those little hills and valleys could grow, ballooning into parallel universes connected to our universe by wormholes, quantum tunnels through space-time. In which case our universe might be one among many.

Throughout all this, Hawking has continued his work despite the devastating deterioration of his body. By 1969, two years after the birth of the first of his three children, he could no longer navigate with a cane and was forced into a wheelchair. Eventually, he came to depend upon full-time nursing care and graduate students who could interpret his every halting mumble. In 1979, upon being elected Lucasian Professor of Mathematics at Cambridge, a post once held by Sir Isaac Newton, he signed his name for the last time. His speech became nearly incomprehensible; then, during an emergency tracheotomy in 1985, he lost the power of speech entirely. It was returned to him with a computerized voice synthesizer attached to his wheelchair.

None of this has removed him from the ordinary crises that flesh is heir to. In 1990, in a remarkably underpublicized divorce, he left his wife, Jane. And on a rainy evening in March 1991, he misjudged the distance of approaching traffic while crossing the street and ended up "in the road with my legs over the remains of the wheelchair." In the accident, he broke his arm, cut his head (which required thirteen stitches), and sustained damage to the computer system that allows him to talk. Despite this, he retains the capability to smile and continues to follow, in his intellectual work if not in his personal life, the injunction Sir Arthur Eddington gave at a lecture in 1928. "I ask you to look both ways," Eddington said. "For the road to a knowledge of the stars leads through the atom; and important knowledge of the atom has been reached through the stars."

In the documentary film about Hawking directed by Errol Morris, Hawking's sister Mary says, "Father was very good at theological debate, so everybody used to argue theology." This seems to be a habit Hawking has never lost. In his writings, Hawking turns repeatedly and ambivalently to the question of, in his sister Mary's words, "the existence or otherwise of God." Often he mocks the idea. He writes humorously of his experiences at the Vatican, where he attended a conference on cosmology in 1981: "At the end of the conference the participants were granted an audience with the pope. He told us that it was all right to study the evolution of the universe after the big bang, but we should not inquire into the big bang itself because that was the moment of Creation and therefore the work of God. I was glad then that he did not know the subject of the talk I had just given at the conference—the possibility that space-time was finite but had no boundary, which means that it had no beginning, no moment of Creation. I had no desire to share the fate of Galileo."

Yet at the same time, Hawking writes that if a unified theory combining the principles of relativity with those of quantum mechanics were ever found, "It should in time be understandable in broad principle by everyone, not just a few scientists. Then we shall all, philosophers, scientists, and just ordinary people, be able to take part in the discussion of the question of why it is that we and the universe exist. If we find the answer to that, it would be the ultimate triumph of human reason—for then we would know the mind of God."

●

Strings

To explain butterflies is unnecessary because everyone has seen them. To explain superstrings is impossible because nobody has seen them. But please do not think I am trying to mystify you.

—Freeman Dyson

 hen a concept's time has come, it appears everywhere. Such is the case with cosmic strings. They are thought to come in two basic sizes—Brobdingnagian cosmic strings and Lilliputian superstrings—and although they are purely speculative, that doesn't stop anyone from describing them.

Cosmic strings are long, massive, vibrating filaments no thicker than a particle. Created a fraction of a fraction of second after the Big Bang, they form loops that twist and curl across millions of light-years. Cosmic strings are made not of ordinary matter but of something so refined they're practically metaphysical. They have been described as "cracks in the structure of space-time," "pure bits of unified force field," "theoretical thin tubes of space-time," and—the most down-to-Earth description—"really thin tubes of false vacuum, with masses of 10^{16} tons per inch." Their incredibly high mass may help account for some of the missing mass—the dark matter—in the universe.

With their immense gravity pulling galaxies and clusters of galaxies into their wake, cosmic strings snake through the universe, lassoing galaxies and possibly creating the foamy mix of galaxy clusters and voids that seems to be the structure of the universe. Eventually, they evaporate or decay. A few especially monstrous cosmic strings might still exist, though there is no proof. But maps showing the distribution of hundreds

KRISHNA OPENS HIS MOUTH

One day, Yaśodā heard that her foster son, Krishna, had eaten dirt. When she scolded him, he proclaimed his innocence, and to prove it, he asked her to look inside his mouth. A section of the *Bhāgavata Purāna*, translated by Wendy Doniger O'Flaherty in *Hindu Myths*, describes what happened next.

She then saw in his mouth the whole eternal universe, and heaven, and the regions of the sky, and the orb of the earth with its mountains, islands, and oceans; she saw the wind, and lightning, and the moon and stars, and the zodiac; and water and fire and air and space itself; she saw the vacillating senses, the mind, the elements, and the three strands of matter. She saw within the body of her son, in his gaping mouth, the whole universe in all its variety, with all the forms of life and time and nature and action and hopes, and her own village, and herself. Then she became afraid and confused. . . .

of thousands of galaxies reveal that galaxies and clusters of galaxies tend to be grouped in streams and chains. Those lines may represent the ancient paths of cosmic strings.

Lilliputian strings, known as superstrings, live in the miniature realm of the subatomic. Superstring theory replaces the fuzzy, pointlike particles of quantum theory with vibrating strings so tiny—perhaps 10^{-33} centimeters across—that, according to physicist Edward Witten, a professor at the Institute for Advanced Study at Princeton, they are "vastly smaller compared to things that you might well think of as unimaginably small." Curled up inside these superstrings are six unknown variables usually thought of as dimensions. "The idea of extra dimensions might sound a little bit strange to anyone who hasn't studied physics," Witten says. "Anyone who has gone into physics professionally will know that there are many things that are a lot stranger than extra dimensions. General relativity is strange, quantum mechanics is strange, anti-matter is strange. All these things are strange but true. Compared to a lot of things that have come true in physics in the past, extra dimensions are not such a radical departure."

Superstring theory, which appears to eliminate some troubling mathematical anomalies that marred other theories, is of great interest to scientists seeking a Theory of Everything—a single, unified theory that works from the subatomic level on up and that explains and unifies the four forces of nature (gravity, the strong nuclear force, the weak nuclear force, and electromagnetism).

None of this has much to do with cosmic strings. But one cannot help but note that although cosmic strings and superstrings are supposedly unrelated, there are parallels between them. They both form loops. They both may help to account for the preponderance of dark matter in the universe, cosmic strings through their mass and superstrings through the hypothesized existence of something known as shadow matter, which can be neither felt nor seen but which could nonetheless be responsible for the existence of an entire, unseen duplicate universe.

It is conceivable that superstrings and cosmic strings may have been formed at the same time and that when the universe inflated during its first second of existence, the newly hatched strings were stretched to supergalactic lengths. Witten notes that "although a superstring representing an elementary particle is incredibly tiny, if you had sufficiently powerful tweezers there is no reason in principle why you couldn't grip one of these things and stretch it, making it larger and larger." Perhaps, in the beginning, the two kinds of strings were one. Or perhaps they don't exist at all.

✳

Why Is the Sky Dark at Night? The Riddle That Won't Go Away

Just why this is a riddle is a riddle. Ask the average person why the sky is dark at night, and you will be told that you've been working too hard. Ask the question in a scientific context, and many generations of astronomers will rigorously work their way through one argument after another in order to confirm again and again a truth everyone knew at the beginning: namely, there aren't enough stars.

The reasoning begins with the assumption that the laws of physics are the same everywhere and that our part of the cosmos is just like any other part. Given that, this must be an infinite universe, because if it weren't (Newton argued), every object would sooner or later attract every other object and the universe would collapse, a victim of the force of gravity. But an infinite universe would have an infinite number of stars. Here, there, and everywhere your line of sight would intersect with the light from a star. At midnight (and at noon, too), the sky would blaze with stars beyond number.

But obviously it doesn't. In 1576, the mathematician Thomas Digges concluded that the explanation for this was that although the universe was infinite and stars were scattered from here to eternity, they became "by reason of their wonderfull distance inuisible vnto us." In short, they were too far away.

Edmond Halley drew a similar conclusion. He imagined the Earth encircled by an infinite series of star-dotted, concentric shells reminiscent of wooden Russian dolls, each of which nests inside another. The more distant the shell was, the fainter the stars would be—but also the more numerous. Halley noted that distant stars that can be seen with a telescope are nonetheless invisible to the naked eye.

Johannes Kepler believed that the sky is dark at night because the universe is surrounded by a dark wall beyond which there is nothing. And Otto von Guericke (1602–86), a German physicist known for his high-profile public experiments demonstrating the properties of a vacuum, concluded that the number of stars is finite but that the void, which we see between and beyond the stars, is infinitely dark and deep.

But even though these astronomers, and others, explored the question, the problem is associated most strongly with the ophthalmologist Heinrich Olbers (1758–1840), a famous amateur astronomer who discovered five comets as well as Pallas and Vesta, two of the first four asteroids. (In recognition of his achievements, asteroid 1002 is named after him, as is a lunar crater.) He wrote about the problem, which is called Olbers' Paradox in his honor. He believed that the sky is dark at night because the dust between the stars absorbs energy from the stars, which keeps the light from reaching us. Unfortunately, as John Herschel showed, that's not the right answer. It's also not the case that mysterious dark stars are blotting out bright stars.

A surprising participant in the debate was Edgar Allan Poe, who believed that the universe pulsed, that the observable universe is only a fraction of everything that's out there, and that we are surrounded, at enormous distances, by the "continuous golden walls of the universe" made of "myriads of the shining bodies that mere number has appeared to blend into unity." They are so far away, Poe wrote in his lengthy essay *Eureka: A Prose Poem,* that the light from these farthest shores has not had sufficient time to reach us, i.e., the stars are not old enough to light the sky. Lord Kelvin (of the absolute-temperature scale) had a similar idea. He thought that even if all the stars were turned on simultaneously, "at no one instant would light be reaching the Earth from more than an excessively small proportion of all the stars." A bright night sky would require the most distant stars to be turned on before the nearby stars, so that their light would have plenty of time to get here, and this, he noted, was improbable.

In the twentieth century, Hermann Bondi, famous for his championship of the Steady State theory, brought renewed interest to the question. Bondi thought the sky was dark at night because the farther away a star is, the more stretched out its light waves become, with the result that light from distant stars would become severely redshifted and essentially weakened.

But as the astronomer Edward Harrison, author of *Darkness at Night,* has shown, that's not why. The reason is that there aren't enough stars. To light the sky from horizon to horizon, day and night, to cover every

NASA

Hot blue stars near the center of the Milky Way Galaxy, photographed in far-ultraviolet light in a 30-minute exposure from the Moon. The large, bright spot is Jupiter.

speck of the celestial sphere with starlight, would require, Harrison asserts, 10 trillion times more stars than now exist.

Or look at it another way. The sky would be awash in light if we could see to a trillion trillion (10^{24}) light-years. But we can't see that far back in time because the universe has only been expanding for 15 billion years or so. It isn't old enough. And in any case, stars have a finite life span, with the shortest lives going to the brightest stars. According to Harrison, if stars live less than 100 billion trillion years, then there simply aren't enough of them. In the real universe, none of the numbers are big enough.

What do we see instead of starlight? Past the stars, past the galaxies, past the quasars, we hit something we might think of as Kepler's Wall. It surrounds us. It is the cosmic background radiation: a souvenir of the Big Bang, when the sky truly did blaze. Since then, the light has dimmed, the temperature has cooled, and the expansion of the universe has turned the fireball of creation into "an infrared gloom invisible to the naked eye," Harrison writes. "Though we see only a wall of darkness, the big bang covers the sky, filling the universe throughout space and time with its afterglow."

One question remains: Why would anyone spend time on a question like this? Hermann Bondi knew the answer. He pointed out that cosmology, which has to do with the universe as a whole, would seem to be immune from observation, as "ivory tower" as you can get. The question of darkness at night was one in which theoretical cosmology presented a plausible scenario—even at night the sky should shine like a curtain of stars—that was clearly negated by observation. Explaining this has required many theories and much rigorously logical analysis. One explanation after another has had to be discarded, and the search has come to seem far from idle. Indeed, according to Bondi, "The discovery of the cosmological significance of the darkness of the night sky made cosmology a science."

UFOs

It seemed to move toward us, then partially away, then return, then depart. It was bluish, reddish, and luminous.

—Jimmy Carter, former president
of the United States

The love of the marvellous is the most dangerous enemy of natural science.

—Eugene M. L. Patrin

ost scientists, it should be noted up front, routinely dismiss anything having to do with UFOs (unidentified flying objects). Like other skeptics who believe that UFOs belong to the province of psychology rather than astronomy, they are unswayed by anecdotal evidence and openly scornful of the questionable methodology of surveys designed to elicit descriptions of UFO encounters—including one that seemed to indicate that 3.7 million Americans had been abducted by aliens.

Nonetheless, believers are legion. Many of them date the inception of modern UFOlogy to July 1947, when, proponents claim, something crashed seventy-five miles northwest of Roswell, New Mexico. A rancher named William Brazel discovered the wreckage, which was inscribed with peculiar hieroglyphic symbols. A few days later he informed the town sheriff, who called Roswell Air Force Base. The base commander announced that the wreckage of a flying disk had been found. Within days that story was retracted and replaced with a standard weather-balloon explanation. Meanwhile, the debris from the crash was collected and taken to Wright-Patterson Air Force Base.

Among that wreckage, UFOlogists claim, were the decomposed remains of four small humanoids. Shortly thereafter, President Harry S. Truman signed a set of top-secret, eyes-only documents, creating an intelligence operation known as Majestic-12.

And that, according to UFO theorists, was only the beginning. Since then, the story has grown, as anyone who has glanced at the headlines on a supermarket tabloid knows. Early stories of contact—an Australian family terrorized by an egg-shaped object, two men in Sweden harassed by small, gray, jelly-like creatures—have given way to more exotic fare. Now, people aren't just harassed: they're kidnapped, a phenomenon brought to widespread public attention by *Communion: A True Story,* Whitley Strieber's 1987 best-seller chronicling his and other people's alleged encounters with alien creatures. These contactees claim to have been given unwanted medical examinations, to have had operations performed on them (often leaving small, geometrically shaped scars), to have had sex with aliens. They include women who claim they were impregnated and had their babies taken from them as part of an ongoing breeding program. They also include people who have presumably repressed these horrific experiences. How can you tell if you might be an abductee in denial? One clue is the realization that you can't account for "lost time."

These abductions don't just happen on deserted roads in remote parts of the country. They also happen in urban areas, where alien spacecraft are said to hover above buildings and abduct people right through windows. "I know it sounds impossible that this could happen in a place like New York City," UFO investigator Budd Hopkins told a reporter, "yet it goes on a lot."

These stories are accompanied by charges of government complicity, cover-up, and disinformation that are downright terrifying. Proponents believe that

* the United States government has collected at least thirty flying saucers and an equal number of alien bodies from three different civilizations. They are in cold storage at Wright-Patterson and at CIA headquarters in Langley, Virginia;

* alien creatures whose digestive systems have degenerated are responsible for worldwide waves of cattle mutilations, and animal and human organs are being stored in an amber-colored liq-

uid in huge underground vats in New Mexico, where they are being processed for alien use;

* the United States and other countries know all about this but have agreed to close their eyes in exchange for alien technology;

* the United States has been invaded by two groups of aliens. The emotionless grays are responsible for cattle mutilations, abductions, and invasive medical procedures that include implanting mind-control devices. The good guys are the whites. They seek peace. But retired Air Force lieutenant colonel Wendelle C. Stevens, writing in *International UFO Library* magazine, explains that it's not so simple and that benevolent gray creatures from two G-type stars in the constellation Reticulum the Net have been unfairly maligned thanks to the wretched customs of other ETs.

And then there are the crop circles—intricate patterns, highly mathematical and often quite beautiful—which have been appearing in English pastures and fields since the late 1970s. People who have claimed responsibility for creating a hoax have been unable to re-create the perfectly formed pictograms. No reasonable explanation has been suggested, and governmental officials, proponents note, don't know how to react.

On the other hand, skeptics point out that UFO stories hook into mythology, folklore, and the human imagination far more than is comfortable. It turns out, for instance, that when people who have not reported a close encounter are hypnotized and asked to imagine one, they come up with the same sorts of details provided by self-proclaimed abductees.

As for the sightings, the great majority of UFO stories—*all* of them, according to most scientists—are traceable to phenomena such as ice crystals, clouds, meteors, lightning, birds, Venus, klieg lights, weather balloons, and the Goodyear blimp. Arthur C. Clarke, science-fiction writer and chancellor of International Space University in Sri Lanka, suggests that the explanation may lie in "strange and surprising meteorological, electrical, or astronomical phenomena still unknown to science." But what about those few instances for which there is no logical explanation? The astronomer J. Allen Hynek (1910–86), who coined the term "close encounters of the third kind" and played a bit part in the movie of the same name, was willing to consider the possibility that UFOs could be involved in such cases, but scientists point out that travel time to other stars is hopelessly long. Even at the speed of light, the nearest star is four and a half years away. And travel at the speed of light isn't possible because the closer you get to that speed, the closer your mass gets to being infinite. Even anti-matter, the *Star Trek* fuel of choice, can't bypass that intergalactic limit. On the other hand, maybe aliens know things we don't. As Clarke has noted, "Any sufficiently advanced technology is indistinguishable from magic."

What to make of all this? It could be that alien creatures with their alien ways really are pursuing our cattle and our women and harvesting our bodies for their own nefarious uses.

Or it could be that a lot of sadly delusional people have concocted a bewildering anthology of genuinely nutty stories. The grays and the whites may be contemporary mythology, brought to you by the same human impulse that led to the Salem witch trials.

Or maybe the government really is involved in a disinformation campaign so vast it makes the Iran-*contra* affair look trivial. In any case, this much is true: One accurate story—one alien skeleton—would be enough. More than enough.

Waiting at the Water Hole: The Search for Extraterrestrial Intelligence

Are we alone? Isaac Newton didn't think so. He wrote in his private papers that just as the land, the sea, and the air were populated, along with "standing waters, vinegar, the bodies and blood of Animals and other juices with innumerable living creatures too small to be seen without the help of magnifying glasses, so may the heavens above be replenished with beings whose nature we do not understand." In the nineteenth century, a few experimentalists began to think of ways to make contact with those beings. One of these was the great prodigy Carl Friedrich Gauss (1777–1855), known as "the Prince of Mathematicians." While still a student, Gauss discovered a way to construct an equilateral 17-sided polygon with a compass and a straight edge—a process the Greeks hadn't figured out. In addition to his mathematical discoveries, he calculated the orbit of Ceres and by age thirty was the director of the Göttingen Observatory.

In his forties he began to ruminate about the inhabitants of the Moon. It occurred to him that one way to signal them would be to create a huge geometrical figure in central Siberia. Dark green pine alternating with green or golden grain in a strict geometrical shape would surely capture their attention. It wasn't the 17-gon, however, that he had in mind but the Pythagorean right triangle. Inscribed on the Earth in sufficient size, it would be obvious to anyone watching that it was not a natural phenomenon.

In the 1840s, the astronomer Joseph von Littrow had a similar notion. He wanted to dig a twenty-mile trench in the Sahara Desert, drench it with kerosene, and set it on fire. The conflagration, he suggested, would be visible to the inhabitants of other worlds. In the late nineteenth century, the idea was raised of building a huge mirror to reflect and focus light in such a way that it would brand the surface of Mars.

The eccentric Croatian-born inventor Nikola Tesla (1856–1943) believed that a better way to contact an alien civilization would be with radio waves. He also believed that the Martians were already sending them our way; Lord Kelvin, who agreed, was convinced that the Martians were specifically trying to contact New York City.

Over half a century later, Philip Morrison and Giuseppi Cocconi wrote the virtual manifesto of the idea that radio waves are the primary form of communication between worlds. Their article "Searching for Interstellar Communications" was published in *Nature* magazine on September 19, 1959. In it, the authors label the specific "channel which must be known to every observer in the universe." It was the radio emission line of neutral hydrogen—1,420 megahertz.

The next year, the astronomer Frank Drake established Project Ozma (named after the princess in the *Wizard of Oz* books), and used that frequency to conduct a radio search of Tau Ceti and Epsilon Eridani, two relatively nearby Sun-like stars. Drake spent 200 hours with the radio telescope at Green Bank, West Virginia, but he detected no organized signals coming from those stars.

Since then, astronomers throughout the world have repeatedly tried to pick up signals from elsewhere. They have investigated hundreds of stars similar to the Sun, both at the hydrogen frequency and at the frequencies of other molecules. Since the mid-1980s, Harvard University's Megachannel Extraterrestrial Assay—funded in part by the filmmaker Steven Spielberg—has scrutinized a wide range of frequencies. These searches have thus far been fruitless.

But they were nothing compared to NASA's SETI (Search for Extraterrestrial Intelligence), a high-resolution microwave survey of the sky inaugurated on October 12, 1992, the 500th anniversary of Columbus' arrival in the Western Hemisphere. Also known as Project Knock Knock, the survey utilizes at least six radio telescopes, including the 1,000-foot metal bowl at Arecibo, Puerto Rico, and it accomplished more in its first thirty seconds than had been done in the field since Frank Drake inaugurated Project Ozma more than three decades before. To date, nary a Klingon has been found, but other discoveries support the notion of life elsewhere. Organic molecules have been found in interstellar space and in comets, and disks of matter have been located around stars, suggesting that the formation of solar systems is common. Life may actually be widespread, and NASA's search, expected to operate for at least ten years, offers the best chance yet of specific proof that we are not alone.

> **S**omewhere in the cosmos . . . along with all the planets inhabited by humanoids, reptiloids, fishoids, walking treeoids, and superintelligent shades of the color blue, there was also a planet entirely given over to ballpoint life forms. And it was to this planet that unattended ballpoints would make their way, slipping away quietly through wormholes in space to a world where they could enjoy a uniquely ballpointoid life-style.
>
> —Douglas Adams,
> *The Hitchhiker's*
> *Guide to the*
> *Galaxy*

But where to look? NASA's search is designed for a short range of microwave frequencies known as the "water hole" because it stretches from the emission line of hydrogen (H) to that of hydroxyl (OH). This is an area disturbed by very little radio noise, and the theory is that other civilizations will also have noticed that. So during the first half of NASA's Microwave Observing Program, radio astronomers will concentrate on those particular frequencies as they conduct a search of 800 Sun-like stars within seventy-five light-years of us. And then, just in case the little monsters are broadcasting on another channel or from another star, the program will include an all-sky survey covering 10 million channels.

Yet if they find nothing, that will prove nothing. The timing has to be right. An alien civilization investigating the Sun over a century ago would have heard nothing from us. A civilization looking right now, but located a hundred or so light-years away, would likewise find nothing. No signals from planet Earth have reached them yet. From their standpoint, we are mute.

But a civilization a few light-years closer would be able to detect the radio signals that have been leaking into the cosmos at the speed of light ever since their original broadcast. An advanced civilization sixty light-years from here should just about now be receiving one of Franklin D. Roosevelt's fireside chats. A civilization thirty-odd light-years away that has just turned its attention in our direction might be receiving news of the Cuban missile crisis—or a rendition of "The Name Game." Since radio waves travel with the speed of light, if they reply immediately, we won't get their response for a generation or more.

So maybe somewhere, on another planet orbiting another star, there lives an unimagined species. Considering the bloody history of contact between alien cultures here on planet Earth, many people are worried that should we ever discover a new world, close encounters of the worst kind are all too predictable. That fear won't stop anyone from looking, nor should it. The search must go on. Yet so far, it has been in vain. At the moment, among all the stars of the galaxy, we know of not a shred of consciousness (not to mention technological know-how) anywhere but on our own small planet. There is so far no reason to doubt that Sir Arthur Eddington was right when he commented, "Not one of the profusion of stars in their myriad clusters looks down on scenes comparable to those which are passing beneath the rays of the Sun."

The Drake Equation

Besides conducting a radio search of two stars, Frank Drake, professor of astronomy and astrophysics at the University of California at Santa Cruz, is known for having turned amorphous hope into a specific formula, $N = R_* f_p n_e f_l f_i f_c L$, which predicts how many advanced civilizations might be willing and able to communicate with us at any given time. The Drake equation is based on the supposition that life arose on Earth through ordinary processes and is likely to arise on other worlds in the same way. "No freak events are required," Drake wrote twenty-eight years after he proposed the equation. "In the Universe, anything that can happen will happen. And often. This is perhaps the Murphy's Law of the Universe. The existence of life, and intelligent, technology-exploiting life, in large quantities, should not be a surprise."

The Drake equation assumes that certain conditions must exist before life can arise. For instance, it seems likely that only stars the size of the Sun could give rise to advanced civilizations. Large stars burn fuel so quickly before expanding into giant stars that there's not enough time for intelligent, technologically sophisti-

"It's the size of things that worries people. No reason for the universe to be so large. It contains more space than I deem absolutely necessary. More time as well. Know who I envy?"

"No."

"Take a guess."

"Don't know."

"Low-gravity creatures," Endor said. "On a low-gravity planet the inhabitants are long, slender and delicate. This is how I think of the Ratnerians. I see them drifting across the terrain, almost ectoplasmically, a race of emanations merely flecked with solid matter. Yes. Beings nearly free of their planet's gravity."

—Don DeLillo, *Ratner's Star*

cated societies to evolve. Small M stars, while long-lived, are so cool that a planet would have to orbit very close to one, in a very narrow band, just in order to maintain a reasonable temperature. Binary stars would produce complicated orbits not suited to the development of life. Thus, solitary stars the size of our Sun offer the best environment for life.

Similarly, not every planet is conducive to life. Some are too close, like Venus; some are too far or too small, like Mars. Only a few are just right, and the margin of tolerance can be surprisingly narrow. For example, if the Earth were only a little farther from the Sun than it is—94 million miles instead of 93 million—the world would be virtually arctic, and life might not have arisen. The Drake equation multiplies a number of such factors in order to come up with a number (N) that suggests how many civilizations might actually be out there, waiting. Those factors are:

* the birthrate for stars similar to the sun (R_*);

* the fraction of stars with planets (f_p);

* the number of planets per solar system on which life could arise (n_e);

* the fraction of those planets on which life does arise (f_l);

* the fraction of those life forms that can be considered intelligent (f_i);

* the fraction of those planets on which intelligent beings are capable of—and desirous of—sending messages into space (f_c);

* the average longevity of technologically advanced civilizations (L). This is the great unknown. We only have ourselves as an example, and although our civilization may be thousands of years old in terms of art, warfare, and general misery, in terms of our ability to send and receive messages from space it is less than a century old.

The Drake equation says that if you multiply these factors together, you get the number of advanced civilizations that might be willing to get in touch with us. Depending upon the values given to the individual terms within the equation, all the factors—except the last—multiplied together equal anything from one tenth to one, which leaves an unavoidable and disturbing implication: The possibility of finding life elsewhere in the Milky Way rests directly on *L,* the average lifetime of a civilization. The longer we—and any other civilization—manage to exist without stumbling back into the Stone Age, the greater the chances that someday we might find each other.

Drake, who has suggested that there might be 10,000 such civilizations, is hopeful of discovering radio signals from outer space by the year 2000.

To Whom It May Concern

On the theory that we shouldn't sit around waiting for other civilizations to call us, the United States in the 1970s launched three specific messages to the stars.

The first message was sent twice: aboard *Pioneer 10,* which was launched on March 2, 1972, and, thirteen months later, aboard *Pioneer 11.* Both spacecraft carried an engraved, gold-anodized aluminum plaque smaller than a piece of loose-leaf paper. Designed by Carl Sagan, Frank Drake, and Linda Salzman Sagan, Sagan's wife at the time, it includes a representation of hydrogen, a map locating Earth in relation to fourteen pulsars, a picture of the solar system with an arrow showing *Pioneer'*s trajectory, several binary numbers, and line drawings of two naked humans standing against a drawing of the spacecraft so that the aliens will know how tall we are. The man stands with one arm at his side and one raised in greeting; the woman, whose long hair is parted in the middle and swept behind her shoulders, holds her arms at her sides. Controversy arose immediately: The picture was pornographic, the woman lacked genitalia, the image was sexist because the man was in an active pose and the woman was in a passive one. Off it went nevertheless on its 80,000-year-long journey to Alpha Centauri. When it arrives, the plaque—which Carl Sagan claims is sturdy enough to survive in space for hundreds of millions of years—should be in fine shape, and the Alpha Centaurians "will learn a great deal about us," writes astrophysicist Rudolf Kippenhahn. "But one thing they will never find out—how we look from the rear. That will remain a mystery to them forever."

The second message to the stars went out on November 16, 1974, from the radio telescope at Arecibo, Puerto Rico. For three minutes, 1,679 yes-no, on-off binary pulses were beamed at globular cluster M13. The number 1,679 was chosen because it is the product of two prime numbers: 23 and 73. Arranged horizontally, in 23 rows of 73 pulses, the collection of pulses looks like nothing in particular. Arranged vertically, in 73 rows of 23, it forms a remarkable pattern.

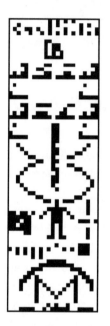

Each component of the design has a meaning. From top to bottom, the images represent the digits 1 through 10 in binary code; the atomic numbers of the elements hydrogen, carbon, nitrogen, oxygen, and phosphorus; information about RNA and DNA, including a picture of the double helix; a picture of a human; binary numbers representative of human height and of the population of the world; below the human figure, a representation of the solar system, with the Sun to the right and the Earth slightly raised; a picture of the Arecibo telescope, and a binary number giving its size. There are perhaps 1 million stars in M13. About 25,000 years from now, if their radio telescopes happen to be pointing in our direction at exactly the right moment, they will receive this message. If, however, they happen to be otherwise occupied during the three minutes allotted to the message, they will miss it forever.

The third message, also produced in duplicate, was launched in 1977 aboard *Voyagers 1* and *2*. Using technology now obsolete even on the home planet, it consisted of a gold-plated copper record attached to each spacecraft in a round container. That container is inscribed with the pulsar diagram from *Pioneer 10*, the representation of hydrogen, and visual instructions on how to use the needle and cartridge to play the record.

Once the aliens figure it out, they will hear greetings in sixty languages; rain; kissing; an automobile changing gears; the song of the humpback whale; Chuck Berry's "Johnny B. Goode"; Bach's Brandenburg Concerto No. 2; part of Mozart's *The Magic Flute;* Senegalese drumming; a Navajo chant; the electrical activity of a woman's eyes, muscles, brain, and heart; and the voice of onetime United Nations secretary general Kurt Waldheim. In addition, there are on the record 115 images of people, places, and the details of life on Earth, ranging from a snowflake to a supermarket.

And that's it. Like notes stuffed into bottles and tossed into the infinite sea, the messages of the space program are slowly drifting toward the stars. They won't arrive anytime soon, though, for astronomical distances are vast. After traveling for 26,262 years, *Voyager 2* will have made it to the Oort cloud, the home of comets. After all those years, it will still be in our solar system.

✪

PART IV

AN ALBUM OF STARS

The appearance of a star (1st magnitude) of exceeding brilliancy dominating by night and day (a new luminous sun generated by the collision and amalgamation in incandescence of two nonluminous exsuns) about the period of the birth of William Shakespeare over delta in the recumbent neversetting constellation of Cassiopeia and of a star (2nd magnitude) of similar origin but of lesser brilliancy which had appeared in and disappeared from the constellation of the Corona Septentrionalis about the period of the birth of Leopold Bloom and of other stars of (presumably) similar origin which had (effectively or presumably) appeared in and disappeared from the constellation of Andromeda about the period of the birth of Stephen Dedalus, and in and from the

AND CONSTELLATIONS

constellation of Auriga some years after the birth and death of Rudolph Bloom, junior, and in and from other constellations some years before or after the birth or death of other persons . . .

—James Joyce, *Ulysses*

Star Names

Still more unfettered,
They left the named
And spoke of the lettered,
The sigmas and taus
Of constellations.

—Robert Frost,
"I Will Sing You One-O"

lthough most stars are known by numbered catalogue identifications, the brightest stars have names. Their origins vary. Some—Sirius, Pollux, Procyon—are Greek. Regulus and Stella Polaris are Latin. Nunki (in Sagittarius) is Babylonian. But the vast majority of star names are Arabic; for when the light of learning went out in Europe during the Middle Ages, responsibility for knowledge of the stars shifted to and was upheld by the Arabs.

In 773, after a traveler from India told the caliph Abu Jafar al-Mansur that he could predict eclipses, the caliph arranged to have a number of astronomy books translated into Arabic. Half a century later, under another caliph, Ptolemy's star catalogue, written more than seven centuries earlier, was translated into Arabic. Other catalogues followed, including one compiled in 1420 by Ulugh Beg (1394–1449). It listed 1,018 stars that the Uzbek prince, the grandson of Tamerlane, had seen from his observatory in Samarkand.

When Europeans finally woke up, they turned to Arabic texts, created new star charts, and mangled Arabic names, with the result that word derivations are not always easy to reconstruct. For instance, some

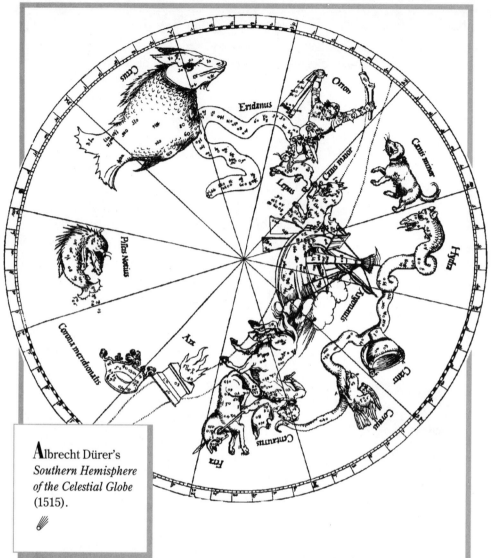

Albrecht Dürer's
*Southern Hemisphere
of the Celestial Globe*
(1515).

Arabs referred to the second-brightest star in Orion as "the hand of the giant"—*yad al-gawzâ*. In 1246, a Parisian translator read the *y* as *b*—an easy mistake to make because in Arabic script the two letters are distinguished by a single dot. *Yad al-gawzâ* became *Bad al-gawzâ,* which in an era of flexible spelling ultimately became Betelgeuse, the name we use today.

In 1603, the Bavarian lawyer and astronomer Johann Bayer (1572–1625) published *Uranometria,* a beautiful, innovative star atlas in which each star was assigned a Greek letter according to its brightness. With a few exceptions (Ursa Major among them), the brightest star in each constellation was given the letter alpha, the next brightest, beta, and so on; the Greek letter was followed by the genitive case (the possessive) of the Latin name of the constellation. Thus Rigel Kentaurus, "the foot of the centaur," also became known as Alpha Centauri.

Bayer's designations are still used today. But there are hundreds of stars in each constellation and only twenty-four letters in the Greek alphabet. Bayer was forced to label some stars with English letters. Soon, other systems arose. John Flamsteed (1646–1719), appointed by King Charles II as England's first astronomer royal, numbered each star according to its position within the constellation. But Flamsteed was a perfectionist, and he refused to authorize publication of his system. Edmond Halley, who had convinced Newton to publish the *Principia Mathematica* but was unable to sway Flamsteed, took matters into his own hands and published Flamsteed's catalogue himself. Flamsteed reacted by burning all the copies he could find. Only in 1725, after his death, did his *Historia Coelestis Britannica* finally become available.

Flamsteed classified 2,682 stars, but of course there are many more. Other systems proliferated. Today, a star could have a name and a Bayer designation; it might have a Flamsteed number, like 61 Cygni or 17 Tauri; it could—like Lalande 21185, the third-closest star to the Sun—be one of the 47,000 stars catalogued by the French astronomer Joseph Jérôme Le Français de Lalande (1732–1807). In the nineteenth century Friedrich Wilhelm August Argelander (1799–1875) of the Bonn Observatory published the last great star catalogue to be created without the use of photography. In the four-volume *Bonner Durchmusterung,* he assigned BD numbers to 324,198 stars, and he also invented a new way of labeling variable stars. Stephen Groombridge (1755–1832), an English merchant and amateur astronomer, catalogued 4,243 circumpolar stars including Groombridge 1830, a fast-moving dim star in Ursa Major. In the early twentieth century, Annie Jump Cannon at the Harvard Observatory was largely responsible for classifying the 250,000 or so stars in the *Henry Draper Catalogue.* There have been many other catalogues since, including the *Guide Star Catalogue* of the Hubble Space Telescope, a computerized list of over 15,169,873 stars and 3.6 million galaxies.

But no matter how important it is to impose a system on the stars, the human need to number seems to be weaker than the desire to name.

While Flamsteed numbered his stars, his would-be patron Edmond Halley named Alpha Canum, the brightest star in Canes Venatici (the Hunting Dogs), after Charles II, England's popular king. Even today, the star Cor Caroli—"the Heart of Charles"—is known by name.

So is Barnard's star in Ophiuchus. It was named after E. E. Barnard, who discovered that this insignificant nearby star, too dim to be seen with the naked eye, is sliding across the sky faster than any other star in the heavens. Van Biesbroeck's star, a red dwarf discovered in 1943, is famous for being dim; Van Maanen's star in Pisces, discovered in 1917, is a white dwarf known for being small; HD 47129 in Monoceros, thought to be the most massive star in the galaxy, is better known as Plaskett's star. And so it goes throughout the celestial sphere. Some stars have names that mean something, like Sualocin and Rotanev in Delphinus the Dolphin. Giuseppi Piazzi, who discovered the first asteroid, gave them those names in his star catalogue of 1814. The names may look vaguely Arabic or Russian, but they're not. They are a secret reference to Piazzi's assistant, one Niccolo Cacciatore, whose name in Latin is Nicolaus Venator—which spelled backward is Sualocin Rotanev. Nor is Cacciatore the only person whose name has been reversed for all eternity. In Cassiopeia one star is sometimes labeled Navi. That's "Ivan" spelled backward—a reference to the middle name of Virgil I. "Gus" Grissom, who died in 1967 when fire swept through his *Apollo* spacecraft.

✶

How Bright Is It?

ou can't tell by looking. A brilliant beacon of light might be an unimpressive but nearby star, while a tremulous wink of light could be a monstrous but very distant sun. Therefore two kinds of magnitude are used to gauge the brightness of a star. Apparent magnitude (m) measures how bright the star looks. It was invented in 134 B.C., when the Greek astronomer Hipparchus saw from his observatory on Rhodes that a new star in Scorpius had appeared in the eternal, unchanging heavens. He couldn't prove it, though, because he lacked an accurate record of how the constellation had appeared previously. To make sure this didn't happen again, he decided to chart the stars. In the process, he rated about a thousand stars according to brightness. Using naked-eye estimates, he placed them into six categories ranging from first magnitude, the smallest category, which included the twenty brightest stars, to sixth magnitude, the largest category and the one with the faintest stars. Thus he established for all time the irritating notion that the brighter a star is, the smaller the number describing its magnitude. "The confusion will disappear immediately if the word 'class' is substituted for 'magnitude,' " Robert Burnham, Jr., counsels in his three-volume *Celestial Handbook*. "Obviously we would expect a 'first-class star' to be brighter than a second or third class star. And the term 'fourth class' already begins to suggest faintness and unimportance."

Hipparchus' basic system, still in use, was calibrated in 1856 by English astronomer Norman Pogson, who observed that first-magnitude stars were 100 times as bright as sixth-magnitude stars and that consequently each magnitude was 2.512 times brighter than the previous one (2.512 is the fifth root of 100). Since then, dimmer stars have been given larger numbers—the Hubble Space Telescope is capable of taking the spectrum of twenty-sixth-magnitude stars—and a few highly luminous objects have been anointed with numbers so small that they extend into the negative (these might be thought of as ruling class). The Sun has an apparent magnitude of –26.72, and Sirius, the brightest star in the night sky, has an apparent magnitude of –1.46.

But apparent magnitude, which only measures how bright a star *looks,* can be misleading. (Nevertheless, when scientists say a star is first- or second-magnitude, they usually are referring to its apparent magnitude.) Distance makes a difference. So the Danish astronomer Ejnar Hertzsprung invented the concept of absolute magnitude (M), which allows stars to be compared according to how bright they would look if they were all at the same distance from Earth. The distance he chose is 10 parsecs, or 32.6 light-years—a little less than the distance of the stars Arcturus in Boötes and Pollux in Gemini, both of which are roughly 36 light-years away. With this scale, Rigel and Betelgeuse in Orion have respective absolute magnitudes of –7.1 and –5.6, while Sirius, which shines more brightly, rates a sorry +1.4. It is the brightest star in our sky only because of its proximity.

As for the Sun, it has an even less impressive absolute magnitude of +4.8. The Sun is an average star in that it lies roughly in the middle in terms of luminosity and mass. But in terms of numbers, the overwhelming majority of stars are smaller, cooler, and less luminous than the Sun, which is a member of the luminous, upper-class minority. Bright stars are exceedingly rare. In our immediate neighborhood, only four stars are equal to or brighter than the Sun. The other forty-nine are fainter and less important—not even fourth-class. And our neighborhood is more or less typical. Throughout the galaxy, about 70 percent of all stars are low-rent, low-luminosity red dwarfs much dimmer than the Sun, and of this starry hoi polloi, we can see not a single example with the naked eye.

✪

The Dog Star and the Pup

The great Overdog.
That heavenly beast
With a star in one eye,
Gives a leap in the east.

—Robert Frost, "Canis Major"

The Dog-star rages!

—Alexander Pope

he most brilliant star in the sky and one of the closest to Earth is Sirius, the Dog Star. Located in Canis Major, a constellation that somewhat resembles a stick figure of a dog, the star has long been feared and revered. Around 3200 B.C., it rose at dawn (an event known as the heliacal rising) on the summer solstice, heralding the annual flooding of the Nile and the beginning of a new year. The ancient Egyptians called it the Star of Isis, represented it hieroglyphically as a dog, and built the Temple of Dendera oriented toward the moment of its rising. The Romans acknowledged the star in late April, when Sirius set at dusk, with a complicated ritual in which a priest burned the entrails of a red dog. Muhammad forbade the worship of Sirius (or of any other star). Its influence was considered baleful; medieval sky watchers thought the "dog days" of August were miserably hot because Sirius had disappeared into the daytime sky and thus presumably was adding its heat to that of the Sun, and Sir Edmund Spenser in *The Shepheardes Calender* linked the star with "plagues and dreerye death."

The possibility that the Dog Star had a companion was first raised in 1752, when Voltaire described the fictional discovery of a satellite of Sirius. In 1844, the German astronomer Friedrich Wilhelm Bessel predicted the existence of just such an object after he noticed that the path of

Sirius was not straight, as a solitary star's would be, but wavy, suggesting that it and an unseen companion were orbiting a common center of gravity. His suppositions turned out to be true, and the star was discovered in 1862 by the telescope maker Alvan Graham Clark. At first, scientists assumed it to be faint and cool. But in 1914, Walter Sydney Adams analyzed its spectrum and proved that Sirius B was the opposite: luminous, hot, surprisingly small, and shockingly dense. Its mass, about equal to that of the Sun's, is squeezed into an area smaller than Neptune. "The message of the Companion of Sirius, when it was decoded, ran: 'I am composed of material 3,000 times denser than anything you have come across; a ton of my material would be a little nugget that you could put in a matchbox,'" Sir Arthur Eddington remembered. "What reply can one make to such a message? The reply that most of us made in 1914 was—'Shut up. Don't talk nonsense.'" Today, Sirius B is the most famous white dwarf in the sky.

The kingly brilliance of Sirius pierced the eye with a steely glitter, the star called Capella was yellow, Aldebaran and Betelgueux shone with a fiery red. To persons standing alone on a hill during a clear midnight such as this, the roll of the world eastward is almost a palpable movement.

—Thomas Hardy,
Far from the Madding Crowd

Sirius has been the source of two mysteries. One concerns its color. Although many writers of antiquity described the star as white, a few, including Cicero, classified it as red, leading some people to question whether it has changed color. That possibility has been rejected in favor of the explanation that when Sirius is scintillating low on the horizon, it flashes many subtle colors.

Stranger still is the discovery announced in the 1930s by a French anthropologist that despite the absence of telescopes or any connection to the usual scientific venues, the Dogon, a West African tribe in Mali, believed as an item of faith what we know to be true: Sirius has a small, heavy companion that orbits it every fifty years (fifty-two years, actually). Various explanations have been offered for this unlikely knowledge. One theory blames extraterrestrials; another puts forth a convoluted mythological analysis involving menstrual blood and rare grains. The mundane likelihood is that astronomically attuned missionaries may have reported to the Dogon the news about the Dog Star and the Pup.

✳

THE BRIGHTEST STARS IN THE SKY

Star	Constellation	Type	Apparent Magnitude	Absolute Magnitude	Distance (in light-years)
Sirius	Canis Major	blue-white main sequence	−1.46	+1.4	8.8
Canopus*	Carina	white supergiant	−0.72	−3.1	100†
Arcturus	Boötes	orange giant	−0.04	−0.2	36
Rigel Kentaurus* (Alpha Centauri)	Centaurus	yellow-white main sequence	0.0	+4.4	4.3
Vega	Lyra	blue-white main sequence	0.03	+0.3	26
Capella	Auriga	yellow giant	0.08	−0.6	42
Rigel	Orion	blue supergiant	0.12	−7.1	910
Procyon	Canis Minor	white subgiant	0.38	+2.6	11
Achernar*	Eridanus	blue subgiant	0.46	−1.6	85
Betelgeuse	Orion	red supergiant	0.50	−5.6	310
Hadar*	Centaurus	blue-white giant	0.61	−5.1	460
Altair	Aquila	white main sequence	0.77	+2.2	17
Aldebaran	Taurus	orange giant	0.85	−0.3	68
Antares	Scorpius	red supergiant	0.96	−4.7	330
Spica	Virgo	blue main sequence	0.98	−3.5	260
Pollux	Gemini	orange giant	1.14	+0.2	36
Fomalhaut	Piscis Austrinus	white main sequence	1.16	+2.0	23
Deneb	Cygnus	white supergiant	1.25	−7.5	1,800
Mimosa*	Crux	blue giant	1.25	−5.0	420
Regulus	Leo	blue main sequence	1.35	−0.6	85

*Too far south to be visible in most of the United States.

†It is a peculiar fact that figures such as these, which ought to be cut-and-dried, sometimes vary wildly. Estimates of Canopus' distance range from 74 light-years to 1,200 light-years, and figures given for its absolute magnitude range from a low of −2.5, to a high of −8.5.

Stella Polaris: The North Star

But I am constant as the northern star,
Of whose true-fix'd and resting quality
There is no fellow in the firmament.
The skies are painted with unnumber'd sparks,
They are all fire and every one doth shine;
But there's but one in all doth hold his place.

—William Shakespeare,
Julius Caesar (III. i. 60–65)

Bright star! Would I were steadfast as thou art.

—John Keats

he ancient Chinese called the forty-ninth-brightest star in the sky "the Great Imperial Ruler of Heaven" and "the Emperor of Emperors" because it is far and away the most useful star in the sky. The Pawnee specified the reason: they called it "the Star That Does Not Walk Around." Known as Stella Polaris, the North Star, the Lodestar, or the Pole Star because the Earth's axis points almost directly at it, it is the outermost star in the tail of the Little Bear, Ursa Minor, and the last star in the handle of the Little Dipper. Celebrated in prose and poetry by Shakespeare, Wordsworth, T. S. Eliot, and scores of others, it is renowned because it doesn't move. Find it now (the last two stars in the Big Dipper point right at it), and for the rest of your life it will always be in the same place: dead north, right above your neighbor's chimney, caught between the branches of a particular sycamore tree. From dusk to dawn, from season to season, it is what Shakespeare called "an ever-fixed mark."

But that's over the short run. In the long run, its constancy is an illusion. As the Earth revolves around the Sun it rotates on its axis around an imaginary pole. But the Earth is not a perfect sphere and its position is not fixed. It swivels on its axis, it wobbles like a top, and over the course of 25,800 years, the imaginary pole shifts direction and inscribes a circle in the sky. (Astronomers call this the precession of the equinoxes.) About 4,600 years ago, the axis pointed toward another North Star, Thuban, in Draco the Dragon. Seven thousand years earlier, Vega was the Pole Star. At the moment, the axis is pointing toward Polaris, but the lineup will be more exact in the year 2102. Five thousand years later, Alpha Cephei, the brightest star in Cepheus, will mark the axis of the world.

Polaris, the North Star.

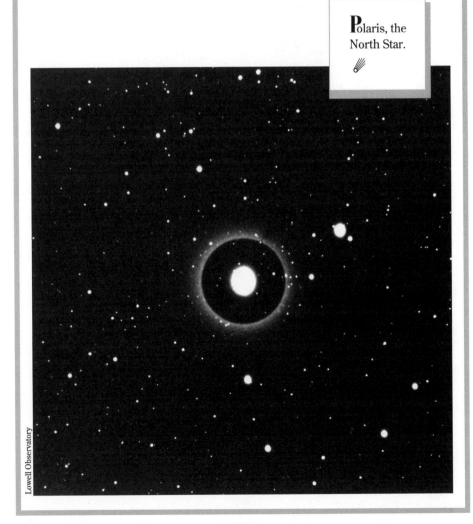

Lowell Observatory

Polaris is inconstant in another way too. A white star with a small, bluish, ninth-magnitude companion, it is a pulsating Cepheid variable. Over a period just shy of four days, its magnitude varies. The change is subtle and difficult to detect. And since the early 1970s, the difference has been decreasing. It may be that one day Stella Polaris will shine with a steadier light. In the meantime, the Earth continues to wobble and nothing is constant.

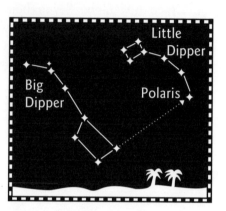

The Sixty-first Star of the Swan and Other Nearby Stars

For over two centuries after the invention of the telescope, astronomers struggled to figure out the distance to a star. The problem was solved in 1838 when the German astronomer Friedrich Wilhelm Bessel (1784–1846) calculated the distance to the inconspicuous fifth-magnitude star 61 Cygni in Cygnus the Swan.

He chose that particular star because it moved rapidly across the sky. To get a sense of its speed, imagine the entire sky, the celestial sphere that surrounds us, divided into 360° of arc. From the horizon to the zenith above your head is 90°; if you stretch your arm out in front of you, your fist should cover around 10°. Each of those degrees is further divided into 60 arc minutes, and each minute is subdivided into 60 arc seconds. Every year, 61 Cygnus covers a distance of 5.22 arc seconds across the sky—an imperceptible amount. At that rate, it takes 61 Cygni about three and a half centuries to cover half a degree, a distance roughly equal to the width of the full moon. In merely human terms, 61 Cygni is standing still. Astronomically speaking, it's moving so fast that it is called "the Flying Star."

Its movement, known as "proper motion," suggested to Bessel that 61 Cygni must be relatively close to the Earth, in which case it should be possible to calculate its distance through the use of parallax. Parallax is the principle invoked when you hold your thumb at arm's length in the classic artist's pose and look around the room shutting first one eye and then the other. When you switch eyes, faraway objects, such as the door, remain stationary, but your thumb seems to jump relative to the background because you're looking at it from two different locations. This makes it possible to figure out distances of nearby objects by calculating how much their position shifts when looked at from two places. By taking the degree of shift (always excruciatingly small in astronomy), measuring the distance between the observation points, and applying the rules of trigonometry, it should be possible to determine distance.

Unfortunately, the stars are so distant that no two locations on Earth are far enough apart to provide a different view. Bessel compensated for this by observing 61 Cygni from opposite ends of the Earth's orbit. By taking measurements at six-month intervals, he created a triangle with a base of around 186 million miles, double the distance between the Earth and the Sun.

Over a three-year period, Bessel repeatedly measured 61 Cygni against the background stars. Using carefully calibrated tools of measurement, telescopes said to be better mounted than the great William Herschel's, and an instrument of his own devising called the heliometer, Bessel discovered that 61 Cygni's parallax was slightly less than one third of an arc second—an infinitesimal amount, a fraction of a fine hair, but sufficient for his purposes. Given the length of the base of the triangle and the angle of parallax, which he found through painstaking measurements, it was a simple enough matter to figure out the distance to the star: 61 Cygni, he calculated, was 10.3 light-years away, slightly short of the current measurement of 11.2 light-years, but close enough to be considered accurate. By figuring out the actual distance to a star (61 Cygni is approximately 219,072,000,000 miles away), Bessel crashed through what Sir John Herschel called "the great and hitherto impossible barrier to our excursions into the sidereal universe—the barrier against which we have chafed so long and so vainly." The true scale of the universe, long hidden in the murky depths of space, swam into view.

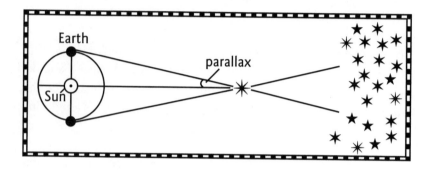

Using this method, astronomers have calculated the distances to the nearest 2,000 or so stars. (Other methods are needed to determine the distances of faraway stars.) Faint but visible, 61 Cygni is a fifth-magnitude binary star consisting of two orange dwarfs revolving around each other every 650 years, possibly accompanied by an insignificant third body thought to be either a very puny star or possibly a planet. 61 Cygni is not the closest star to Earth—fourteen stars, excluding the Sun, are closer, although ten of those are too faint to be seen. Nor is it the star with the fastest proper motion; that honor belongs to Barnard's star. 61 Cygni is the star that put the first notch on the measure of the universe.

☆

The Nearest Stars to Earth

Star	Constellation	Type	Apparent Magnitude	Absolute Magnitude	Distance
The Sun		yellow main sequence	−26.72	+4.8	8 light-minutes
Proxima Centauri	Centaurus	red dwarf	11.05	+15.4	4.2 ly (light-years)
Alpha Centauri	Centaurus	yellow-white main sequence	0.0	+4.4	4.3 ly
Alpha Centauri B	Centaurus	orange main sequence	1.33	+5.71	4.3 ly
Barnard's star	Ophiuchus	red dwarf	9.54	+13.22	5.9 ly
Wolf 359	Leo	red dwarf	13.53	+16.65	7.6 ly
Lalande 21185	Ursa Major	red dwarf	7.50	+10.5	8.1 ly
Luyten 726-8A	Cetus	red dwarf	12.52	+15.46	8.4 ly
Luyten 726-8B	Cetus	red dwarf	13.02	+15.96	8.4 ly
Sirius A	Canis Major	blue-white main sequence	−1.46	+1.4	8.8 ly
Sirius B	Canis Major	white dwarf	8.3	+11.2	8.8 ly
Ross 154	Sagittarius	red dwarf	10.45	+13.14	9.4 ly

* Proxima Centauri, the second star on this list, is one of the least luminous stars known. The nearest star to us (other than the Sun), it revolves at a considerable distance from its parent pair of Alpha Centauri A and Alpha Centauri B, whose distance was calculated by Thomas Henderson (1798–1844). In the 1830s, Henderson returned to Scotland from a stint as director of the observatory at the Cape of Good Hope, studied the data he had collected about Alpha

The double star Alpha Centauri A and Alpha Centauri B. Together with Proxima Centauri, not shown in this photograph, they constitute the nearest star system to our own.

Centauri, and determined its parallax and hence its distance. This accomplishment garnered less acclaim than he might have hoped only because by the time it was published in 1839, Friedrich Bessel had already determined the distance to 61 Cygni.

* Barnard's star, discovered by E. E. Barnard in 1916, has the largest proper motion of any star in the sky: it hurries across the cosmos, moving twice as far in a year, or in a century, as 61 Cygni.

* Wolf 359 is the faintest red dwarf yet discovered.

* Lalande 21185 is the brightest red dwarf.

* Luyten 726-8A, also a red dwarf, is part of a binary system; its companion, 726-8B, is a famous flare star that once brightened 160 times in twenty seconds.

✳ Sirius A is the brightest star in the sky; Sirius B, its tiny white dwarf companion, is known as the Pup.

✳ And Ross 154 in Sagittarius, the last star on this list, is just another red dwarf of no particular significance, a sort of low-end Everystar.

F. W. Bessel and the Personal Equation

Friedrich Wilhelm Bessel (1784–1846) entered the import business as an accountant at age fifteen. He soon left the profession behind—but not the pursuit of exactitude it required and for which he is known. He became an astronomer, after studying mathematics, geography, and languages, by writing a paper about Halley's Comet and mailing it to Heinrich Olbers, the doctor and comet hunter famous for wondering why the sky is dark at night. Olbers got Bessel's paper published and introduced him to Johann Hieronymus Schröter, who hired the twenty-year-old Bessel as an assistant at his private observatory. By the time he was twenty-six years old, Bessel had been appointed director of the Königsberg Observatory by King Frederick William III of Prussia. He maintained that post for the rest of his life.

Bessel's eternal concern was measurement. In addition to measuring the parallax, and hence the distance, of 61 Cygni, he measured the diameter of Mercury (he was accurate to within ten miles) and compiled a star catalogue for which he made an estimated 75,000 observations and pinpointed the positions of 31,895 stars. In the process, he developed a concept concerning mechanical instruments. They were prone to inexactitude, he said, for however careful the design, in real life the instrument will fail to achieve its intended level of precision. Something will be off: not quite horizontal, not quite smooth, not perfectly calibrated. These tiny errors diminish the instrument's precision, and the astronomer must correct for them. "Every instrument in this way is made twice," he said at a public lecture. "Once in the workshop of the artisan, in brass and steel, and then again by the astronomer on paper, by means of the list of necessary corrections which he derives by his investigation."

Another correction also had to be made, one that he noticed while making the measurements necessary to catalogue all those stars. The process involved marking the moment when the star first appeared to touch the thin lines within the eyepiece of the telescope. His assistant, engaged in the same tedious task, inevitably registered that moment a second later than Bessel did. This led Bessel to formulate the notion that accurate observations are compromised by "personal errors" according to which observers are always a little early or, more commonly, a little late in record- ing the timing of an event. The amount by which they are off tends to be consistent; hence, it can be accounted for. He called it the personal equation.

✫

F. W. Bessel
(1784–1846).

Zodiac Tales

very year as the Sun, the Moon, and the planets drift across the heavens, they pass through the same strip of sky in front of the twelve constellations we call the zodiac. As early as 3500 B.C., the Sumerians recognized Taurus, Leo, and Scorpius, the most dramatic constellations among them. The other constellations were added slowly (although Ophiuchus the Serpent Bearer and Cetus the Whale, both of which intrude into that same band of the cosmos, were never incorporated into the zodiac at all). By 1600 B.C., the Chaldeans—a people so interested in prognostication that "Chaldean" and "astrologer" were once considered synonymous—were studying a zodiac very similar to our own, and by 400 B.C., the zodiac as we know it was firmly established, along with much of the mythology. Every culture has told stories about the stars. The tales told herein are primarily from the Greeks.

♈ Aries ♈

People born under the sign of Aries have powerful personalities, astrologers say. The constellation, on the other hand, is strictly second-rate, its chief feature being a flat triangle of second- and fourth-magnitude stars.

The story of Aries begins with King Athamas of Boeotia and the two children he had with his first wife. The marriage didn't work out, and his second marriage also had problems: His new wife, Ino, hated her stepchildren so much that she wanted to kill them.

The German astronomer known as Regiomontanus (1436–76) published tables of planetary motion, astrology books, and a revision of Ptolemy's *Almagest*. A 1496 edition of that book, published twenty years after he died of the plague, included this frontispiece. It pictures Regiomontanus and Ptolemy, his crowned head bent over a book, sitting beneath an armillary sphere.

There was also famine in the land. So the king sent a messenger to consult the Delphic oracle. He awaited a reply, but Ino intercepted the courier and substituted a message saying that Athamas should sacrifice his son, Phrixus. Like the biblical Abraham, Athamas prepared to do this.

Fortunately, at the last moment, his first wife, Nephele, sent a winged ram whose fleece was like gold. Phrixus climbed onto the ram with his sister, Helle, who was also in danger, and flew to the Black Sea. As they soared over the water, Helle lost her grip and tumbled into the narrow channel between Europe and Asia (it was for a long time known as the Hellespont in her honor). Phrixus arrived safely, sacrificed the ram, and hung the soon-to-be fabled Golden Fleece from a tree, where it was guarded by a dragon.

A related tale says that Neptune, afraid he might be rejected by the woman he desired, once changed both himself and his object of desire into sheep; the ram of Aries, with the Golden Fleece, was born of that union.

♉ Taurus ♉

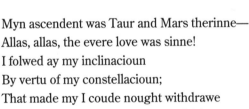

Myn ascendent was Taur and Mars therinne—
Allas, allas, the evere love was sinne!
I folwed ay my inclinacioun
By vertu of my constellacioun;
That made my I coude nought withdrawe
My chambre of Venus from a good felawe.

    —Geoffrey Chaucer,
      "The Wife of Bath's Tale," *Canterbury Tales*

If Aries the Ram is an unimpressive constellation, Taurus the Bull is a great one. Because ancient mythologies are replete with bull legends, it's not clear which bull Taurus represents. It might be the Bull of Osiris, a sort of Dalai Lama among bulls, recognizable by certain physical signs and chosen in childhood to be the vessel containing the soul of the great god. Eventually, the bull was sacrificed, whereupon a new Bull of Osiris was found.

Or Taurus may represent Zeus, who once turned himself into a snow-white bull so as to more readily abduct the Phoenician princess Europa. Once she climbed onto his back, he swam across the Mediterranean to Crete, where he seduced her. They had three sons. Among his gifts to her was a dog, which later became the constellation Canis Major.

Or Taurus may have been the Minotaur, born on Crete with the head of a bull and the body of a man. In the constellation, only the head and shoulders of the bull are visible.

> **C**hris Callinan
>
> *What is the parallax of the subsolar ecliptic of Aldebaran?*
>
> *Bloom*
>
> *Pleased to hear from you, Chris. K. 11.*
>
> —James Joyce, *Ulysses*

Among the notable features in Taurus are two star clusters—the Hyades, in the face of the bull, and the nearby Pleiades—and the bright star Aldebaran, which marks the eye of the bull. A red-orange giant with a diameter thirty-six times that of the Sun, Aldebaran trails across the sky after the Pleiades, rising in the same place one hour later. (Thus its

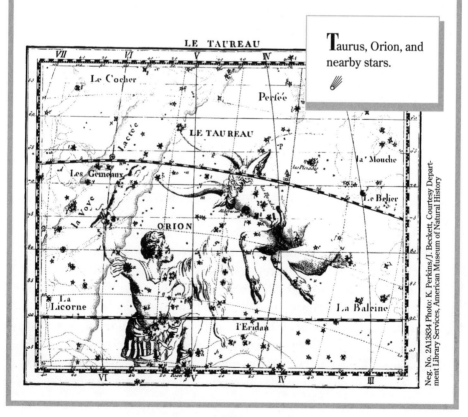

Taurus, Orion, and nearby stars.

Neg. No. 2A13834 Photo: K. Perkins/J. Beckett, Courtesy Department Library Services, American Museum of Natural History

Arabic name, which means "the follower.") According to a Polynesian story, the most brilliant star in all creation once lived in this part of the sky, but it couldn't help bragging about its beauty. This eventually irritated the Maori god Tane so much that he threw Aldebaran at it, causing the star to shatter. The Pleiades are all that remain of that bright light.

Taurus is also the home of the Crab Nebula, the fuzzy remains of the supernova explosion seen in the year 1054.

♊ Gemini ♊

Castor and Pollux (Polydeuces to the Greeks) were brothers who sailed with Jason and the Argonauts in search of the Golden Fleece. Hatched from a giant egg, they may have been twins, conceived when Leda, the queen of Sparta, mated with Zeus in the form of a swan.

Or perhaps they were half brothers, sired by different fathers on the same day. According to this version of the story, Pollux and Helen of Troy were the offspring of Zeus, and were therefore immortal. Castor and his sister, Clytemnestra, were the children of Tyndareus, the king of Sparta, and thus were fated to die. The mortal Castor and his immortal brother were nonetheless inseparable.

One day the brothers attended a wedding and decided to kidnap the brides. This led to a feud that resulted in the death of both grooms and Castor. Grief-stricken, Pollux begged Zeus to let him share his immortality with his brother. Zeus responded by placing the brothers together in the skies. But their time has to be split; for every day on Olympus, they must spend a day in Hades.

The Australian aboriginal people say that during the Dreamtime, the Moon, whose name is Kulu, chased a group of women so relentlessly that finally two lizard men, known collectively as Wati-kutjara, fatally wounded him with their boomerangs, which turned his face white and bloodless. The women were turned into the Pleiades, and the lizard men became the constellation Gemini.

Pollux, the seventeenth-brightest star in the sky, is a yellow-orange giant about thirty-six light-years away from us. Castor, the twenty-third-brightest star, is actually a system of six stars grouped into three binary pairs forty-six light-years away.

Also in Gemini is NGC 2392, a planetary nebula with a white dwarf at its center. It was discovered by William Herschel and is known as the Eskimo Nebula because it resembles a face surrounded by a furry hood. Robert Burnham, Jr., accurately notes that it "irresistibly suggests the classic and unforgettable features of W. C. Fields."

Another unusual object in Gemini is the mysterious pulsar Geminga, known since 1972 as a powerful source of gamma rays. Located only 195 light-years away, it was created in a supernova 340,000 years ago, an explosion that destroyed perhaps 20 percent of the Earth's ozone layer, flooded our Stone Age planet with ultraviolet radiation, and may have helped form the gas bubble within which the Sun and other stars are found. Although Geminga is in Gemini now, it moves across the sky so rapidly that at the time of the explosion, it was probably in Orion.

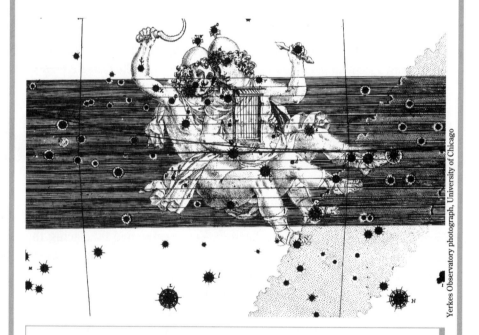

Yerkes Observatory photograph, University of Chicago

Gemini from Bayer's *Uranometria*. The star Castor, labeled α, is in the forehead of the twin at the right, while Pollux (β) is in the jaw of the other twin. Below the dark strip representing the zodiac are the bright stars Procyon (L) in Canis Minor, and Betelgeuse (H), which marks Orion's shoulder. The dotted diagonal band is the Milky Way.

♋ Cancer the Crab ♋

Hera hated Hercules even before he was born. When he was just a baby, she tried to kill him by putting two snakes in his bed. Later on, when Hercules fought the many-headed Hydra, she rooted for the monster. In the midst of that battle, a large crab emerged from the swamp, bit Hercules on the foot, and was crushed to death by the demigod. In recognition of its attempt to defeat her enemy, Hera raised the crab into the celestial sphere. But since the crab had failed in its task, she withheld any bright stars. As a result, Cancer is the most inconspicuous constellation in the zodiac.

It does have one notable feature: the hazy star cluster M44, known as Praesepe, or the Beehive Cluster. With the unaided eye, it looks like a faint smudge, but as Galileo discovered in 1610, it is an amalgamation of many stars. Galileo saw 36 stars in the cluster, and at least 350 have been counted since. It is about 530 light-years from Earth.

♌ Leo the Lion ♌

One of the labors of Hercules was to kill the Nemean lion, which fell from the Moon in the shape of a meteor and landed in Nemea, where it threatened the countryside and harassed the people. No weapon could pierce its hide, but Hercules blocked off one of the openings to the lion's cave, cornered the animal, clubbed it, and choked it to death. Afterward, he stripped off the lion's hide and wore it himself.

Leo actually resembles the animal after which it is named. A right triangle of stars outlines the back legs, where a cluster of spiral galaxies can be found. The front of the constellation, like a giant backward

question mark, defines the head, mane, and front legs. At the base of the question mark is Regulus, the heart of the lion.

Regulus, a name this star acquired in the fifteenth century meaning "Little King," is a large bluish star with a small orange companion that is itself a binary star. Immediately above Regulus is Eta Leonis, a huge white star. Comparing these stars offers a striking contrast. Regulus looks much brighter than Eta Leonis only because it is closer. Regulus is 85 light-years away, whereas Eta Leonis is perhaps 2,000 light-years away. But Eta Leonis is so large that, if the Sun were the size of a pea, Regulus would be the size of, say, a Ping-Pong ball, and Eta Leonis would have the diameter of a hula hoop. If Eta Leonis were at the same distance as Regulus, it would outshine every star in the sky several times over; if it were as near as Sirius, it would be visible by day. But since Eta Leonis is 2,000 light-years away, it is only an unimpressive, third-magnitude dot of light.

♍ Virgo ♍

There have been, according to the Roman poet Ovid (43 B.C.–A.D. 17), four ages of mankind, each worse than the previous one. First came the utopian Golden Age, when mankind lived off fruit and "acorns that fell from the spreading oak of Jupiter." Next came the Silver Age, when "the scorched air glowed with parching heat and ice hung frozen by the winds" and "mankind lived in houses." The Bronze Age saw the arrival of "a race harsher in nature and quicker to take up rough weapons, but nevertheless not wicked people." And then came our own age, the Iron Age, when "men lived by plundering. Friends were not safe from friends, nor relatives from relatives, and even brotherly love was rare." So unpleasant was it that the gods and goddesses left. The last deity to abandon the planet Earth was Astraea, the daughter of Aurora.

Another version of this story says that the last deity to bail out was Tyche, the little-known goddess of good fortune who was given ruler-ship over dice (she is sometimes referred to as Dike). When the Iron Age came, bringing war and crime, she ascended to the heavens, where she is thought of as the goddess of justice because of her proximity to the scales of Libra. Tyche and Astraea are both represented in the sky by the constellation Virgo.

But then Virgo, the only constellation of the zodiac identified as a woman, has been associated with virtually every female goddess you can name, from Ishtar to Urania, the Muse of astronomy. Virgo is also asso-ciated with Persephone, who spends six months underground with her

M104. Arguably the most beautiful galaxy in the sky, M104 is edge-on to the Earth, with a dark lane of dust and gas that runs across the equator like the narrow brim of a luminous hat, as a result of which M104 is known as the Sombrero Galaxy. It is thought to be somewhere between a spiral galaxy, like the Milky Way, and an elliptical one.

abductor, Pluto, and six months above ground with her mother, Demeter, the goddess of grain; and with Erigone, the dutiful daughter of the mortal Icarius, who was given the gift of wine. He carted the wine from town to town, but because the people thought it was toxic, they killed him. When Erigone learned of her father's murder, she wept inconsolably and was consequently turned into a constellation. (Icarius is sometimes identified with Boötes the Herdsman; the wagon in which he carried his wine is linked with Ursa Minor, which is occasionally called "the Lesser Wagon.")

The second-largest constellation (Hydra is first), Virgo has a single bright star—Spica, a first-magnitude star whose name signifies a spike of wheat. Spica is famous for being easy to find. Once you've located the Big Dipper, follow the curve of its handle to Arcturus—arc to Arcturus—and just continue along the curve until you spy Spica. It's a long, graceful line that flows over half the night sky.

Through the stars of Virgo can also be found the Virgo Cluster, the nearest major cluster of galaxies. An agglomeration of as many as 3,000 galaxies including 250 large ones, it is about 70 million light-years away—far beyond the Milky Way. In the sky, the Virgo Cluster covers a patch fourteen times the size of the Moon. In the universe, it stretches out over 10 million light-years.

♎ Libra ♎

Large yet inconspicuous, Libra is the only constellation in the zodiac that involves no living being. Four thousand years ago, the Sumerians thought of it as "the balance of Heaven," because on the first day of autumn, when the day and the night were equal in length, the Sun rose in this sign. Two thousand years later, the Romans and the Egyptians still saw the constellation in a similar way.

To the Greeks, Libra was the chariot used by Pluto to bring Persephone to the Underworld, but they also connected Libra with the majestic constellation Scorpius by defining its two brightest stars as the southern and northern claws of the scorpion. Their names in Arabic are Zubenelgenubi ("southern claw") and Zubeneschamali ("northern claw").

Zubeneschamali is a white star, but it has also been described as pale emerald green. Although it does not shine especially brightly to us, in the third century B.C., Eratosthenes of Cyrene found it more luminous than Antares, the brilliant red star at the heart of Scorpius, suggesting that Zubeneschamali was brighter in those days, that Antares has changed, or that Eratosthenes, famous for calculating the circumference of the Earth, was not paying attention.

♏ Scorpius ♏

To the Chinese, this constellation was a dragon. To the Polynesians, it was the fishhook used by the trickster god Maui to pull up islands from the ocean floor. The Greeks saw it as a giant scorpion, and it's easy to understand where that image came from. On a summer's night, when Scorpius seems to crawl out from under the horizon, it looks the way it's supposed to look, and the stories told about it clearly reflect its position in the sky:

Orion, the great hunter, bragged that he could not be harmed by any animal, whereupon Hera sent the scorpion to teach him a lesson. It stung him fatally, but before he died, he smashed it with his club.

Or maybe he threatened to kill all the animals, and Gaia sent the scorpion to kill him.

Or maybe he tried to rape Artemis, who picked up a giant scorpion and killed him with it.

In any case, all versions of the story end the same way, with Orion and the scorpion placed in opposite parts of the sky. When one rises, the other sets.

Scorpius from Bayer's *Uranometria* (1655). The bright star Antares is in the middle of the constellation.

Its most notable feature is the bright red supergiant Antares, the heart of the scorpion—an analogy emphasized by the Arabic names of the stars immediately above and below it, which are known as "*al niyat*," the arteries.

At the southern tip of South America, in Tierra del Fuego, these three stars are linked in a different way. Antares is the god Kwonyipe, the stars above and below are his two wives, and on the opposite side of the sky is his adversary, the bright star we call Canopus.

Like the planet Mars, who was known to the Greeks as Ares, this star is reddish; indeed, the name "Antares" means "Anti-Ares," or rival of Mars. Yet despite its obvious redness, Antares, like Zubeneschamali, is also known for its greenness. Nineteenth-century observers noticed that "when the star was watched intently, especially with an instrument of adequate power, a peculiar green light was found to force itself persistently into view," the popular-science writer Mary Proctor wrote in 1924. "Suspicion was at last aroused among observers to the effect that the star must have a green companion which had so far escaped detection."

It was sighted in 1845 or 1846 by Professor O. M. Mitchell, director of the Cincinnati Observatory. "As this was the first noteworthy achievement of the Cincinnati telescope," Proctor wrote, "the discovery was a source of considerable gratification."

Unfortunately, it was not a discovery, for the small sixth-magnitude companion star was first seen in 1819 by a Viennese professor. As for its color, that seems to be in the eye of the beholder, for this little star has been described as pale yellow, very blue, purple, and green—which may simply be an optical illusion created by its contrast to red Antares. Still, it's pleasant to mull over John Herschel's vision of life near Antares: "Imagination fails to conceive the charming contrasts and graceful vicissitudes of a red and green day, alternating with light or with darkness, in the planetary systems belonging to these suns."

✒ Sagittarius ✒

It takes little effort to imagine Scorpius as a scorpion, and its identification as such dates back 5,000 years. But it's not easy to picture Sagittarius as an archer. A more vivid image is that of a teapot, with a handle, a knob on top, and a spout pouring briskly into the Milky Way. On a sum-

National Optical Astronomy Observatories

The Trifid Nebula, M20, in Sagittarius. Dark trails of dust and gas are silhouetted against clouds glowing from the radiation of unseen stars.

mer night, when Scorpius has climbed above the horizon, Sagittarius the Teapot follows and is easy to identify. It marks the center of the galaxy, where the Milky Way is thickest and brightest. Star clusters and nebulae abound, including the Lagoon Nebula, the strangely lobed Trifid Nebula, and the Omega Nebula, where young stars are being born among glowing hydrogen clouds.

Three thousand years ago in India, this constellation was seen as a horse; the Babylonians associated it with Nergal, the king of war; to the Chinese, it was a tiger; and the Arabs saw it as a flock of ostriches, with the stars near the spout as "the going ostriches," the stars in the handle as "the returning ostriches," and the star that marks the teapot's lid as "the keeper," who watches over both groups as they travel to and from the Milky Way.

In classical mythology, Sagittarius was a centaur, half horse and half man. According to one story, Ixion, notorious for having killed his own father-in-law, tried to seduce Hera. After she told her husband, Zeus, about Ixion's advances, the great god set a trap by creating a cloud that resembled his wife in every way. Ixion mated with this cloud, and from that union came the centaurs, a rowdy and lascivious group.

Another centaur was Chiron, the friend of Apollo and the tutor of Jason, Asclepius, and Achilles. Chiron was born as a result of the duplicitous union of his father, Cronos, and a sea nymph named Phylira. Because Cronos didn't want his wife, Rhea, to find out about his liaison, he coupled with Phylira in the shape of a horse, and in consequence Phylira gave birth to Chiron. As the son of a god, Chiron was immortal, but once, when the other centaurs and Hercules got into a drunken mêlée, Chiron, an innocent bystander, was wounded in the knee by one of Hercules' poisoned arrows. In intense pain but unable to die, he begged to be allowed to bestow his immortality on someone else. His wish was granted. Prometheus, who had stolen fire from the gods to give to mankind and been punished by having his liver continually eaten by an eagle, became immortal, and Chiron was lifted into the heavens. Exactly where remains in dispute. Some say that Chiron is Sagittarius; others maintain that by the time Chiron ascended into the skies, most of the good spots were taken, and so he became the constellation Centaurus, in the southern sky, while the more bellicose Sagittarius, shown aggressively pulling back a bowstring and aiming an arrow at Scorpius, is one of the sons of Ixion.

♑ Capricornus ♑

One of the dimmest constellations of the zodiac, Capricornus is poor in galaxies and deep-sky objects. It does, however, have a curious mythological symbol. Like Sagittarius, which is half man and half horse, Capricornus is an amalgam: half goat and half fish. This image, which appears on Babylonian tablets, is identified with a number of figures, including the horned god Pan, who was sitting by the Nile one day with a few other gods when the monster Typhon suddenly appeared. Terrified by the monster's thunderous voice and hundred snake heads, the gods changed shapes, turning themselves into animals in order to escape. But Pan was frozen in indecision. At the last possible moment, he jumped into the river and was transformed so instantly that his body wasn't even entirely in the water yet. His submerged lower half became a fish while his upper half, still above the water, remained a goat.

〰 Aquarius 〰

Aquarius, a man carrying a water jug, is probably a Babylonian figure connected with early flood myths. The Egyptians linked it to the annual flooding of the Nile, and it is surrounded in the sky by other watery constellations, including Cetus the Whale and Pisces. No particular myths seem to be connected with this figure, although Aquarius is often identified as the boy Ganymede, who was chosen by Zeus to be his cupbearer and was either abducted by the gods, as Homer claims, or lifted into the firmament by Aquila the Eagle, who is nearby.

Aquarius is the home of two great planetary nebulae. Planetary nebulae are formed when a dying star ejects a shell of luminous gas. Often the gas maintains its spherical form even as it expands outward. The largest and nearest example is the Helix Nebula (NGC 7293), which clearly shows a glowing circular ring with a single star at the exact center.

Also in Aquarius is the Saturn Nebula (NGC 7009), which received its name because it vaguely resembles the planet.

Hannah Berman

The Saturn Planetary Nebula, in Aquarius.

The brightest stars in Aquarius are Sadalmelik, Sadalsud, and Sadachbia, which can be translated as "lucky one of the king," "luckiest of the lucky," and "lucky star of hidden things" or "lucky star of the tents."

♓ Pisces ♓

> Now follow me, we should be getting on;
> the Fish are shimmering over the horizon . . .
> —Dante Alighieri, *Inferno*

Like Pan, Aphrodite and her son Eros (known to the Romans as Venus and Cupid) were once threatened by the monster Typhon. They too escaped by leaping into the river, where they were transformed into fish. To make certain they didn't lose each other in the watery deep, they took a length of string and tied their tails together. The fourth-magnitude binary blue-white star known as Al-rescha, from the Arabic *al risha* ("the cord"), represents a knot in the string. It is the brightest star in the constellation.

Pisces is notable for another reason. Over 2,000 years ago, the signs of the zodiac, as astrologers know them, roughly corresponded to the constellations recognized by astronomers. (However, the constellations are not equal in size, whereas the signs of the zodiac have traditionally been assigned equal space on the calendar.) On March 21, the first day of Aries, the Sun entered the same swath of sky inhabited by that constellation; a month later, it could be found among the stars of Taurus. But because the Earth wobbles on its axis, the dates astrologers use no longer correspond to the astronomical reality. And so it is that on the first day of spring, once the first day of Aries, the Sun can now be found traveling through the stars of Pisces. But that too is an impermanent situation. Around the year 2500 the Sun will enter the constellation of the Water Bearer on the first day of spring, and the Age of Aquarius will begin.

> **T**he great cusp—green
> equinox and turning,
> dreaming fishes to young ram,
> watersleep to firewaking,
> bears down on us.
>
> —Thomas Pynchon,
> *Gravity's Rainbow*

The Ten Brightest Stars in the Zodiac

Some constellations in the zodiac are dramatic presences in the night sky and some are not. It all depends on the presence of first- and second-magnitude stars, and they are not distributed equally. Of the fifty brightest stars in the sky, three are in Scorpius, two are in Taurus, two are in Gemini, and one each is in Leo, Virgo, and Sagittarius. The other constellations of the zodiac, worthy though they may be in other respects, can claim not a single bright star.

Star	Constellation	Type	Apparent Magnitude	Distance (in light-years)
Aldebaran	Taurus	Orange giant	0.9	68
Antares	Scorpius	Red supergiant	1.0	330
Spica	Virgo	Blue main sequence	1.0	260
Pollux	Gemini	Orange giant	1.1	36
Regulus	Leo	Blue main sequence	1.4	85
Castor	Gemini	Blue/white main sequence	1.6	46
Shaula	Scorpius	Blue subgiant or main sequence	1.6	275
El Nath	Taurus	Blue giant	1.7	130
Kaus Australis	Sagittarius	Blue subgiant	1.9	85
Sargas	Scorpius	White supergiant	1.9	915

A Drama of the Northern Skies: Perseus, Cassiopeia, Cepheus, Andromeda, Pegasus, and Cetus

The great mythological drama of the northern skies, visible most fully in autumn, began when King Acrisius was told that his daughter Danaë would bear a son who would eventually kill him. Intent on avoiding this fate, he imprisoned her in a tower, but Zeus slipped into her cell in the shape of a shower of gold and she became pregnant. When the baby Perseus was born, Acrisius locked him and his mother into a chest and pushed it into the sea. They drifted ashore on the island of Seriphos, where a fisherman named Dictys found them and raised Perseus as his own child. Time passed. Then one day the brother of Dictys, King Polydectes, decided he wanted to marry Danaë. She rejected the offer.

When Perseus, grown up by now, came to his mother's defense, Polydectes angrily ordered him to bring back the head of one of the three Gorgons: Medusa, a beautiful young woman whose hair had been turned into a nest of snakes. So terrible was the sight that anyone looking at her was turned to stone. In consequence, she lived in a sculpture garden, surrounded by the statues of men and beasts who had glanced her way. Because the danger was so great, the gods offered Perseus protection in the form of various magical objects. Hermes gave him winged sandals and an iron sword; the nymphs gave him a helmet of invisibility and a special pouch for the head; Athena gave him a shiny bronze shield. By looking at Medusa's reflection in the shield, Perseus was able to slice off

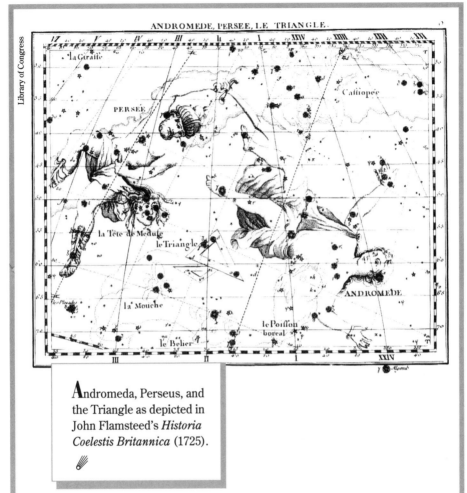

Andromeda, Perseus, and the Triangle as depicted in John Flamsteed's *Historia Coelestis Britannica* (1725).

her head without gazing directly at her. He dropped the head into the pouch, then donned the helmet and disappeared from sight, thereby protecting himself from the attacks of her immortal sister Gorgons. Drops of Medusa's blood fell into the ocean, mixed with sea foam, and formed Pegasus. Perseus mounted the winged horse and flew away.

On the way home, he chanced to see Andromeda chained to a rock by the sea. The lovely daughter of Cepheus and Cassiopeia, the king and queen of Ethiopia, Andromeda was in peril as a result of her mother's vanity. Cassiopeia had bragged that she was more beautiful than the Nereids, the sea nymphs. Insulted, the Nereids asked Poseidon, the god of the sea, to punish her. He did this by sending down a giant flood and a vicious sea monster, Cetus. When Cepheus consulted an oracle about how to rid himself of this curse, he was told to sacrifice his daughter. He chained her to a rock along the shore.

The tide was rolling in when Perseus flew by. The moment he saw Andromeda "he breathed the fire of love," according to Ovid. He was promised Andromeda's hand in exchange for a vow to kill the monster, which he promptly did with his sword. Then, to protect the head of Medusa, he carefully placed it in a nest of seaweed, which immediately turned to stone—and that's how coral was created.

Afterward, Perseus discovered that Cepheus had been less than scrupulously honest, for he had already promised his daughter's hand to his own brother, Phineus. The wedding feast of Perseus and Andromeda was in progress when Phineus showed up with his troops to claim the bride. But Perseus brandished the head of Medusa, turning Phineus to stone. A similar fate awaited the Titan Atlas, who was turned into a mountain of stone upon which the heavens rest, and Polydectes, who expressed doubt that Perseus had secured the head at all and was consequently forced to look at the evidence. Eventually, Perseus gave the head to Athena, who placed it on her shield.

Later on, while attending funeral games held in honor of a Thessalonian king, Perseus threw a discus that spun into the stands and killed a spectator. By happenstance, the spectator was his grandfather, old King Acrisius. The prophecy had been fulfilled.

In the sky, Cassiopeia is easy to find. Old star atlases depict the constellation as a queen sitting on her throne, but it forms a figure more easily described as a *W* or an *M,* depending upon the time of year. The constellation rotates around the North Star, with the result that half the time, Cassiopeia is in a humiliating, upside-down position—a form of punishment, according to early Greek and Roman writers. Of course, all the circumpolar constellations rotate around the sky in precisely the same way, but for reasons we can only guess at, this upside-down position figures only in the mythology of the beautiful, vain queen Cassiopeia.

Cepheus is nearby. Although usually pictured in star atlases as a king, the constellation more clearly resembles a house drawn by a five-year-old, with the peak of the roof nearest Polaris. Pegasus is recognizable as a giant square. Andromeda springs from one corner of the square and stretches toward Perseus, where the eclipsing binary star Algol represents the terrible blinking eye of Medusa.

The Demon Star

Old celestial atlases show Perseus, sword in hand, holding aloft the head of the snake-haired Medusa. Her blinking eye is the variable star Algol. The name, derived from the Arabic *alghul*, means just what it sounds like: the ghoul, or demon. The Chinese name for the star is even worse: Tseih She, or Piled-Up Corpses.

The talented young astronomer John Goodricke (1764–86) discovered the star's variability. A deaf-mute from birth, Goodricke was born in Holland to an English family and was educated in England, where his best friend's father inspired him to study the heavens. One night shortly after his nineteenth birthday, Goodricke saw that Algol, normally a second-magnitude star, was so faint that it looked like a fourth-magnitude star.

Now, Goodricke was not the first to notice this; Geminiano Montanari of Bologna had observed the same phenomenon in 1667, but hadn't pursued it. After five months of careful observation, Goodricke discerned the star's pattern. After 2 days, 20 hours, 48 minutes, and 56 seconds, it suddenly dims. Ten hours later, it brightens again.

The cause of this behavior, Goodricke reasoned, was that Algol had an invisible companion. As the companion star orbited in front of Algol (from the point of view of the observer on Earth), it blocked the light, creating an eclipse. The rest of the time, the light was constant.

This interpretation is correct. Although it looks like a dot of light, Algol—like most stars—belongs to a family. The two main stars, Algol A and Algol B, orbit each other closely at a distance of about 6.5 million miles—about $\frac{1}{14}$ the distance between the Sun and the Earth and close enough so that the stars pull each other out of shape. Algol A is so bright that in comparison the slightly larger Algol B—which is more luminous than the Sun—looks dark. When this "dark" star passes in front of its companion, the luminosity of the system plummets, but when the bright star eclipses the "dark" one, the luminosity takes only a tiny dip.

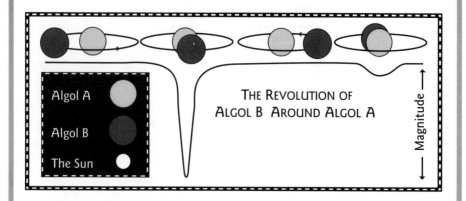

Algol A

Algol B

The Sun

THE REVOLUTION OF
ALGOL B AROUND ALGOL A

Magnitude

Less than a hundred years after Goodricke's untimely death from pneumonia, Algol was found to be not just a binary but a multiple star. Fifty million miles from the primary pair is a third star, which revolves around them every twenty-two months or so. A fourth star, even more distant, takes 188 years to circle the binary pair.

Goodricke also made other discoveries, including Beta Lyrae, an eclipsing binary made up of a blue giant, a red giant, and a river of gas flowing between them, and Delta Cephei, a pulsating star so important that an entire category of stars, called Cepheid variables, has been named in its honor. Goodricke's genius was recognized immediately. At age nineteen, he won the Royal Society's Copley Medal, and at age twenty-one, he was elected to that organization. His sudden death two weeks later was blamed on exposure to the cold night air.

The Wonderful Star

Although most stars shine steadily, an amazing number do not. Among them are variable stars that swell and contract so rapidly that they seem to blink; stars that exhale huge portions of their mass in plumes of nebulosity; stars whose brightness is dimmed by sooty clouds of carbon in the stellar atmosphere; stars that explode. Some stars are so misshapen that they look brighter when their long side is facing us and dimmer when we're watching the short end; some stars flare up after siphoning matter from a companion star; some are regularly eclipsed by their partners.

The first variable star was discovered on August 13, 1596, when David Fabricius (1564–1617) noticed a star he'd never seen in Cetus the Whale. A Lutheran minister and active amateur astronomer in Germany, he counted among his friends Johannes Kepler and Tycho Brahe, who had recorded a supernova in Cassiopeia a quarter of a century before. So Fabricius knew that stars could change. Maybe he even thought that Omicron Ceti—a designation that would be given the star by Johann Bayer in 1603—was the same kind of star that had so excited Tycho in 1572.

But it was not. Unlike Tycho's star, a supernova that shone brilliantly and then dwindled in luminosity, Omicron Ceti blinks on and off, fading in and out of visibility on an eleven-month schedule. In 1638, after years of observation, the schedule of its changes was established, and in 1662, Johannes Hevelius wrote a book about it called *Historiola mirae stellae* (*A Short History of the Wonderful Star*). Ever since then, the variable star in Cetus has been known as Mira, the wonderful star.

A red giant so puffed-up and diaphanous that it practically isn't there, Mira is one of the biggest, coolest stars in the sky. Over a period averaging 331 days, it swells from a star not quite as bright as the Sun into one that's 250 times brighter. Then, having blown off gusts of gas and dust, it contracts. At its dimmest, it is a tenth-magnitude star; at its bulging maximum, it is usually a second-magnitude star, although once, in 1779, William Herschel saw it shine as brightly as Aldebaran, a first-magnitude star. By the end of the eighteenth century, only a dozen Mira-type variables were known, but today astronomers have identified literally thousands, including one that is known to vary over a period of 3,000 days. Despite their large population, not one has ever reached a maximum as bright as the wonderful original.

*

Orion: The California of the Sky

Orion plunges like a drunken hunter
over the Mohawk Trail   a parallelogram
slashed with two cuts of steel . . .
— Adrienne Rich, "The Spirit of Place"

"You know Orion always comes up sideways."
— Robert Frost, "The Star Splitter"

f you had to choose a quintessential constellation, Orion would be the one. It's got everything: a recognizable shape, dramatic mythology, brilliant stars, splendiferous nebulae. Camille Flammarion (1842–1925), a French astronomer who believed in vegetation on the Moon and intelligent life on Mars, called it "the California of the Sky."

In Greek mythology, Orion is a hunter holding a club and a shield as protection against Taurus the Bull. His dogs, Canis Major (with the Dog Star, Sirius) and Canis Minor (with the bright star Procyon), follow right behind him, rising shortly after the great hunter himself, while Orion's mortal enemy, Scorpius, glimmers on the opposite end of the sky. In ancient China, Orion and Scorpius represented two brothers who fought so much they had to be separated.

According to another story, Orion once attacked the Pleiad Merope, who was the daughter of the king of Chios and the granddaughter of Dionysus. When her father found out about it, he blinded Orion. An oracle told Orion that if he faced the Sun at dawn, his vision would be restored. So the hunter paid a visit to the ugly blacksmith god Hephaestus, who offered the assistance of his servant Cedalion. Cedalion climbed onto Orion's shoulders and directed him toward the dawn. When the Sun peeked above the horizon, Orion could see once more. (A painting of this scene by the seventeenth-century French artist Nicolas Poussin hangs in New York City's Metropolitan Museum of Art.)

In Hindu mythology, this constellation represents the god Prajapati, who lusted after Rohini, one of his twenty-seven daughters, all of whom were married to the Moon. Prajapati made the star Sirius so angry that Sirius aimed an arrow at him and shot him in the side. In the sky, Rohini, represented by the star Aldebaran, rises first. Prajapati/Orion follows, with the three stars in the middle representing the arrow. Sirius rises last, still chasing Prajapati.

In some places the three stars in the belt rate their own stories. In Greenland, they are three seal hunters adrift at sea. In Australia, they are a trio of young men captivated by the music of the Pleiades. And in Argentina, they are known as the three Marías. Claude Lévi-Strauss recounts a South American story about three brothers. One was married; the other two—one good-looking, one not—were bachelors. One day the handsome bachelor made the ugly bachelor climb a tree to get some seeds. When his brother was high in the branches, the handsome bachelor speared him with a stake, killed him, and then chopped off his legs and tossed them into the water to feed the fishes. The soul of the ugly brother ascended to the heavens and became the middle star in the belt. The legs changed into catfish and rose into the sky to become the stars on either side of the belt. The handsome brother was transformed into the planet Venus, and the married brother became the star Sirius.

Of the thirty-one brightest stars in the sky, five are in Orion. The brightest is Rigel, derived from *Ryl*, the Arabic word for "foot." A brilliant blue-white supergiant 910 light-years away, it is as bright as 57,000 Suns and its diameter is nineteen times that of the Sun. Sometime during the next 20 million years, Rigel is expected to erupt in a supernova.

Betelgeuse will explode even sooner—probably within a few hundred thousand years, which in an astronomical sense qualifies as any moment now. A pulsating orange variable supergiant with a mass about twenty times that of the Sun, Betelgeuse is spread out over an area whose diameter is about 800 times the Sun's, and is made of gases so thin that it is essentially a vacuum. Because it is so large, astronomers have been able to map it according to temperature. Its coldest area: the north pole.

Other bright stars include Bellatrix, which means "woman warrior," and the stars of Orion's belt: Mintaka, Alnilam, and the triple star Alnitak, Arabic names meaning, respectively, "the belt," "a belt of pearls," and "the girdle."

Dangling from these three stars like a sword from a belt is a series of wonders that includes the Orion Nebula, an incandescent cloud about 20,000 light-years across that William Herschel described as "an unformed fiery mist, the chaotic material of future suns." Home to 1,000 or so hot young stars, all created within the last million years or so, the Orion Nebula is a star-forming area embedded within a giant molecular cloud known to contain water, carbon dioxide, and other familiar substances. Ingredients like those inevitably give rise to a certain sort of speculation. "Astronomers who once dismissed as pure imagination Fred Hoyle's science-fiction novel *The Black Cloud*, in which an interstellar cloud turns out to be alive, are now willing to discuss seriously the question of whether life could originate in interstellar clouds," write astronomers Donald Goldsmith and Nathan Cohen. "We may yet find—though this must still be called a long shot—that complex molecules in the Orion Nebula have assembled themselves to the point that primitive forms of life (to speculate mildly) exist there."

✳

The Pleiades

Borax, Borax, Borax
Borax the Dinosaur slounges thru
fronds under Pleiades. . . .
—Allen Ginsberg, "Iron Horse"

startlingly rich collection of lore and legend is connected to the Pleiades, a jewel-like cluster of stars about 415 light-years from Earth. Glittering in Taurus near the shoulder of the bull, for many ancient cultures this knot of stars marked the turning of the seasons. In spring, it rose at dawn to denote the time for sowing (or, in ancient Greece, the start of the navigational season). In autumn, it rose at sunset in time for the harvest (or the rainy season—a legend recounted in the Talmud connects the Pleiades with Noah's flood). And in November, it reached its highest point in the sky, whereupon the Australian aborigines would stage a corroboree in celebration of the new year and in honor of these stars. The Pleiades have been linked to almost every memorial festival, including Halloween, All Souls' Day, the Mexican Day of the Dead, and an elaborate ritual celebrated by the Aztecs, whose fifty-two-year calendar reached its climax with a human sacrifice scheduled for the moment when the Pleiades were overhead at midnight. (The last such ceremony was held in 1507 in Tenochtitlán, a city razed to the ground fourteen years later by Hernando Cortés.)

Stories about the Pleiades often describe scenes of escape. For instance, the Monache Indians of central California tell a story about six women who discovered onions. Despite their husbands' displeasure they couldn't stop eating them, and when their husbands went out to hunt cougar, the women returned to the fields to indulge their newfound appetite; one brought her baby with her. At the end of the day, when the men came home empty-handed, they blamed their wives for their failure and refused to sleep with them. The women kept on eating the onions anyway, and eventually it dawned on them that they were just as happy without the men. So they decided to escape. The oldest wife threw a rope into the air and uttered a magic word, whereupon the six women (and the baby) all climbed the rope into the skies and became the Pleiades.

When the husbands saw this, they tried to follow, using a rope of their own. They are now visible in the six bright stars that mark the head and horns of Taurus the Bull.

In a story told by the Onondaga Indians of the Iroquois Confederacy, the Pleiades were a group of children who escaped from their parents every day and went to a lake to dance. One day an old man in a feathered cloak appeared to them and warned them not to continue. But they redoubled their efforts and danced with such abandon that their bodies

Some of the stars in the Pleiades.

National Optical Astronomy Observatories

were lifted off the ground and they began to dance on air. Their parents could only stand below and watch them ascend. When the chief called out to them, one child looked down and became a falling star. The other children continued rising into the heavens, where they became the Pleiades.

In Greek mythology, the Pleiades are the daughters of Atlas and Pleione. Orion became enamored of them and chased them so relentlessly that Zeus helped them escape by turning them at first into doves and then into stars. Another story says that they killed themselves after their half sisters the Hyades, grief-stricken by the death of their brother, were turned into stars.

No convincing explanation exists for why so many stories describe seven Pleiades even though most observers nowadays can see only six bright stars. Similarly, it is unclear why so many stories are told about a missing Pleiad. Perhaps the star Pleione, a rapidly spinning fifth-magnitude star and a known variable, was once much brighter than it is now. In any case, the stories are legion. In the Monache story told above, the baby represents the dimmest of the Pleiades. In Greek mythology, the missing Pleiad is Merope, who, unlike her sisters, mated not with a god but a mortal, and a miserable one at that: Sisyphus, who was condemned to push a boulder up a hill again and again for all eternity. According to Ovid, this was so humiliating that she hid her face in shame. The Roman author Hyginus claimed that she hung her head and became a comet with her hair streaming out behind her. Either way, she disappeared.

Or perhaps the missing Pleiad was Electra, who turned her face away so that she wouldn't have to watch Troy burn. Or Celaeno, who was struck by lightning.

The Pleiades are often seen as a group of women or children who have ascended to heaven. But the term has also occasionally been applied to groups of learned men on Earth. In the sixth century B.C., the Seven Wise Men of ancient Greece (one of them was the astronomer Thales) were known as the Philosophical Pleiad, and various groupings throughout the centuries followed, including an Alexandrian Greek Literary Pleiad of the third century B.C., a sixteenth-century French septet of poets called (by one of its own members) *la grande Pléiade*, and a group of post–Revolutionary War poets known as the Pleiades of Connecticut.

Although literally hundreds of stars have been counted in this cluster, only nine have proper names—the seven sisters; their mother, Pleione;

and their father, Atlas, who carried the weight of the world on his shoulders. They were formed together and are so young that the bright, easily visible B-type stars are still wrapped in the fuzzy wisps of blue nebula in which they were born as recently as 20 million to 50 million years ago—many millennia after the death of the dinosaurs. Thus, despite the quotation at the top of this section, neither hadrosaurus nor velociraptor ever slounged beneath the Pleiades.

Ursa Major
and the Big Dipper

My father compounded with my mother
under the Dragon's Tail, and my nativity
was under Ursa Major, so that it follows, I
am rough and lecherous.

— William Shakespeare,
King Lear (I. ii. 124–27)

he stars in Ursa Major, and especially the seven stars we call the Big Dipper, have not been universally perceived as a bear. The ancient Chinese saw them as the god of literature and his bureaucratic attendants. The Egyptians imagined the leg of a bull; the Sioux saw a skunk; the Lapps saw a reindeer, and the Arabs pictured these stars as a funeral bier followed by three daughters in mourning. Throughout Europe, these stars, which revolve around the sky like the hour hand around the clock, were thought of as a plow or wagon variously attributed to, among others, Odin, King Arthur, and Charlemagne.

But the Algonquin, the Zuni, the Eskimo, and the ancient Greeks saw these stars as a bear. The Greek story, like many other classical myths, concerns the extramarital activities of the god Zeus. It seems he was interested in a nymph named Callisto, but she had decided to devote her life—and virginity—to Artemis, the goddess of the hunt and of the moon. So the great god transformed himself into the image of Artemis and in this guise raped Callisto, who nine months later gave birth to a baby named Arcas. Because she had, however unwillingly, broken her vow of chastity, Artemis and Callisto's fellow nymphs shunned her, and Hera,

Neg. No. 2A13112, photo by Larry Brown. Courtesy
Department Library Services, American Museum of
Natural History

Donati's Comet of 1858.
The star near the head of the
comet is Arcturus, and the
Big Dipper is clearly visible
to the right.

Zeus' distressed wife, turned Callisto into a bear. She wandered the earth
for years until one day she spied her son Arcas. Forgetting her bestial
form, she ran to hug him; horrified, he drew an arrow to shoot.

At that exact moment, Zeus appeared. He picked them both up,
changed Arcas into a bear, and flung them into the heavens. He threw
them so hard that their tails stretched—which is why Ursa Major and
Ursa Minor have much longer tails than have ever been seen on a real
bear.

Dissatisfied, Hera asked two gods of the ocean (according to Ovid) to
"forbid these bearlike creatures in the stars to wade your waters." This
was accomplished: At the latitude of Athens two or three thousand years
ago, neither bear was ever low enough in the sky to touch the water.

Lowell Observatory

Mizar and Alcor, the double star in the Big Dipper.

Native American stories also view these stars as bears. A Northwest tale describes what happened when Coyote heard that two strange animals had been seen in the sky. First, he convinced the five Wolf brothers to investigate with him. Next, he shot an arrow into the sky, and then another arrow, which plunged into the shaft of the first, and then a third, which stuck to the second, and in this way he constructed a long chain of arrows that stretched from the Earth to the sky. Coyote and the Wolf brothers climbed up this ladder of arrows and reached the sky, where Coyote discovered that the two animals were grizzly bears. He promptly left, tearing down the arrows one by one as he climbed back to Earth. To-day, the two grizzly bears are still in the sky, followed by the five Wolves (plus a dog belonging to one of them), all stranded in heaven by their erstwhile friend.

Another story is told by the Iroquois and the Micmac of Canada, who imagined a bear chased by birds. The stars in the ladle of the Big Dipper represent the bear. The first star in the handle represents a robin; the second star, which is actually a double star, is a chickadee carrying a pot in which to stew the bear; and the last star in the handle and four bright stars in the constellation Boötes represent other birds. The chase begins in spring, when the bear is high in the sky, and as the year progresses, the bear circles the North Star, coming closer and closer to the horizon, with the birds in pursuit. In the autumn, when Ursa Major hovers closely above the horizon, the four birds in Boötes (which now dips below the horizon) fly south, forgoing the chase. But the robin aims an arrow at the bear and wounds it. Blood spurts everywhere, staining the robin's breast red and coloring the leaves of the trees. The bear dies, but its spirit remains, and in the spring, when the world comes back to life, the bear emerges from hibernation and the chase begins anew.

Ursa Major is much larger than the Big Dipper, which is actually only a part—an asterism—of the larger constellation. The Big Dipper represents the tail and the hindquarters of the Great Bear. Three of its stars are especially celebrated. The two far stars in the ladle, known as Dubhe (Arabic for "bear") and Merak, are well known because a line drawn through them points almost directly to the North Star. The middle star in the handle is famous because it is the most obvious multiple star in the sky. Mizar, the brighter of the pair, is a second-magnitude star, while Alcor is a dimmer fourth-magnitude star. (In one of the Native American stories told above, Alcor represents the dog belonging to one of the Wolf brothers; in the other story, Alcor symbolizes the pot in which the chickadee intends to cook the bear.) English stargazers saw the two stars as a horse and rider, and since Alcor and Mizar comprise the middle star in the handle, they sometimes referred to Alcor as "Jack on the Middle Horse." A thirteenth-century Persian writer described the double star as a test of good vision, but assuming your glasses have an up-to-date correction, its components are not impossible to see.

Despite appearances, the stars in the Big Dipper are not members of the same group, and the pattern they form is not permanent. The middle five stars move together, unaccompanied by the stars on either end, which move in different directions. A hundred thousand years from now, the shape will have changed. The deep bowl will be shallower, the gently curving handle will be bent, and the most recognizable constellation in the sky will look less like a soup ladle and more like a battered snow shovel. It will resemble a bear as much as it ever has.

◆

Defunct Constellations

n aura of classical inevitability clings to the eighty-eight recognized constellations. But don't be fooled. Although forty-eight were cited by Ptolemy and hence can rightly be considered classical, the remainder have been invented in recent centuries. Coma Berenices was added by Tycho Brahe in 1590. Johann Bayer added eleven constellations, all pictured in his ornate celestial atlas of 1603. The Southern Cross was added in 1679. In 1690, Elisabeth Hevelius, widow of the Danzig astronomer Johannes Hevelius, produced a catalogue with eleven new constellations including the Lizard (Lacerta) and Sobieski's Shield, now known as Scutum.

But many constellations struck out. Among the constellations that have been tossed out of the skies are:

* Antinoüs. Not a mythological figure but a real one, Antinoüs was sailing down the Nile in A.D. 130 with his lover, the emperor Hadrian, when he drowned, either by accident or as a result of a suicidal decision to fulfill an oracle's prediction and save Hadrian by sacrificing himself. Hadrian founded a city, Antinoöpolis, in his lover's memory, and two decades later Ptolemy included Antinoüs as a part of Aquila. It was pictured on a star globe of 1551, listed as a constellation by Tycho Brahe . . . and ultimately eliminated from the celestial city;

* Taurus Poniatowski (Poniatowski's Bull). Invented in 1777 to honor Stanislaw II Augustus Poniatowski, the king of Poland, it has been subsumed in Ophiuchus;

* a number of technological constellations proposed by Johann Bode, including Herschel's Telescope (Telescopium Hersche-

AN ALBUM OF STARS AND CONSTELLATIONS

lii), the Balloon (Globus Aerpstaticus), the Electrical Machine (Machina Electrica), and the Printing Press (Officiana Typographica). These rejected constellations might sound too mundane to deserve inclusion, but keep in mind: the heavens are full of such objects. Many of this sort were suggested in 1752 by Nicolas Louis de Lacaille, whose maps of the southern skies included, among other constellations still in use, the Telescope, the Microscope, the Sculptor (originally the Sculptor's Apparatus), the Furnace (Fornax), the Octant (Octans), the Clock (Horologium), the Air Pump (Antlia), and a faint constellation originally named the Rhomboidal Net (Reticulus Rhomboidalis), now simply called Reticulum;

* Noctua the Owl, which disappeared from sky maps, along with its predecessor, Turdus Solitarius (the Solitary Thrush);

* the Tigris and Jordan Rivers, which simply never caught on;

* Lilium the Lily and Cerberus (a three-headed snake—not a dog), both rejected in the seventeenth century;

* Tarandus (the Reindeer), a doomed constellation of the eighteenth century;

* Felis the Cat, a constellation recommended by Joseph Jérôme Le Français de Lalande ("I am very fond of cats," he explained), depicted in *Uranographia,* Bode's 1801 star atlas, and rejected for reasons incomprehensible even today.

Finally, although most commentators simply hoped to add occasional favored constellations to those already accepted, a few had more grandiose ideas. In 1688, for instance, Erhard Weigel hoped to redesign the skies using a heraldic motif honoring the ruling families of Europe. But no attempt was more ambitious than that of Julius Schiller, a committed Catholic who, shortly before his death, rearranged the stars into . . .

* The Christian Constellations. Schiller got a little help from Johann Bayer, but Bayer died in 1625, and Schiller died in 1627, and responsibility for the celestial catechism passed to Jakob Bartsch, Kepler's son-in-law, who made sure the new designs

Neg. No. 314279. Courtesy Department Library
Services, American Museum of Natural History

Saints in the sky—an idea
that never caught on.

were successfully published. In this scheme, Aries became Saint
Peter, Taurus became Saint Andrew, Orion metamorphosed into
Joseph, Perseus was renamed Saint Paul, Cassiopeia turned into
Mary Magdalen, Ursa Major became Saint Peter's Boat, and on
and on.

None of these designations survived.

Neg. No. 313166, photo by Hugh S. Rice. Courtesy
Department Library Services, American Museum of
Natural History

A nineteenth-century engraving showing four constellations of the zodiac.

THE ORIGINAL FORTY-EIGHT

Of Ptolemy's original forty-eight constellations, all but one are still in use today.

Andromeda

Aquarius (the Water Bearer)

Aquila (the Eagle)

Ara (the Altar)

Argo Navis* (the Ship *Argo*)

Aries (the Ram)

Auriga (the Charioteer)

Boötes (the Herdsman)

Cancer (the Crab)

Canis Major (the Great Dog)

Canis Minor (the Little Dog)

Capricornus (the Sea-Goat)

Cassiopeia (the Queen)

Centaurus (the Centaur)

Cepheus (the King)

Cetus (the Whale)

Corona Australis (the Southern Crown)

Corona Borealis (the Northern Crown)

Corvus (the Crow)

Crater (the Cup)

Cygnus (the Swan)

Delphinus (the Dolphin)

Draco (the Dragon)

Equuleus (the Foal)

Eridanus (the River)

Gemini (the Twins)

Hercules

Hydra (the Watersnake)

Leo (the Lion)

Lepus (the Hare)

Libra (the Balance)

Lupus (the Wolf)

Lyra (the Lyre)

Ophiuchus (the Serpent Bearer)

Orion

Pegasus (the Flying Horse)

Perseus

Piscis (the Fish)

Piscis Austrinus (the Southern Fish)

Sagitta (the Arrow)

Sagittarius (the Archer)

Scorpius (the Scorpion)

Serpens (the Serpent)

Taurus (the Bull)

Triangulum (the Triangle)

Ursa Major (the Great Bear)

Ursa Minor (the Little Bear)

Virgo (the Virgin)

Yerkes Observatory photograph, University of Chicago

*Argo Navis—the Argonauts' ship—is no longer used. It was subdivided in the 1750s into its constituent parts: Carina (the Keel), Puppis (the Poop Deck), Pyxis (the Compass), and Vela (the Sails).

Celestial Art

It will be clear that putting little white dots on a blue-black surface is not enough.

—Vincent van Gogh

hy don't artists pay closer attention to the sky? Architects and decorators do, as the history of the celestial ceiling indicates. A Chinese Han dynasty tomb over 2,000 years old is adorned with a pastel sky map complete with the Milky Way and the Big Dipper; the ancient Greeks decorated concave ceilings with celestial designs; Byzantine mosaics in Ravenna are adazzle with hundreds of golden stars against a vault of midnight blue; a ceiling in the Farnesina in Rome portrays constellations in the horoscope of Roman banker Agostino Chigi, born December 1, 1466; a similar ceiling in the J. Pierpont Morgan Library in New York City emphasizes zodiac constellations important to the financier; a few blocks away, Grand Central Station is famous for the inadvertently reversed constellations on the high ceiling above its main concourse. It's even possible to inscribe the constellations on your bedroom ceiling with a glow-in-the-dark kit.

But stroll through an art museum looking for stars and you are bound to be disappointed. You will see skies populated by angelic orchestras, clouds, doves, and, on occasion, rainbows, the Sun, the full moon, and the new moon (although the lopsided gibbous moon is a rare sight). But a starry sky? Orion? The Milky Way? These ordinary miracles, visible almost every night of our lives, have been painted so infrequently, it's shocking. Artists, evidently, don't look up.

Fortunately, there are exceptions:

* the unknown needlework artists who created the Bayeux Tapestry and included in it a picture of Halley's Comet of 1066;

* the Florentine artist Giotto (1266–1337), who painted Halley's Comet in a fresco on the walls of Padua's Scrovegni Chapel;

* the Limbourg brothers, Paul, Herman, and Jean (or Jannequin), who created the sumptuous miniatures in the fifteenth-century illuminated manuscript known as the *Très Riches Heures* of Jean, Duke of Berry. In addition to painting decorative stars in the astrological semicircles that crowned their monthly calendar scenes, they painted a star-spattered sky, complete with meteors; a gloomy scene at the death of Christ showing the Sun and the Moon both in eclipse; and, in a small medallion, a puzzled astronomer watching the celestial signs;

* Albrecht Dürer (1471–1528), who made detailed engravings of two celestial maps. In 1515, a mathematician put the stars in the proper places and Dürer drew the mythological figures, which were subsequently copied by, among others, Johannes Bayer in his influential celestial atlas, *Uranometria*. Later, Dürer published forty-eight drawings of individual constellations for a star catalogue based on Ptolemy;

* Matthias Grünewald (c. 1475–1528), who painted rainbows around Christ and a sprinkling of stars in one painting. In another, a small Crucifixion scene on display at the National Gallery of Art in Washington, D.C., he painted a star-spotted sky with an eclipsed Sun, now thought to be inspired by a solar eclipse visible in Germany on October 1, 1502;

* Tintoretto (1518–94), a Venetian painter who knew something about celestial mythology. *The Origin of the Milky Way,* now at the National Gallery in London, shows Zeus holding the baby Hercules up to Hera so that he can nurse; milk and stars are spilling from her breast. Patches of starry sky also appear among the angels in *Paradise,* painted in about 1588;

* Adam Elsheimer (1578–1610), who painted his *Flight into Egypt* in 1609, shortly after the invention of the telescope. It shows a

gibbous moon, constellations, and a star-studded Milky Way, suggesting that even in Rome, Elsheimer had heard the news about Galileo's discoveries. True, his constellations are not in the right place relative to the Moon, but he nonetheless tried to create a realistic sky;

* Donato Creti (1671–1749), an Italian artist who created a series of landscapes, now in the Vatican Collection, showing stargazers with and without telescopes observing, among other phenomena, the gibbous moon, Jupiter and its moons, the ringed Saturn, and a glowing comet plunging toward the horizon at dusk;

* Thomas Cole (1801–48), an American who painted highly romantic oils. His *Landscape (Moonlight)* depicts a glowing full moon against a background of carefully placed stars so subtly rendered that one or two of them have pale reddish casts;

* Jean-François Millet (1814–75), who did not paint stars but did look at the sky, painting a full-color rainbow in one painting and a gibbous moon in another;

* Edouard Manet (1832–83), who occasionally sprinkled stars across his skies, as in the brooding *Moonlight over Boulogne Harbor, 1869*;

* Henri Rousseau (1844–1910), who dotted his skies with biplanes, blimps, balloons, suns, moons, and, in his famous *Sleeping Gypsy,* painted in 1897, a few isolated stars;

* Wassily Kandinsky (1866–1944), who once included a photograph of a globular cluster in a book he'd written. His paintings include *Several Circles* (1926) and others that look like foreshortened solar systems in motion;

* František Kupka (1871–1957), a Spiritualist whose abstract paintings, including *The First Step* and *Disks of Newton,* seem to illustrate heavenly bodies in motion;

* Diego Rivera (1886–1957), who revealed an acquaintance with astronomy in the ill-fated fresco mural he designed for Rockefeller Center. The mural was almost completed when Rivera's leftist leanings, symbolized by a portrait of Lenin on the mural, released a flood of hysterical publicity ("RIVERA PERPETUATES

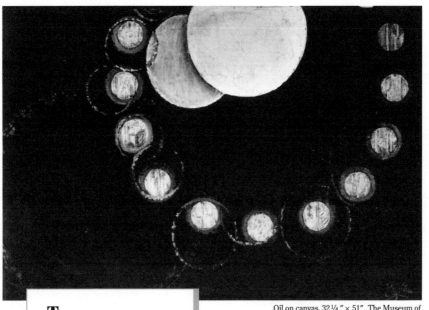

The *First Step*, by
František Kupka (1910–13;
dated on painting 1909).

Oil on canvas, 32 ¼″ × 51″. The Museum of
Modern Art, New York. Hillman Periodicals Fund

SCENES OF COMMUNIST ACTIVITY FOR R.C.A. WALLS—AND ROCKEFELLER
FOOTS BILL," read a headline from the *World Telegram*). Rivera
was ousted from the building, the Rockefellers withdrew their
support, and in February 1934, the mural was literally chopped
off the walls. After its destruction, Rivera reconstructed the art-
work in Mexico City. The mural shows "Man" at the crossroads
of two ellipses, one representing biological organisms and one
showing the cosmos, complete with stars, planets, solar systems, comets, nebulae, spiral galaxies, and a "hammer and
sickle" star;

* Joseph Cornell (1903–73), who often used celestial imagery in
  his collaged box constructions. Other twentieth-century artists
  who have occasionally used celestial imagery include Maxfield
  Parrish, Paul Klee, Joan Miró, Georgia O'Keeffe, and, more re-

cently, Billy Al Bengston, Vija Celmins, and Betye Saar. The conceptual artist James Turrell is notable for using the celestial vault—the actual stars—as the focus of his transcendent earthworks located in the Arizona desert.

But above and beyond those artists, one painter in the last thousand years deserves applause for looking at the sky and noticing that stars form patterns and that these patterns—the constellations—are part of the general landscape. Vincent van Gogh's *Starry Night, on the Rhône,* painted in the incredibly productive year of 1888, clearly depicts the Big Dipper hanging over the river. *Café Terrace, Night, on the Place du Forum* shimmers under the constellation Aquarius. And his famous *Starry Night* (1889) may represent, according to the UCLA art historian Albert Boime, the constellation Aries (the sign of the zodiac under which

The Starry Night, by Vincent van Gogh (1889).

Oil on canvas, 29″ × 36 ¼″ (73.7 × 92.1 cm). Collection, The Museum of Modern Art, New York. Acquired through the Lillie P. Bliss Bequest

Van Gogh was born). In these paintings, stars are not dots but orbs, not decorative devices but massive globes with the sky swirling about them.

It's not just that Van Gogh looked at the stars. They were part of his mental landscape, and he envisioned them even when he was painting interior scenes such as his portrait of the poet Eugène Boch. He wrote, "Behind the head, instead of painting the ordinary wall of the mean room, I paint infinity, a plain background of the richest, intensest blue that I can contrive, and by this simple combination of the bright head against the rich blue background, I get a mysterious effect, like a star in the depths of an azure sky."

He thought about stars: how to paint them, what they meant. "To look at the stars always makes me dream as simply as I dream over the black dots of a map representing towns and villages," he wrote to his brother Theo. "Why, I ask myself, should the shining dots of the sky not be as accessible as the black dots on the map of France? If we take the train to get to Tarascon or Rouen, we take death to reach a star. One thing undoubtedly true in this reasoning is this: that while we are alive we cannot get to a star, any more than when we are dead we can take the train. So it seems to me possible that cholera, gravel, phthisis, and cancer are the celestial means of locomotion, just as steamboats, omnibuses, and railways are the terrestrial means. To die quietly of old age would be to go there on foot." In the work of Vincent van Gogh, beleaguered soul, the stars fill the sky.

What spectacle confronted them when they, first the host, then the guest, emerged silently, doubly dark, from obscurity by a passage from the rere of the house into the penumbra of the garden?

The heaventree of stars hung with humid nightblue fruit.

—James Joyce, *Ulysses*

An Essential Glossary

Absolute magnitude—a measure of how bright a star would look if it were located ten parsecs, or 32.6 light-years, from Earth.

Accretion disk—a disk of gas and other matter orbiting around a central star or black hole.

Anti-matter—matter made up of anti-particles, which have the same mass as ordinary particles but an opposite charge. Thus, the electron has a negative charge, while the anti-electron, or positron, has a positive charge.

Apparent magnitude—a measure of how bright a star looks.

Asteroids—small chunks of rock and metal orbiting the Sun; also known as minor planets.

Astronomical unit (AU)—the average distance between the Earth and the Sun: about 93 million miles.

Astrophysics—the principles of physics as applied to celestial objects, be they black holes or elementary particles.

Barred spiral galaxy—a spiral galaxy with a bar of stars across the nucleus.

Big Bang—the moment approximately 10 billion to 20 billion years ago when space began to expand from a hot, dense state into the universe as it exists today.

Big Crunch—the ultimate end of the universe if there is enough matter to halt the outward expansion of galaxies and cause a collapse.

Binary star—two stars orbiting around a common center of mass.

Black dwarf—the stellar cinder that remains after a white dwarf has ceased to shine.

Black hole—an object, generally a completely collapsed star, with such strong gravity that nothing can escape it, including light.

Blueshift—the change in wavelength created when an object is approaching the observer. As the object approaches, waves of radiation from that object are pushed together or shortened, shifting them toward the blue end of the spectrum. (See **Doppler shift.**)

Bok globule—a dense, roundish, dark nebula that may hide a protostar.

Brown dwarf—a small, low-mass object not quite hot enough in its core to produce nuclear fusion and become a star.

Cassini's Division—the seemingly empty gap between Saturn's main rings, discovered in 1675. It is actually populated by many small ringlets.

Central bulge—the ball of stars surrounding the nucleus of a spiral galaxy.

Cepheid variable—a yellow, supergiant, variable star whose absolute magnitude is directly proportional to the rate at which its brightness changes; useful as a standard candle for measuring distances in the universe.

Chandrasekhar limit—the maximum mass for a white dwarf star, about 1.4 solar masses.

Chondrite—a stony meteorite studded with small, round silicate globules.

Coma—the thin gaseous halo surrounding the nucleus of a comet.

Comet—a small clump of ice and dust in orbit around the Sun. While passing near the Sun, some of the comet's ices melt and are vaporized to gas, which is swept by the solar wind into a long tail.

Constellation—a collection of stars that appear to make a pattern in the sky.

Corona—the outermost parts of the Sun's atmosphere, visible during a total solar eclipse.

Cosmic background radiation—radiation created in the Big Bang that now permeates the universe. Once having the temperature of a star, that radiation has been redshifted into the microwave area and is now extremely cool.

Cosmological constant—a quantity added by Einstein to his equations in order to achieve a stable solution—a universe that neither expands nor contracts.

Cosmology—the study of the evolution and structure of the universe.

Dark matter—unknown, unseen matter thought to exist by virtue of its gravitational force and believed to constitute at least 90 percent of the gravitational mass of the universe.

Dark nebula—a dark interstellar cloud of gas and dust that blots out the light of more distant stars.

Doppler shift—the apparent change in wavelength of light (or other forms of radiation) that occurs as a result of relative motion between two objects. When an object is approaching the observer, its waves are essentially pushed together, making them shorter and of higher frequency; when an object is moving away, its

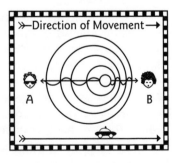

Direction of Movement →

A   B

waves are essentially stretched, making them longer and of lower frequency; named after the Austrian physicist Christian Johann Doppler (1803–53), who explained why train whistles sound higher as they approach and lower as they recede. A related process causes light that is approaching to shift toward the blue end of the spectrum, where wavelengths are shorter, while light that is moving away shifts toward the red end, where longer wavelengths are found.

Eclipsing binary—a pair of stars aligned in such a way, relative to the Earth, that one star regularly moves in front of and eclipses the other.

Ecliptic—the apparent yearly path of the Sun against the stars of the zodiac.

Einstein ring—an optical effect created when light heading toward us from a remote galaxy is bent by an intervening galaxy in such a way that it forms, under ideal conditions, a circle.

Electromagnetic spectrum—the full range of wavelengths, ranging from radio waves, which are the longest, through microwaves, infrared radiation, visible light, ultraviolet radiation, X-rays, and gamma rays, which are the shortest.

Electron—a light subatomic particle with a negative charge.

Ellipse—a closed, oval-shaped figure mathematically created by slicing at a slant through a circular cone with a flat plane.

Elliptical galaxy—a galaxy with an elliptical shape, no spiral arms, and a population primarily made up of old stars with little gas and dust.

Escape velocity—the minimum speed needed to escape from a star or planet. To leave the Earth behind, you need to achieve an escape velocity of 7 miles per second (25,200 miles per hour).

Event horizon—the border surrounding a black hole where gravity becomes so strong that nothing, not even light, can escape.

Fusion—the nuclear process whereby two atomic nuclei combine to form a heavier nucleus plus a substantial release of energy.

Galaxy—a collection of millions or billions of stars all bound together by gravity. Our own Milky Way Galaxy, with its hundreds of billions of stars, is approximately 80,000 to 100,000 light-years in diameter.

Gamma rays—the most-high-energy form of electromagnetic radiation, with the shortest wavelengths.

Globular cluster—a sphere of stars numbering anywhere between a few hundred thousand stars and a few million; usually found in the galactic halo.

Gravitational lens—an optical effect caused when light from a distant source is interrupted in its journey toward us by an intervening galaxy that curves space (according to Einstein's theory of general relativity) and consequently bends the light in such a way as to produce multiple mirror images of itself.

Gravity—the force of attraction between any two objects; one of the four basic forces of the universe (the others being the strong force, the weak force, and the electromagnetic force).

Hertzsprung-Russell (H-R) diagram—a chart in which stars are placed according to absolute magnitude (or luminosity) and color or temperature.

Hubble constant—the rate at which the universe is expanding, based on the proportion between the distances and the recessional speeds of faraway galaxies.

Hubble's law—Edwin Powell Hubble's description of the expansion of the universe, stating that the more distant a galaxy is, the faster it is receding.

Inflation—a theory that suggests that for a fleeting moment immediately after the Big Bang, the universe expanded at an extremely rapid rate.

Infrared light—radiation with wavelengths slightly longer than that of visible light (and hence not visible to the naked eye).

Light-year—the distance light travels in a year: approximately 5.9 trillion miles.

Local Group—the small cluster of galaxies that includes the Andromeda Galaxy and the Milky Way.

Magellanic Clouds—two small companion galaxies to the Milky Way, visible from the southern hemisphere.

Main Sequence—the major period of a star's life during which it sustains itself by converting hydrogen into helium; also the diagonal strip on the Hertzsprung-Russell diagram extending from hot, bright stars in the upper left-hand corner to cool, dim stars in the lower right.

Mass—the total amount of matter in an object. Note that an object has mass even if it is weightless, as an astronaut orbiting the Earth would be.

Meteor—the shooting star produced when a particle, often dust from the path of a comet, streaks through Earth's atmosphere and is heated by friction.

Meteorite—a chunk of material, often from an asteroid, large enough to survive a fall through the atmosphere.

Microwaves—electromagnetic radiation with wavelengths longer than infrared radiation but shorter than radio waves; also considered to be short-wavelength radio waves.

Milky Way—our galaxy; also the plane of the galaxy that is visible as a faint river of stars across the sky.

Nebula—a diffuse molecular cloud of interstellar dust and gas. Within some nebulae, stars are being born. Other nebulae, such as the Crab Nebula, are created when a dying star explodes in a supernova. The term "spiral nebula," once used to describe such objects as the Andromeda Galaxy, is no longer used, having been supplanted by the term "spiral galaxy."

Neutrino—an elementary particle believed to have little or no mass. Neutrinos do not interact with anything and can pass right through us.

Neutron star—a small, dense star approximately the size of Manhattan, made almost entirely of tightly packed neutrons. A neutron star is created in a supernova.

Nova—a star that suddenly brightens as much as a thousand times, probably caused when a white dwarf in a binary pair pulls material from its companion star.

Oort cloud—a spherical cloud of comets believed to surround the solar system.

Parallax—the apparent shift of an object's position caused by looking at it from two different places. It is easily demonstrated by holding up a finger and looking at it first with one eye shut and then with the other. The finger will seem to shift against the background because of the slightly different locations of the eyes. In astronomy, different positions are provided by the Earth at opposite points in its orbit around the Sun.

Parsec—a unit of distance equal to 3.26 light-years, or approximately 19 trillion miles, a figure chosen because a star at that distance would have a parallax of one arcsecond.

Photosphere—the visible surface layer of the Sun.

Planetary nebula—a luminous shell of gas expelled from a dying star on its journey from red giant to white dwarf. Through a small telescope, a planetary nebula looks round and greenish, like a planet. Otherwise, it has no connection with planets.

Proper motion—a star's apparent movement—always very slight—across the sky, caused by the star's orbiting the galaxy.

Protostar—a collapsing mass of dust and gas that is becoming a star.

Pulsar—a rapidly rotating neutron star that emits radio waves.

Quark—an elementary particle that combines to make up others, including protons and neutrons, each of which consists of three quarks.

Quasar—a very remote, extremely luminous object that appears starlike and shines 100 times brighter than an ordinary galaxy; thought to be the nucleus of an active galaxy with a central black hole.

Radio galaxy—an active galaxy that emits large amounts of radio waves.

Radio waves—the form of electromagnetic radiation with the longest wavelengths.

Red giant—a large, cool, luminous star created when a star leaves the main sequence and expands.

Redshift—a change in wavelength produced when an object is receding, which causes the waves of radiation from that object to become lengthened, shifting them toward the red end of the spectrum. A large redshift implies that the object is receding quickly, and hence is very distant. (See **Doppler shift**.)

Reflecting telescope—a telescope using a curved mirror to collect and focus light.

Refracting telescope—a telescope using a glass lens to collect and focus light.

Retrograde motion—the apparent westward drift of a planet, opposite to the usual eastward motion of the planets.

RR Lyrae variable—giant variable stars with periods of less than a day, usually found in globular clusters and other old stellar systems; used to measure clusters.

Singularity—a point of infinite density at the center of a black hole.

Solar system—the Sun and everything that orbits around it, including the planets and their moons, the asteroids, and the comets.

Spiral galaxy—a galaxy with a central bulge of old stars surrounded by a flat dish of younger stars, dust, and gas arranged in spiral arms; often found in a halo of dark matter.

Standard candle—a celestial object whose intrinsic brightness is known, which makes it useful for determining distance.

Strong force—the force that binds quarks together and joins protons and neutrons into nuclei; one of the four fundamental forces of nature.

Sunspot—a cool area of magnetic disturbance on the surface of the Sun.

Supercluster—a cluster of galactic clusters.

Supergiant—a very large, extremely luminous star.

Supernova—an exploding star that blows off most of its mass and increases in brightness so enormously that it can briefly outshine its entire galaxy and can look like a new star in the sky.

Theory of relativity—Einstein's theory. Special relativity, published in 1905, states that the speed of light is a constant, the same for all observers, and that energy and matter are equivalent ($E = mc^2$). The general theory of relativity, published in 1916, links gravity with the curvature of space and the passage of time.

Ultraviolet radiation—electromagnetic radiation with wavelengths shorter than visible light but longer than X-rays. Ultraviolet radiation does not penetrate the Earth's atmosphere and hence must be measured from space.

Variable star—a star whose brightness varies.

Weak nuclear force—one of the four basic forces, responsible for radioactive decay.

White dwarf—a small, dense star forms at the end of an ordinary star's life when its nuclear fuel has been exhausted.

Wormhole—a theoretical narrow tunnel leading from a black hole into another universe.

X-rays—a high-energy form of electromagnetic radiation with wavelengths shorter than ultraviolet rays but longer than gamma rays.

Selected Bibliography

Abell, George O. *The Realm of the Universe*. Philadelphia: Saunders College Publishing, 1984.

Allen, Richard Hinckley. *Star Names: Their Lore and Meaning*. New York: Dover Publications, 1963.

Ashbrook, Joseph. *The Astronomical Scrapbook: Skywatchers, Pioneers, and Seekers in Astronomy*. Cambridge, Mass.: Sky Publishing Corporation and Cambridge University Press, 1984.

Asimov, Isaac. *Asimov's Biographical Encyclopedia of Science and Technology* (new revised edition). Garden City, N.Y.: Doubleday, 1972.

Bartusiak, Marcia. *Thursday's Universe*. New York: Times Books, 1986.

Beatty, J. Kelly, and Andrew Chaikin, eds. *The New Solar System*. Cambridge, Mass.: Sky Publishing Corporation and Cambridge University Press, 1990.

Benford, Timothy B., and Brian Wilkes. *The Space Program Quiz & Fact Book*. New York: Harper and Row, 1985.

Bertotti, B., R. Balbinot, S. Bergia, and A. Messina, eds. *Modern Cosmology in Retrospect*. Cambridge, England: Cambridge University Press, 1990.

Blum, Howard. *Out There: The Government's Secret Quest for Extraterrestrials*. New York: Pocket Books, 1990.

Boslough, John. *Stephen Hawking's Universe*. New York: Avon, 1989.

Bova, Ben, and Byron Preiss, eds. *First Contact: The Search for Extraterrestrial Intelligence*. New York: Penguin, 1990.

Bucky, Peter A., in collaboration with Allen Weakland. *The Private Albert Einstein*. Kansas City, Mo.: Andrews and McMeel, 1992.

Burke, John G. *Cosmic Debris: Meteorites in History.* Berkeley: University of California Press, 1986.

Burnham, Robert, Jr. *Burnham's Celestial Handbook: An Observer's Guide to the Universe Beyond the Solar System.* New York: Dover Publications, 1978.

Burrows, William E. *Exploring Space: Voyages in the Solar System and Beyond.* New York: Random House, 1990.

Chaisson, Eric. *Relatively Speaking: Relativity, Black Holes, and the Fate of the Universe.* New York: W. W. Norton, 1988.

Chandrasekhar, S. *Truth and Beauty: Aesthetics and Motivations in Science.* Chicago: University of Chicago Press, 1987.

Chapman, Clark R. *Planets of Rock and Ice: From Mercury to the Moons of Saturn.* New York: Charles Scribner's Sons, 1982.

Chartrand, Mark. *The Audubon Society Field Guide to the Night Sky.* New York: Alfred A. Knopf, 1991.

Christianson, Gale E. *This Wild Abyss: The Story of the Men Who Made Modern Astronomy.* New York: Free Press, 1978.

Cornell, James, ed. *Bubbles, Voids and Bumps in Time: The New Cosmology.* Cambridge, England: Cambridge University Press, 1989.

Cronin, Vincent. *The View from Planet Earth.* New York: Morrow, 1981.

Davidson, Norman. *Astronomy and the Imagination.* London: Routledge and Kegan Paul, 1985.

Davies, John K. *Cosmic Impact.* New York: St. Martin's Press, 1986.

Davies, P.C.W., and J. Brown. *Superstrings: A Theory of Everything.* Cambridge, England: Cambridge University Press, 1988.

Davis, Joel. *Journey to the Center of Our Galaxy: A Voyage in Space and Time.* Chicago: Contemporary Books, 1991.

Ferguson, Kitty. *Stephen Hawking: Quest for a Theory of Everything.* New York: Bantam, 1992.

Ferris, Timothy. *Coming of Age in the Milky Way.* New York: Morrow, 1988.

———. *Galaxies.* New York: Stewart, Tabori and Chang, 1982.

———. *The Red Limit: The Search for the Edge of the Universe.* New York: Quill, 1983.

Ferris, Timothy, ed. *The World Treasury of Physics, Astronomy, and Mathematics.* Boston: Little, Brown, 1991.

Flaste, Richard, Holcomb Noble, Walter Sullivan, and John Noble Wilford. *The New York Times Guide to the Return of Halley's Comet.* New York: Times Books, 1985.

Friedman, Herbert. *The Astronomer's Universe: Stars, Galaxies, and Cosmos.* New York: W. W. Norton, 1990.

Gamow, George. *The Great Physicists from Galileo to Einstein.* New York: Dover Publications, 1961.

Gingerich, Owen. *The Great Copernicus Chase.* Cambridge, Mass.: Sky Publishing Corporation and Cambridge University Press, 1992.

Goldsmith, Donald. *The Astronomers.* New York: St. Martin's Press, 1991.

Goldsmith, Donald, and Nathan Cohen. *Mysteries of the Milky Way.* Chicago: Contemporary Books, 1991.

Greenstein, George. *Frozen Star.* New York: Freundlich Books, 1983.

Gribbin, John. *In Search of the Big Bang.* New York: Bantam, 1986.

Grossinger, Richard. *The Night Sky: The Science and Anthropology of the Stars and Planets.* Los Angeles: Jeremy P. Tarcher, 1988.

Harrison, Edward R. *Cosmology: The Science of the Universe.* Cambridge, England: Cambridge University Press, 1981.

———. *Darkness at Night: A Riddle of the Universe.* Cambridge, Mass.: Harvard University Press, 1987.

Hartmann, William K. *Astronomy: The Cosmic Journey.* Belmont, Calif.: Wadsworth, 1982.

Hawking, Stephen. *Black Holes and Baby Universes and Other Essays.* New York: Bantam, 1993.

———. *A Brief History of Time: From the Big Bang to Black Holes.* New York: Bantam, 1988.

Hawking, Stephen W., ed. *Stephen Hawking's A Brief History of Time: A Reader's Companion* (prepared by Gene Stone). New York: Bantam, 1992.

Hellemans, Alexander, and Bryan Bunch. *The Timetables of Science: A Chronology of the Most Important People and Events in the History of Science.* New York: Simon and Schuster, 1988.

Hermann, Dieter B. *The History of Astronomy from Herschel to Hertzsprung.* Translated and revised by Kevin Krisciunas. Cambridge, England: Cambridge University Press, 1973.

Hockey, Thomas A. *The Book of the Moon.* New York: Prentice-Hall, 1986.

Hoskin, Michael A. *William Herschel and the Construction of the Heavens: An Analysis of the Work of the Founder of Sidereal Science.* New York: W. W. Norton, 1963.

Hoyle, Fred. *Astronomy: A History of Man's Investigation of the Universe.* Garden City, N.Y.: Doubleday, 1962.

Jastrow, Robert. *Red Giants and White Dwarfs.* New York: W. W. Norton, 1990.

Jones, Bessie Zaban, and Lyle Gifford Boyd. *The Harvard College Observatory: The First Four Directorships, 1839–1919.* Cambridge, Mass.: Harvard University Press, 1971.

Kaufmann, William J., III. *Universe.* New York: W. H. Freeman, 1991.

Kippenhahn, Rudolf. *Bound to the Sun: The Story of Planets, Moons, and Comets.* Translated by Storm Dunlop. New York: W. H. Freeman, 1990.

Koestler, Arthur. *The Sleepwalkers: A History of Man's Changing Vision of the Universe.* London: Penguin, 1959.

Krupp, E. C. *Beyond the Blue Horizon: Myths and Legends of the Sun, Moon, Stars and Planets.* New York: HarperCollins, 1991.

Lang, Kenneth R., and Charles A. Whitney. *Wanderers in Space: Exploration and Discovery in the Solar System.* Cambridge, England: Cambridge University Press, 1991.

Lear, John. *Kepler's Dream.* Translated by Patricia Frueh Kirkwood. Berkeley: University of California Press, 1965.

Ley, Willy. *Watchers of the Skies.* New York: Viking, 1963.

Lightman, Alan. *Time for the Stars: Astronomy in the 1990s.* New York: Viking, 1992.

Lightman, Alan, and Roberta Brawer. *Origins: The Lives and Works of Modern Cosmologists.* Cambridge, Mass.: Harvard University Press, 1990.

Littman, Mark. *Planets Beyond: Discovering the Outer Solar System.* New York: John Wiley and Sons, 1988.

Mason, Stephen F. *A History of the Sciences.* New York: Macmillan, 1962.

Maurer, Richard. *Junk in Space.* New York: Simon and Schuster, 1990.

Monroe, Jean Guard, and Ray A. Williamson. *They Dance in the Sky: Native American Star Myths.* Boston: Houghton Mifflin, 1987.

Moore, Patrick. *The Amateur Astronomer.* New York: W. W. Norton, 1990.

———. *Fireside Astronomy: An Anecdotal Tour Through the History and Lore of Astronomy.* New York: John Wiley and Sons, 1992.

———. *Passion for Astronomy.* New York: W. W. Norton, 1992.

Morrison, David, and Tobias Owen. *The Planetary System.* Reading, Mass.: Addison-Wesley, 1988.

Motz, Lloyd, and Carol Nathanson. *The Constellations: An Enthusiast's Guide to the Night Sky.* New York: Doubleday, 1988.

Munitz, Milton K., ed. *Theories of the Universe: From Babylonian Myth to Modern Science.* New York: Free Press, 1957.

Murray, Bruce. *Journey Into Space: The First Thirty Years of Space Exploration.* New York: W. W. Norton, 1989.

Nininger, Harvey H. *Find a Falling Star.* New York: Paul S. Ericksson, 1972.

Olcott, William Tyler. *Star Lore of All Ages.* New York: G. P. Putnam's Sons, 1911.

Overbye, Dennis. *Lonely Hearts of the Cosmos.* New York: HarperCollins, 1991.

Pais, Abraham. *"Subtle Is the Lord . . .": The Science and the Life of Albert Einstein.* Oxford and New York: Oxford University Press, 1982.

Pannekoek, A. *A History of Astronomy.* New York: Dover Publications, 1989.

Parker, Barry. *Creation: The Story of the Origin and Evolution of the Universe.* New York: Plenum Press, 1988.

Payne-Gaposchkin, Cecilia. *An Autobiography and Other Recollections.* Edited by Katherine Haramundanis. Cambridge: Cambridge University Press, 1984.

Preston, Richard. *First Light: The Search for the Edge of the Universe.* New York: New American Library, 1987.

Proctor, Mary. *Evenings with the Stars.* London: Cassell and Company, 1924.

Raymo, Chet. *The Soul of the Night: An Astronomical Pilgrimage.* New York: Prentice-Hall, 1985.

Regis, Ed. *Who Got Einstein's Office?: Eccentricity and Genius at the Institute for Advanced Study.* Reading, Mass.: Addison-Wesley, 1987.

Richardson, Robert S. *The Star Lovers*. New York: Macmillan, 1967.

Ridpath, Ian. *Star Tales*. New York: Universe Books, 1988.

Ronan, Colin A. *The Natural History of the Universe*. New York: Macmillan, 1991.

Room, Adrian. *Dictionary of Astronomical Names*. London: Routledge and Kegan Paul, 1988.

Rowan-Robinson, Michael. *Our Universe: An Armchair Guide*. New York: W. H. Freeman, 1990.

Sagan, Carl. *Cosmos*. New York: Random House, 1980.

Schweighauser, Charles A. *Astronomy from A to Z: A Dictionary of Celestial Objects and Ideas*. Springfield, Ill.: Sangamon State University, 1991.

Shapley, Harlow. *Through Rugged Ways to the Stars*. New York: Charles Scribner's Sons, 1969.

Sheehan, William. *Planets and Perception: Telescopic Views and Interpretations, 1609–1909*. Tucson: University of Arizona Press, 1988.

Staal, Julius D. W. *The New Patterns in the Sky: Myths and Legends of the Stars*. Blacksburg, Va.: McDonald and Woodward, 1988.

Tombaugh, Clyde W., and Patrick Moore. *Out of the Darkness: The Planet Pluto*. Harrisburg, Pa.: Stackpole Books, 1980.

Trefil, James S. *Space, Time, Infinity: The Smithsonian View of the Universe*. New York: Pantheon, 1985.

———. *The Dark Side of the Universe: A Scientist Explores the Mysteries of the Cosmos*. New York: Doubleday, 1988.

Wali, Kameshwar C. *Chandra: A Biography of S. Chandrasekhar*. Chicago: University of Chicago Press, 1984.

Whitney, Charles A. *The Discovery of Our Galaxy*. Ames: Iowa State University Press, 1988.

Wilford, John Noble. *Mars Beckons*. New York: Alfred A. Knopf, 1990.

Wilson, Colin. *Starseekers*. London: Hodder and Stoughton, 1980.

Yeomans, Donald K. *Comets: A Chronological History of Observation, Science, Myth, and Folklore*. New York: John Wiley and Sons, 1991.

Zeilik, Michael. *Astronomy: The Evolving Universe*. New York: John Wiley and Sons, 1991.

Star Charts

The latitude of these charts is 34°N, but they can be used throughout the continental United States. Hold the chart vertically and turn it so the direction you are facing shows at the bottom.

Star charts from
Griffith Observer,
Griffith Observatory,
Los Angeles

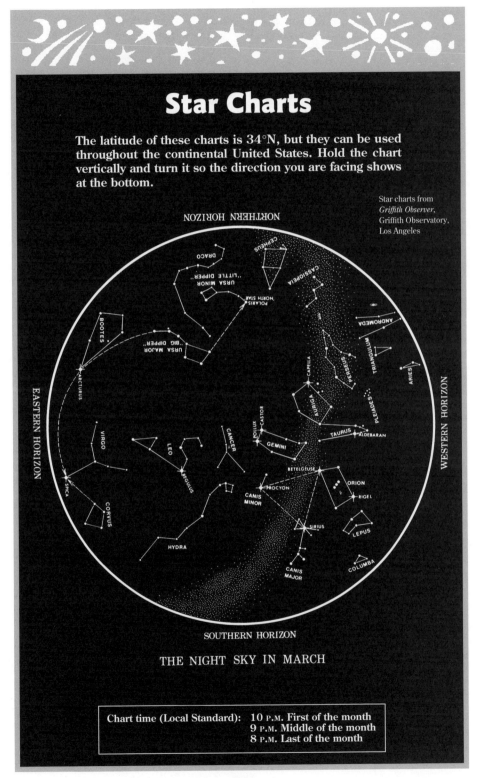

THE NIGHT SKY IN MARCH

Chart time (Local Standard):   10 P.M. First of the month
9 P.M. Middle of the month
8 P.M. Last of the month

445

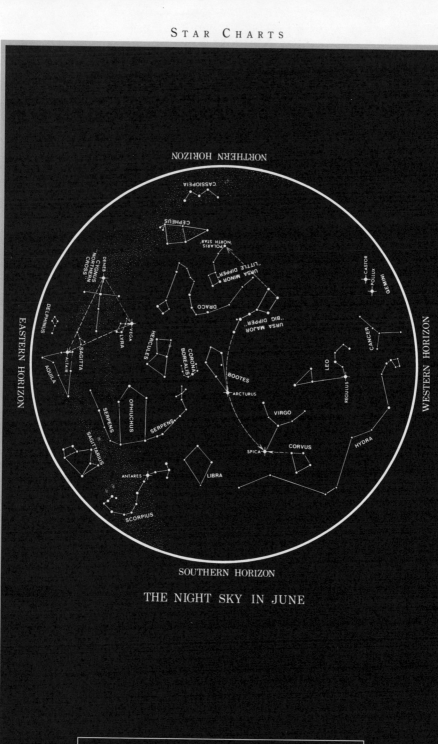

THE NIGHT SKY IN JUNE

Chart time (Local Standard): 10 P.M. First of the month
                               9 P.M. Middle of the month
                               8 P.M. Last of the month

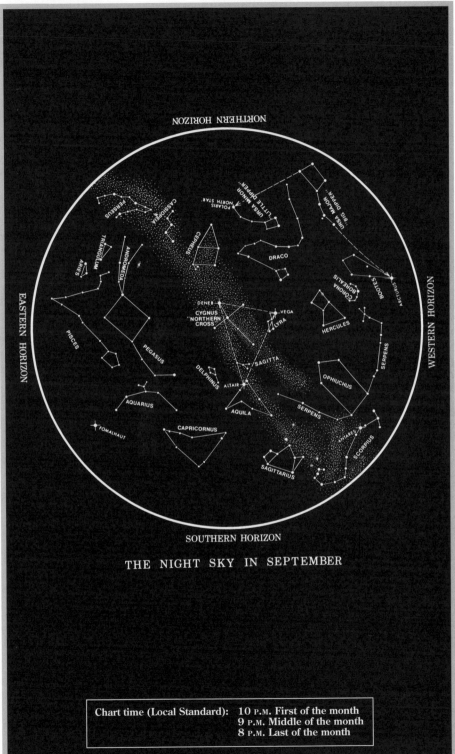

THE NIGHT SKY IN SEPTEMBER

Chart time (Local Standard): 10 P.M. First of the month
9 P.M. Middle of the month
8 P.M. Last of the month

447

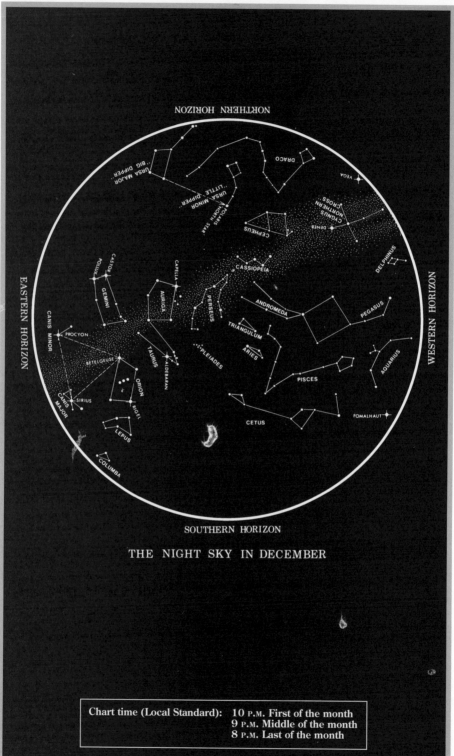

THE NIGHT SKY IN DECEMBER

Chart time (Local Standard):   10 P.M. First of the month
                                9 P.M. Middle of the month
                                8 P.M. Last of the month

448

Index